Unmasking the Methyl Code: A History of Protein Methylation Research

Mathew

Table of Contents

Chapter 1: Introduction

1.1.   Protein methylation: a tale slowly unravelling over decades

N-methylation of lysine (K) and arginine (R) sidechains (Figure 1.1) is now a widely studied post-translational modification (PTM), however it took half a century after its discovery for the significance to be understood and appreciated.[1] The first report on side chain methylation dates to 1959 when Ambler and Rees showed the presence of methylation on flagellar proteins *of Salmonella typhimurium*.[2] The hint that methylation was installed post-translationally came two years later in 1961 when presence or absence of methylation was shown to be correlated to a gene other than that encoding flagellin.[3] The relationship was firmly established by Kim and Paik in 1965 who identified *S*-adenosylmethionine (SAM, Figure 1.1 red box) as the source of the methyl group and showed that the modified tRNAs alone were not able to install methyllysine at the intended position during protein synthesis.[4] This suggested that the gene responsible for installation of methyllysine was encoding for an enzyme that used SAM as a substrate. We now call such enzymes methyltransferases, and further classify them as protein lysine methyltransferase (PKMT) or protein arginine methyltransferase (PRMT).[5]

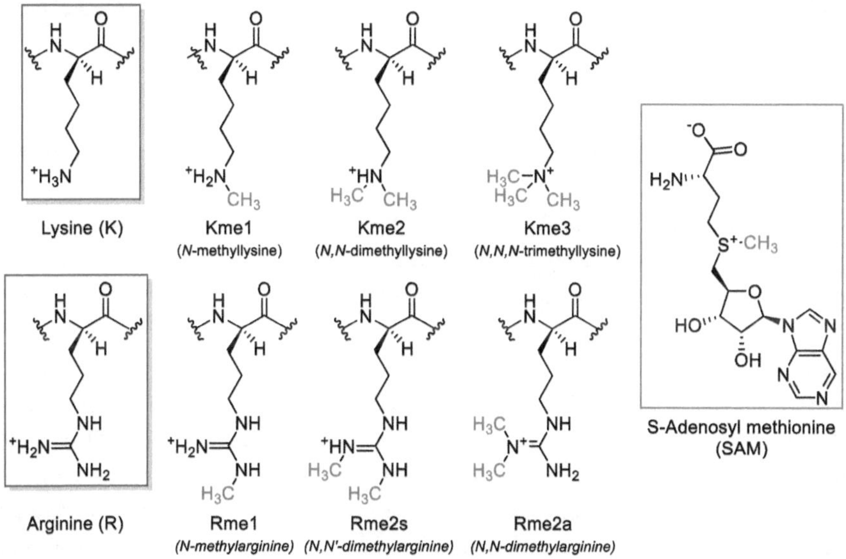

Lysine (K)

Kme1
(*N*-methyllysine)

Kme2
(*N,N*-dimethyllysine)

Kme3
(*N,N,N*-trimethyllysine)

S-Adenosyl methionine
(SAM)

Arginine (R)

Rme1
(*N*-methylarginine)

Rme2s
(*N,N'-dimethylarginine*)

Rme2a
(*N,N-dimethylarginine*)

Figure 1.1. *N*-methylation states of lysine and arginine found *in vivo*. Unmodified amino acids are shown in blue boxes. Each methyl group is installed sequentially by methyltransferase enzymes using SAM (red box) as substrate.

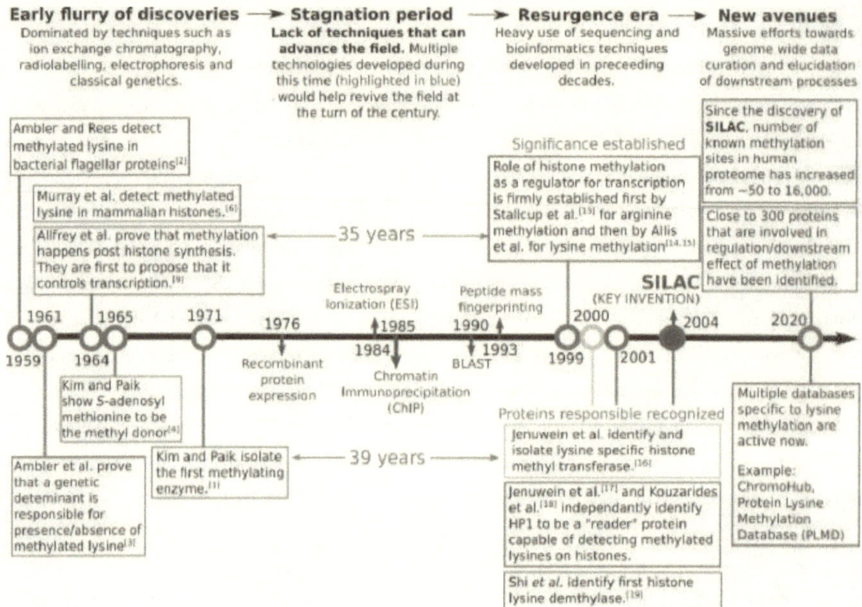

Figure 1.2. A timeline of important discoveries related to K/R methylation. Boxes indicated important milestones while blue text refers to invention of techniques that would later drive the field forward.[1]

Despite early exciting discoveries, the study of *N*-methylation of lysine and arginine was a dormant field for much of latter half of 20th century (Figure 1.2). The prime reason for this was lack of suitable techniques that could unambiguously detect *N*-methylated amino acids without destroying the peptide backbone. Research, at this time, was heavily driven by radioisotope labelling, classical genetics and chromatographic techniques. For example, Ambler and Rees detected *N*-methylated lysine by hydrolyzing flagellar proteins of a bacteria and separating individual amino acids using ion exchange and 2D paper chromatography.[2] They found a small shoulder on a lysine peak which, upon elemental analysis, proved to be *N*-methyllysine. Murray et al. used similar techniques to detect *N*-methyllysine in calf thymus histone.[6] Only one from the three possible *N*-methyllysines (Figure 1.1) is mentioned in these early papers likely due to coelution of higher methylated lysines with non-methylated lysine in standard ion-exchange chromatography procedures. Detection of di- and tri-methyllysines came almost a decade later.[7,8] The other big issue was ascribing functions to these methylation events. Clever molecular biology experiments did prove that methylation was an enzyme dependent and controlled process, not just a lysine substitute, but their significance was not properly understood.[4,9] Because the standard protocols for amino acid analysis at that time involved hydrolysis of peptide backbone, the site-specific knowledge about methylated lysines and their role was never uncovered during this era.

The field became stagnant since the late 1970s and would have to wait till the early 2000s for its revival (Figure 1.2). The complex nature of *N*-methylation needed more sensitive and less destructive techniques to be invented in several subfields of biochemistry (which happened throughout the last quarter of the 20th century). Techniques like recombinant protein expression made it possible for scientists to obtain proteins easily in preparative scale, and study their properties.[10] The invention of electrospray ionization (ESI) mass spectrometry and peptide mass fingerprinting (a decade apart) removed the need to completely hydrolyze the protein to individual amino acids and allowed determination of the precise positions of methylated lysine.[11] Chromatin immunoprecipitation (ChIP) and related techniques helped determine the heterogeneity of methylated lysine on the chromatin and provide insight towards chromatin regulation.[12] New bioinformatics tools (like BLAST) provided a way to compare the now growing database of genome and proteome sequences and, for the first time, allowed the researcher to predict the function of proteins without isolating them by comparing them to known enzymes.

The advantage provided from these techniques is obvious in the next era of study of K/R methylation. First, proper roles for K/R methylated protein were determined by various researchers in processes like intracellular transport[13] and transcription regulation[14,15] (Allfrey's hypothesis[9] from 1964). This research was heavily dependent on new techniques like ChIP, two-hybrid screens and bioinformatics. Second, multiple enzymes responsible for K/R methylation were identified and isolated. Once again, growing data on protein sequences and modern bioinformatics practices made it possible to associate functions to otherwise uncharacterized proteins, isolate them, and then test the hypothesis. Methylating enzymes (called writer proteins) were first to be identified, while proteins that detect methylated lysine and start the downstream process (called reader proteins) were second.[16–18] Perhaps the most important was the discovery of the first demethylase enzyme (later called eraser proteins) by Shi et al. in 2004.[19] The identification of an eraser protein proved a long-standing hypothesis that *N*-methylation is a dynamic event and opened the door for new research avenues. All these ground-breaking results were published within a 5-year period (Figure 1.2) and completely turned a 40-year-old doctrine upside down.

The study of *N*-methylation of K/R residues has since become a focal point for studies on chromatin structure[20], gene regulation[21,22], developmental biology[23,24], prognosis of diseases[25] and as a potential target for drugs[26,27]. With the discovery of SILAC, a mass spectrometry dependent semi quantitative technique for detection of protein and PTMs, the identity and location of methylated K/R residues can now be rapidly determined.[28,29] It is now understood that writer and eraser proteins alter the methylation state of these K/R residues depending on presence/absence of a chemical stimulus. Reader proteins are responsible for detecting these changes and triggering downstream processes by recruiting/activating appropriate effector proteins which can activate or repress gene products at both transcriptional and translational stages.[22,30] This allows the cell to respond to changes in its environment in a controlled and careful manner. The number of known K/R residues that are methylated *in vivo* has increased from 50 to >16,000 in last two decades.[1] However, only few of these residues are well studied, once again, partially due to their ease of detection and established phenotype and physiological roles.

1.2.   Roles of *N*-methylated lysine/arginine inside cells

1.2.1. N-methylation of histone

The first report of methylated lysine in higher organisms came from the work of Kenneth Murray in mammalian histone sample in 1964.[6] Since then, lysine and arginine methylation has been most studied in the context of histones.[30] These are octameric proteins that have K/R-rich unstructured tails on each subunit that are frequently subjected to *N*-methylation (Figure 1.3).[31] The prevailing theory about histones suggests that the abundance of K/R residues gives histones a positive charge at cellular pH and helps in binding to negatively charged DNA.[32] The Histone-DNA complex is called a nucleosome and is the chief component of chromatin.[33] A nucleosome is made of DNA wrapped around a core of 8 histones — two each of four different histone subtypes (Figure 1.3).[34] The tail and core of each histone subunit harbors sites for multiple PTMs (13 different types) including *N*-methylation of K/R (>40 different sites).[35,36] More new sites are being discovered every year. Combined with the fact that there are millions of nucleosomes inside the nucleus, likely (semi)independently modified, the complexity of chromatin becomes enormous.

The degree of methylation and its location on histone tails serves as an important signal for various transcriptional factors to regulate gene transcription.[21,22,37] This crucial result was underappreciated for a long time in the study of K/R-methylation because the enzymes involved were difficult to study. Some early work had linked K/R methylation to RNA synthesis and established importance of SAM for methylation inside cells.[4,9] But it was only after the discovery of the first demethylase enzyme that the multifaceted role of *N*-methylation in gene regulation started emerging.[38] Before this discovery, it was thought that only the synthesis of new histones and degradation of old histones was responsible for the removal of *N*-methylated side chains.[39] This was due to the fact that the half-life of methylated K/R were found to be almost same as histones. The discovery of demethylases added dynamics into the field.[40] Genome wide studies on the distribution of methylation marks have suggested the intricate involvement of

K/R methylation in regulation of transcription levels.[41,42] In general, N-methylation increases the positive charge on histones and increases the strength of DNA-histone complex while N-acetylation does the opposite by neutralizing charge. Patterns of N-methylation and acetylation combined with other PTMs lead to remodelling of chromatin where DNA is either exposed or condensed.[20] A tightly condensed DNA-histone complex represses transcription because the protein assembly required to initiate the process cannot dock properly. Loosening of the nucleosome gives access to docking sites and results in increased transcription.

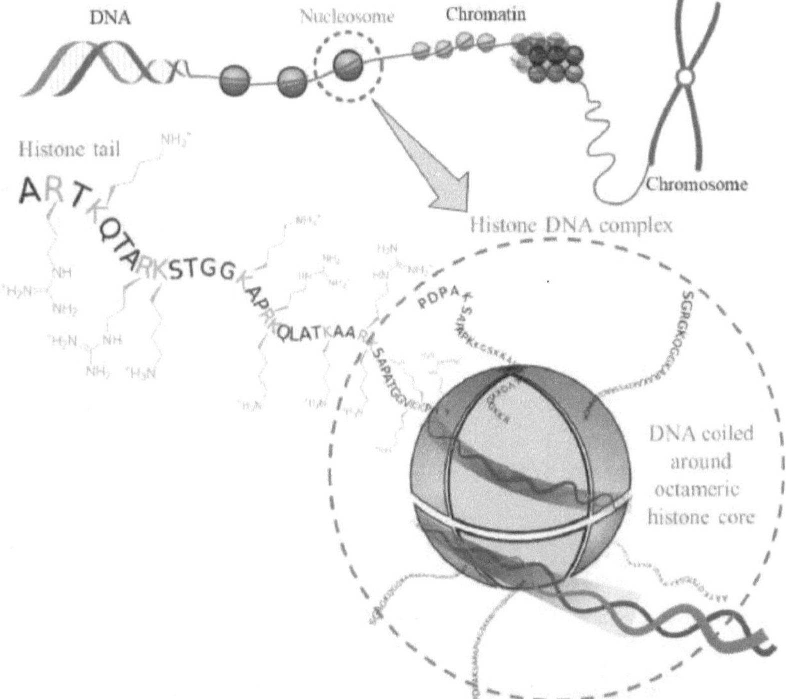

Figure 1.3. A schematic representation of chromatin structure and hierarchy found *in vivo*. DNA is coiled on an octameric histone protein, which then further coils into complex structures that eventually condense into a chromosome. The key feature of DNA-histone complexes, called the nucleosome, is the unstructured tail that protrudes from the complex. These tails can be methylated at multiple lysine and arginine residues (highlighted in green) giving rise to a staggering number of combinations.

Methylation patterns on histone tails are so prominent that they have now been proposed as a determinant in cell cycle[43], cell fate[44] and differentiation[38], and are now dubbed in the literature as the "histone code".[45,46] Distinct methylation marks on histone tails, formed and controlled due to reversible (multi)methylation of multiple K/R residues, attract different transcription factors and lead to distinct phenotypes.[47–49]. For example, methylations of the fourth lysine of histone 3 (H3K4), H3K36 and H3K79 are correlated with increase in transcription, while H3K9, H3K37 and H4K20 methylations do the opposite.[50] Monomethylation of H3K4 (H3K4me1) is closely correlated with enhancer elements on a gene while H3K4me3 is usually found closer to promoter regions.[51] H3K4me2, the least studied form of H3K4 methylation, is now considered to be involved in transcriptional memory.[52]

1.2.2. *N-methylation of non-histone proteins*

It is increasingly clear that methylation also occurs on non-histone proteins.[30,53–55] In fact, multiple methyltransferase and demethylase enzymes originally thought to act only on histones have been shown to methylate non-histone proteins at >4000 different K/R residues.[56] For example, SETD7 is an enzyme known to monomethylate H3K4. It is now known to methylated up to 30 non-histone proteins including kinases, transcription initiators and membrane proteins.[56–58] G9a, another methyltransferase, is known to methylate 17 non-histone proteins, including itself, in addition to methylating H3K9.[56,59] The diversity of functions of these proteins prove that *N*-methylation is not restricted to any one domain and that it plays important roles in cell signaling cascades in every compartment of cellular machinery. In addition, crosstalk between different PTMs and methylation in cytosolic proteins has been shown to control protein fate and activity.[60–62] These examples make it clear that the role of *N*-methylation has only begun to be understood. The ever-growing list of methylation sites, promiscuity of methyltransferase enzymes and significance of their substrate proteins suggests that they are central to multiple biochemical pathways. While the bulk of this work is focused on histone methylation, a choice made because of abundance of literature on the topic, the goal, methods and significance of each Chapter can be easily applied to non-histone methylation sites.

1.2.3. *Studies on methylated amino acids can help determine the prognosis of multiple diseases.*

With the ever-increasing list of proteins that contain *N*-methylation sites, it should be no surprise that mistakes in installation/removal of methyl groups from lysine/arginine side chains have been correlated to various diseases. An online database PTMD (PTMs that are associated with human Diseases) list 42 proteins where aberrant *N*-(de)methylation of lysine/arginine has been associated with different diseases in humans.[63] p53, a tumor suppressor protein also known as "the guardian of the genome", is a good example to illustrate this point. Its function has been shown to be affected by lysine methylation at multiple sites, both agonistically (K372) and antagonistically (K382 and K373).[64] Because of p53's well established role in multiple cancers, it is easy to see how a wrong methylation event can lead to irreparable DNA damage. In fact, p53 is found to be compromised in human embryonal carcinoma cell lines due to a combined monomethylation at K370 and K382.[65] In the absence of p53-directed DNA repair processes, the cancer cells proliferate quickly and dangerously. Histone methylation also exerts a tight control over transcription and as such its mismanagement has been implicated in multiple diseases including cancers, neurodegenerative diseases like Alzheimer's disease, neurodevelopmental disorders and Immune-related diseases like HIV.[66–69]

A big motivation to associate K/R methylation to a disease phenotype is the idea that they can be used in prognosis and most importantly to measure individual response to treatment. Cancer recurrence is, once again, the best example to illustrate this. Cancer is a collection of diseases and response to treatments varies from person to person. Global changes of different methylation marks have been correlated with survival rates and recurrence in difference cancers.[70–73] Detection of changes in histone methylation marks parallel to treatment has been suggested to predict a patient's prognosis.[25,74] This is easier said than done because of difficulties associated with confident detection of specific methylation marks. Both sides are equally affected. On the laboratory side, the discovery and association of methylation marks with a phenotype has been slow. The usual workflow requires multiple independent experiments which are all individually time consuming. On the clinical side, applying the existing knowledge in diagnosis is non-existent yet. Improvement in the current situation requires new methodologies to be developed that specifically target K/R methylation.

1.3.   The Basics of Methyl Proteomics

1.3.1.   Two very different approaches

Proteomics is a broad field that encompasses the discovery and cataloguing of all proteins as well as studying their synthesis, role and fate in the functioning of an organism.[75] The study of PTMs also falls under the realm of proteomics. The field is ever-growing, complex and requires an interdisciplinary approach that combines analytical chemistry, biochemistry, bioinformatics and chemical biology. It is estimated that there are little over 20,000 unmodified proteins in our body. To illustrate the complexity PTMs bring to the field, let's consider the fact that each protein can be modified by multiple PTMs. This not only includes *N*-methylation but >300 other known PTMs.[76] Each combination of PTMs on one protein has a different role in the cellular machinery. Even if we assume that only 3 on/off PTM sites exist on every protein, we can have 8 different PTM forms of one protein that can exist inside the cell, increasing the number from 20,000 to 160,000. We know of proteins such as histones that have >30 PTM sites on them, so the real number is much bigger — probably in the hundreds of thousands.[77]

The field of proteomics has developed in to two broadly defined subfields: a) discovery-based proteomics, and b) targeted proteomics.

    a. Discovery-based proteomics aims to find new and unknown PTM sites on a given protein, inside an organelle, cell, or in the entire proteome.[78] The end goal of discovery-based proteomics is to catalog, usually in curated electronic databases (UniProt, PIR etc.), all possible variants of any protein. The field is heavily dependent on techniques like Peptide Mass Fingerprinting (PMF, Section 1.3.2.)

    b. Targeted PTM proteomics is the study of already known PTM sites and modifications. The aim is to elucidate the mechanism of installation and removal of the PTM and all its downstream effects on gene regulation/protein-protein interaction.[79] Samples for targeted proteomics are often focused on a small set of purified proteins. Selective reaction monitoring (SRM) or selective ion monitoring (SIM) are the common techniques used to confidently track individual biomolecules.[80]

Despite fundamental differences in their approach, a common theme to both fields is the heavy use of mass spectrometry (MS). PMF, SIM and SRM are all MS-based techniques. The finding in this book has direct significance towards discovery-based proteomics and as such PMF will be discussed in more details below. For a more detailed discussion on Targeted proteomics and related MS-based techniques, readers are referred to Reference 79.

1.3.2. Mass spectrometry and its use in proteomics

Before MS-based methods became popular, peptide/protein sequencing was done using Edman degradation.[81] The procedure involved selectively cleaving one amino acid from the protein (starting from the N-terminus), identifying it using chromatography or electrophoresis and then repeating until the entire sequence is deciphered. Despite mass spectrometers existing since the early 1920s, their use in studying biopolymers like proteins only became possible after Electrospray Ionization-MS (ESI-MS) was introduced as a soft ionization method in 1980s.[82]

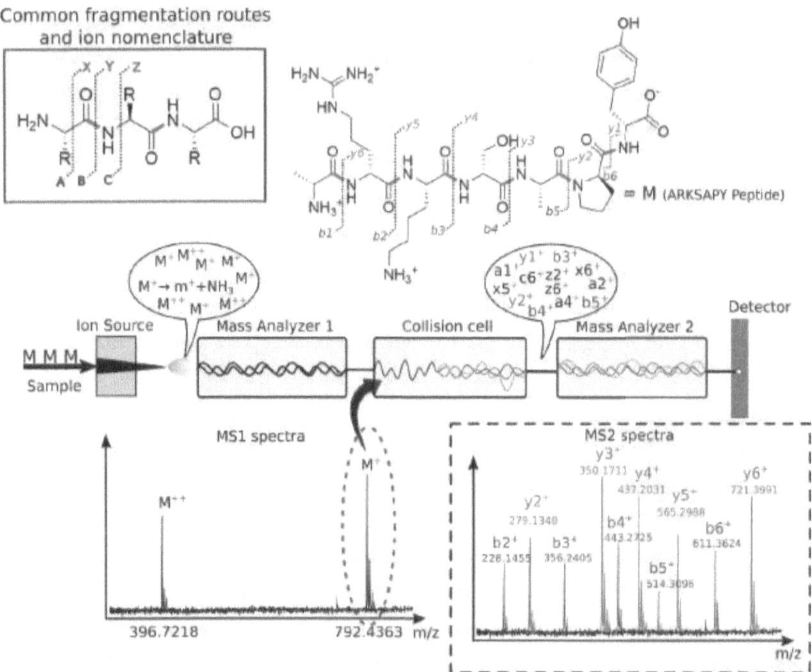

Figure 1.4. Workflow for Tandem MS (or MS^n) based peptide sequencing (*de novo* sequencing). Common fragmentation routes and ion nomenclature are shown in blue box. An example of fragmentation pattern for the peptide ARKSAPY is shown in dotted box. For clarity, only b and y ions, generated due to fragmentation of bold red bonds, are shown.

Mass spectrometry has since become synonymous with proteomics, and trends in proteomics have followed the advancements in MS very closely.[83] It has several advantages over the now obsolete Edman degradation. It can detect small quantities of analytes (femto- to picomoles) using only a tiny amount of sample (micro- to nanoliters).[84,85] Modern high-resolution MS can also differentiate between two very different analytes with very similar masses (For example, Kacetyl and Kme3, Δm/z = 0.04 amu). Most importantly, amide bonds of a peptide can be fragmented very predictably inside a MS and the daughter ions measured again in a subsequent experiment (an MS/MS or MS^n or Tandem MS experiment, Figure 1.4.).[86] In a typical experiment, a peptide is first ionized (using ESI) and the resulting ions are scanned (MS1). Suitable ions (high intensity, low charge) are then isolated and diverted into a collision cell where they are bombarded with neutral atoms like helium or argon. This leads to fragmentation of the parent ion into daughter ions (Figure 1.4). The most common fragmentation routes of peptides are highlighted in the blue box in Figure 1.4. A second mass analyzer scans the daughter ions and generates the fragmentation spectra (MS2). Each fragment helps deduce position of one amino acid in the parent peptide.[87] For example, in Figure 1.4, $y6^+$ is 71.03 amu less than M^+ which can only be due to loss of an alanine. Subsequentially, $y5^+$ is 227.1375 amu less than M^+ and must contain the fragment lost in $y6^+$ too. Therefore, it must be due to an AR fragment breaking from the M^+ and so the second amino acid in the sequence is arginine. Many algorithms have been developed to automate this analysis (*de novo* peptide sequencing).[88] If two peptides are made up of the same amino acids but have different sequences, they will have same molecular weight but a different fragmentation pattern. Any PTM (or their combination) and its exact position in the sequence can be identified based on the change in molecular weight of fragments.[89]

Peptide sequencing using tandem MS (Figure 1.4) is best suited for sequencing peptides between 8-15 amino acids in length.[90] While intact protein fragmentation approaches do exist (a top-down approach), they are only utilized on pure proteins or simple mixtures due to instrumental limitations.[91] The most common method in current science is a bottom-up approach which involves initial proteolysis of protein or mixture (like a cell lysate) which is then injected into an UPLC that feeds into a MS (Figure 1.5). This allows for in-line fractionation of peptides before MS. Each peak from the UPLC is then subjected to tandem MS as shown above. Another key difference lies in the analysis of fragmentation spectra (MS2). *De novo* sequencing is a successful strategy but becomes increasingly slow as the complexity of sample increases. Modern methods use fast database search algorithms to match MS2 with candidate peptides.[92] These databases are generated by *in silico* proteolysis of all possible proteins that can produced by an organism's genome. They also contain information about all possible fragment ions that can be generated by these peptides in tandem MS. Peptides are scored based on their similarity to experimental MS2 and the best candidate is identified. Finally, the identification of a given protein in the pre-proteolysis sample is inferred by tracing the origin of peptide in the entire proteome (Peptide Mass Fingerprinting, PMF).[93] Confident identification requires verification of multiple characteristic peptides that map back to one protein. Quantitative versions of PMF using isotope labelling have been developed and are now a common protocol in most proteomics labs.[94]

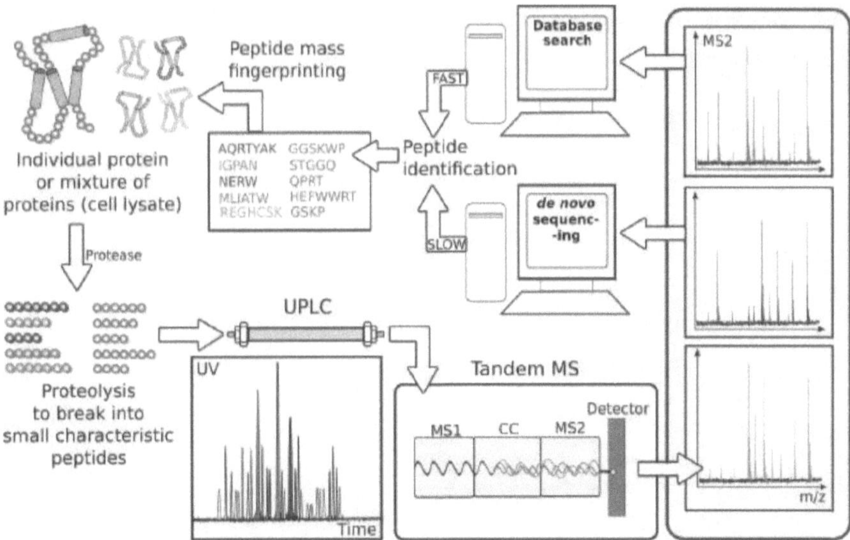

Figure 1.5. Workflow for peptide mass fingerprinting (PMF), a routine method used in proteomics for protein/PTM profiling. UPLC – Ultra High-Pressure Liquid Chromatography.

1.3.3. Needle in a haystack

Even though PMF requires a minuscule sample, the methodology still suffers from averaging issues, noisy backgrounds, and signal suppression and interference (Figure 1.6). The prime reason for this is the fact that proteins are expressed in dissimilar proportions inside a cell. The dynamic range of proteins measured in a cellular proteome can be as a high as 10 orders of magnitude.[95–97] Inside cells, ribosomal, cytoskeletal, and membrane proteins are present in high copy number while proteins like transcription factors, enzymes etc., are present in much lower amounts.[98] Large dynamic ranges are also observed for proteins that are secreted outside from the cells (secretome). For example, the highest blood plasma concentration of interleukin 6 measured so far is 11.5 pg/mL[99] which pales in comparison to albumin, which can be as high as 35 mg/mL[100]. Consequently, the most highly studied PTMs tend to be ones that are most abundant (phosphorylation, ubiquitylation etc.), or present on highly abundant proteins.[101] N-methylation of histone sidechains is no different. As discussed in section 1.2., most N-methylation sites are studied in the context of histones which are present in high copy numbers inside the nucleus, while the appreciation for non-histone K/R methylation is only recent. Many of the catalogues of PTMs only have a handful of publications on them because: a) they are present is such low concentration that it is difficult to study them accurately, b) their role in cellular machinery is not known, and c) there are no good techniques that target them specifically.

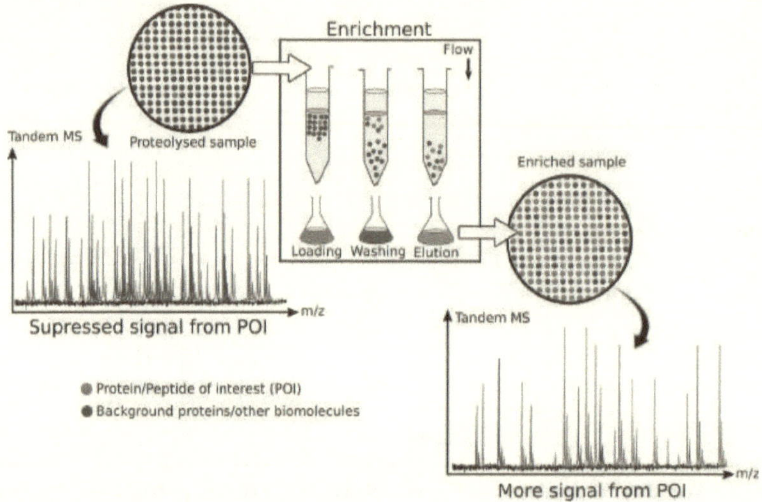

Figure 1.6. A typical enrichment step (shown in black box) used in proteomics improves overall signal from Protein (or peptide) of Interest (POI).

1.3.4. Common methods for enrichment

N-methylated peptides makeup a small fraction of total peptides in a typical proteolyzed sample. Since there are likely thousands of different methylated peptides, the individual copy number is even lower. Mass spectrometry on this sample would rarely be able to identify methylated peptides with confidence because most of the signal would be from non-methylated peptides. The bulk of the non-methylated fraction must thus be removed before UPLC-MS/MS to better increase the chances of detection (Figure 1.6). This problem is common to all PTM proteomics and the steps to alleviate this is called enrichment.[102–104]

Many enrichment techniques are set up as some form of chromatography.[105] Installation of a PTM brings a change in physiochemical property which must be exploited to separate PTM peptides from unmodified peptides. Ion exchange chromatography,[106] hydrophilic interaction chromatography,[107] size-exclusion[108] and electrophoresis[109] have all been used to this effect.[104,110] The use of immobilized proteins as a stationary phase to enrich their respective substrates (affinity enrichment) has also been successful.[103,110] Proteins have evolved to distinguish their substrate with high specificity in complex cellular environments. Immobilized proteins can trap molecules that bear their substrate at any position. In a typical setup, proteins are immobilized on a solid support (yellow phase in Figure 1.6) which is then loaded with sample. The immobilized proteins bind their target substrate (the analyte) stronger than other molecules. A washing step removes all weakly bound material. Retained analytes can then be eluted by either denaturing the protein or by outcompeting the analyte (Figure 1.6). This fraction would contain increased levels of target analyte which improves the signal strength. Antibody-based enrichment (Immunoaffinity purification (IAP), Figure 1.7) is an extension to this idea and is widely used technique.[110–115] Their prevalence is due to the fact the new antibodies specific for different antigens can be easily raised inside the immune system of animals[116] or cultured cell lines.[117] This opens opportunities to enrich targets that are not a natural substrate of any known protein. Consequently, a lot of research and development has gone into production, adaptation and commercialization of IAP kits and researchers have plenty of options to choose from.[118,119]

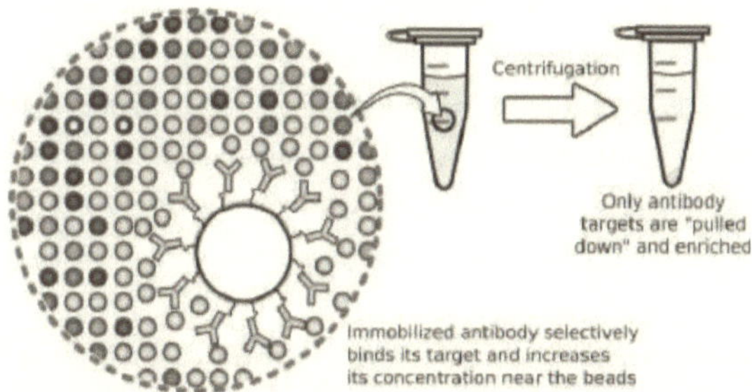

Figure 1.7. Scheme for a pull-down assay or Immunoaffinity Purification (IAP). Immobilized antibodies (Y shaped) on solid support (typically on beads or resin) are used as the stationary phase. Centrifugation drags molecules that bind to the antibody-bead conjugate to a pellet.

While antibodies use non-covalent forces to identify and distinguish their guests, chemical reactivity can also be exploited for enrichment. The perfect example of this methodology can be found in phosphoproteomics, which relies on techniques like immobilized metal affinity chromatography (IMAC) and metal oxide affinity chromatography (MOAC).[105] Both techniques exploit the ligand-like properties of phosphate. TiO_2 or ZnO_2, typically used in MOAC, and resins containing chelated Fe^{3+}, Ga^{3+} or Zn^{2+}, used in IMAC, have one thing in common – they can form strong dative bonds between phosphate (ligand) and their metal center (Figure 1.8).[120] This is used to trap all the phosphorylated peptides on a solid substrate while non-phosphorylated peptides flow through. Phosphorylated peptides can then be eluted by breaking the bonds between phosphate and metal either by simply changing the pH or adding a competing agent like dihydroxy-benzoic acid.[121] These techniques are so efficient and reliable for phosphopeptide enrichment that they are now a common practice in almost every proteomics facility worldwide.

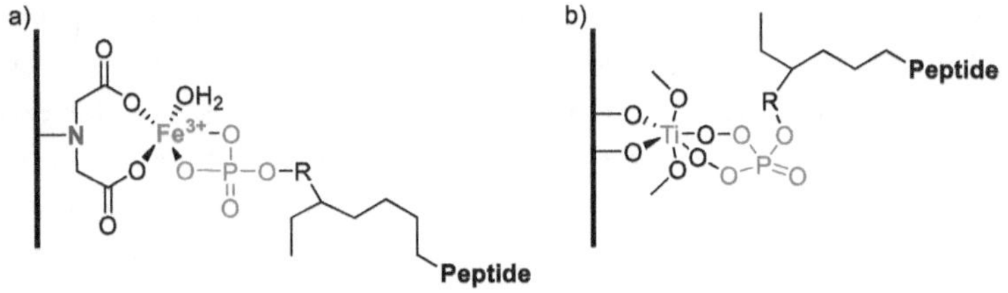

Figure 1.8. Mechanism for entrapment of phosphopeptide using a) IMAC and b) MOAC. Bold line represents solid support on which the metal organic complex is anchored.

1.3.5. *Problems and pitfalls of enrichment protocols in N-methyl proteomics*

Existing strategies for enrichment of *N*-methylated peptides are inadequate. *N*-methylation is the smallest known modification, and it doesn't significantly alter the chemical reactivity of the lysine or arginine side chain. The pK_a of the lysine or arginine side chain also doesn't change significantly upon methylation. All methylated states except trimethyllysine are H-bond donors and nucleophilic (including unmethylated K and R). Unlike phosphorylation, *N*-methylation does not confer any special reactivity to the side chain. This makes it difficult to develop an enrichment technique analogous to IMAC and MOAC.

Antibodies against methylated lysine have been produced[122–124] but major concerns have been raised recently over their performance.[125–128] In particular, Rothbart et al. screened more than 100 commercially available antibodies against peptides with varying sequences using fluorescence peptide arrays and found them to be inconsistent and to have high off-target binding.[126] They catalogued their results in an interactive web platform for easy access by the community (www.histoneantibodies.com). Figure 1.9 shows an exemplary set of data for an anti-H3K9me3 antibody from this catalog. The green bar represents its response towards a target containing peptide (with no additional modification). Multiple off target hits that do not contain the targeted K9me3 sequence (grey bars) were detected by the antibody. More importantly, yellow and blue bars represent response from target peptide with additional modification at neighboring positions (like phosphorylation, acetylation etc.). Similar off-target binding is shown by almost every antibody in their catalog and some antibodies show stronger response for completely unmodified peptides than for the PTM peptides that they are supposed to be targeting.

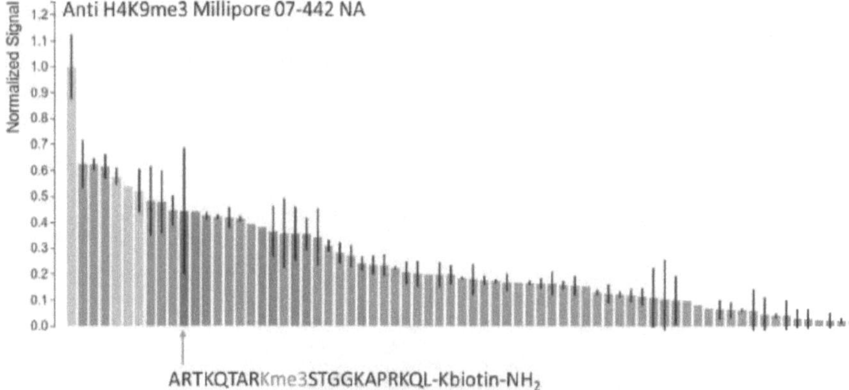

Figure 1.9. Performance of a commercial anti H3K9me3 antibody in a peptide array screen, taken from www.histoneantibodies.com.[126] **Green bar represents the response against target peptide of the antibody. Yellow and blue bars are K9me3 containing peptide with other modifications also present. Grey bars are false positives that do not contain K9me3 at all.**

The data from Figure 1.9 (and from in general) is worrying as it suggests that modifications on neighboring residues significantly affect antibody-PTM interaction. Fuchs et al. also reached a similar conclusion, independently, a few years prior.[127] From a molecular recognition perspective, such sensitive behavior is to be expected when raising antibodies against a small target like Kme3 (or any other methylated residue). Antibodies are big proteins and their binding surface is too large to efficiently sequester small targets like Kme3 without being affected by nearby residues. Figure 1.10 illustrates this problem in a crystal structure obtained recently by Hattori et al.[124] Their in-house developed anti-K9me3 antibody was co-crystalized with target peptide. The structure clearly shows Kme3 sidechain to be deeply embedded in the active site of antibody. But it also demonstrates engagement of peptide backbone and, more importantly, strong interaction of neighboring sidechains of lysine, threonine, alanine and serine (highlighted grey in Figure 1.10) with the binding surface. Any modification on these amino acid positions will strongly affect antibodies' performance. It is this inability to cleanly sequester Kme3 sidechains without engaging the neighboring groups that gives rise to the binding profile shown above (Figure 1.9). This antibody is highly specific to this target peptide, which the authors note, and should aptly be called anti-ARTKQTARKme3ST rather than "anti-H3K9me3."

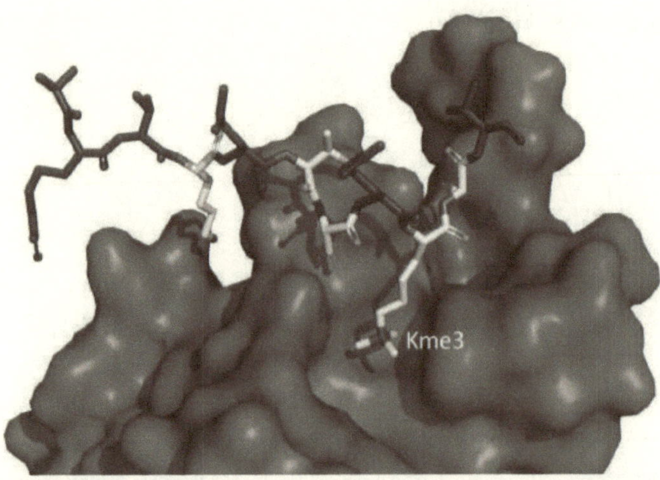

Figure 1.10. Crystal structure of an anti-H3K9me3 antibody with a target peptide (ARTKQTARKme3ST), PDB 4YHP.[124] The red area denotes the active site of Fab region of antibody while the peptide backbone is highlighted in blue. Deeply engaging side chains are coloured grey.

The inherent difficulties in raising sequence-independent, Kme3-selective (pan-Kme3) antibodies makes them less attractive as an enrichment agent. The use of currently available antibodies in such endeavors will result in a biased enrichment and fail to show true abundance and distribution of methylation marks. Multiple authors have raised concerns over the continued use of such antibodies in chromatin research and have urged the community to do a more vigilant characterization before committing to a commercial supplier.[129–132] Similar concerns over long-term storage, low lot-to-lot reproducibility, and unintended off-target binding of antibodies have also been raised by researchers outside the PTM proteomics community.[133,134] Still, global proteome analysis utilizing IAP protocols for methylated peptides have been reported with most authors using a combination of different antibodies to obtain greater sequence coverage.[54,135–140] In at least two such reports authors have explicitly commented on the undesirable sequence-specific behavior of the antibodies used.[54,135] Bremang et al. also noted that anti-methyllysine antibodies fare much worse that anti-methylarginine antibodies when it comes to reproducibility.[54]

1.3.6. *Novel approaches towards N-methyl peptide enrichment*

In the light of the drawbacks discussed in section 1.3.5., multiple groups have made attempts to advocate alternate strategies for *N*-methyl peptide enrichment.[141,142] These can be divided into: a) chromatography-based, b) affinity-based, and c) chemical proteomics.

1.3.6.1. *Chromatography-based methods*

Uhlmann et al. compared the prospect of three different chromatography techniques viz Hydrophilic Interaction Liquid Chromatography (HILIC), Strong Cation Exchange chromatography (SCX) and Isoelectric Focusing (IEF) for enrichment of methylated arginine (Rme) containing peptides.[143] Among these three, HILIC turned out to be an outright winner, identifying 215 methylated peptides compared to 39 by SCX and 55 by IEF from the same sample. Chenxi Jia and coworkers have also advocated the use of HILIC but identified possible interference from coeluting glycosylated peptide. Their modified workflow, named DOMAIN, suggests an additional deglycosylation step (using enzyme PNGase) post proteolysis as a solution.[144,145] Similary, Mingliang Ye and coworkers identified that the failure of SCX was due to coelution of a) long highly positively charge peptides and b) histidine containing peptides. Their two-pronged solution involved, first, proteolysis using multiple proteases simultaneously (Trypsin, Lys-C and Arg-C) to remove long highly charged peptides and, second, performing SCX at high pH (>9) at which only methylated peptides remain positively charged.[146,147] Curiously, this strategy worked well for methylarginine containing peptides but didn't enrich many methyllysine containing peptides.

1.3.6.2. *Affinity-based methods*

Fundamentally, affinity-based enrichment methods are like chromatography-based methods both in theory and instrumentation. The difference, however, lies in the fact that affinity-based methods partition analytes based on presence/absence of a chemical group rather than physico-chemical properties. This "affinity-based" partition is achieved by immobilizing hosts or agents (like antibodies) capable of binding targeted chemical group on a solid support. In the absence of pan-specific antibodies, Or Gozani and coworkers have put forward an alternate affinity purification strategy. They observed that 3xMBT domain of protein L3MBTL1 binds mono- and di-methyllysine with little substrate specificity.[148,149] Their protocol utilizes this broad specificity to enrich mono- and di-methyl peptide using a pull-down method analogous to IAP (Figure 1.7).[150,151] Hundreds of proteins were enriched using this strategy but only 26 methylation sites were confirmed. Their efforts were severely hampered by the fact the 3xMBT domain can only bind to proteins and not peptides. This forced Gozani and coworkers to enrich their sample before proteolysis, unlike a typical proteomics workflow (Figure 1.5). This allowed multiple non methylated protein, that form complexes with methylated proteins, to also be enriched and reduced the overall efficiency. Another similar demonstration by Liu et al. uses immobilized HP1β as the bait protein.[57] This, however, lacks the broad specificity like 3xMBT domain and can only be used to enrich substrates of HP1β.

1.3.6.3. *Chemical proteomics methods*

In another attempt to utilize SCX, Ning et al. cleverly eliminated the positive charge on unmethylated lysine and arginine by reacting them with orthophthalaldehyde and malondialdehyde, respectively, which they called a charge-suppression strategy (Figure 1.11).[152] The resulting mixture can be purified on SCX where methylated lysine containing peptides (which can't react with orthophthalaldehyde and are still positively charged) and methylated arginine containing peptides (which can't react with malondialdehyde and are still positively charged) are retained on the column. This strategy allowed for significant enrichment of both methyllysine and methylarginine containing peptides.

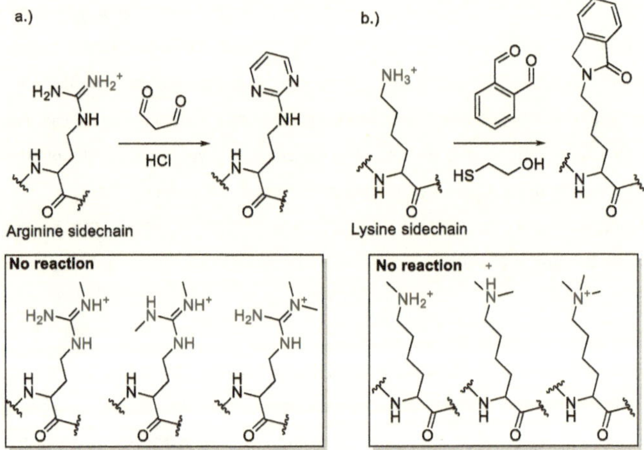

Figure 1.11. A charge suppression strategy by Ning et al. removes positive charge from unmethylated a.) arginine and b.) lysine and free *N*-terminus (not shown) of peptides. Methylarginines and methyllysines do not react and retain their charge.

A biorthogonal chemistry-based method has also been suggested. This strategy became possible after Luo and coworkers found that allyl-SAM, an analogue of SAM, is well tolerated by multiple wild-type protein methyltransferases (PMT).[153,154] Growing SAM-auxotrophic cells in presence pf allyl-SAM leads to *in vivo* labelling of methylation sites with allyl groups (Figure 1.12).[155] Unlike methyl group, allyl groups can partake in chemical reactions and the authors choose an oxidative Heck reaction to attach biotin fragments to labelled sites. A simple streptavidin based pull-down assay resulted in significant enrichment and identification of 167 protein methylation candidates.[155] This strategy labels *O*- and *S*-methylation in addition to *N*-methylation site and is much more tailored towards global methylation enrichment. However, swapping methyl group for allyl group is bound to miss some substrates that don't tolerate the allyl substitution, and also to bring some changes in protein-protein interactions. While the methylation sites detected are true, this strategy cannot be used to study a biochemical process whose levels depend on processing of methylated residue by proteins.

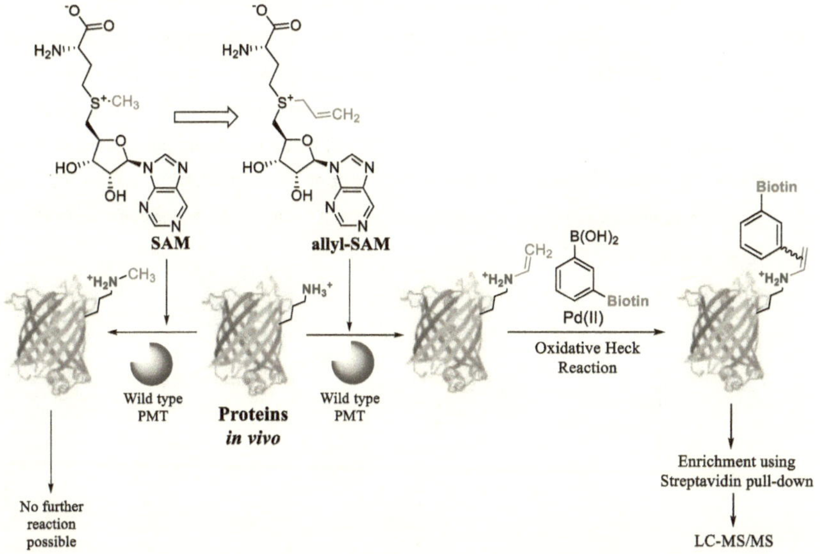

Figure 1.12. An *in vivo* labelling strategy for global discovery of methylated sites that utilizes allyl-SAM analogues. Allyl-labelled proteins can be further modified with biotin using biorthogonal chemistry and enriched using streptavidin pull-down.

1.3.6.4. *Shortcomings and workarounds*

Almost all methods discussed above, except for affinity enrichment, involve additional chemical/biochemical reaction of some sorts on the sample. While they are clever and elegant tricks, each extra chemical reaction will have its own bias. Additionally, no chemical reaction is hundred percent efficient and there will always be missed sites due to incomplete labelling. Affinity enrichments are better in this regard but are lacking is qualities that can readily integrate them into proteomics workflow without change (*vide supra*). A reagent that can enrichment post-proteolysis samples without any further chemical/biochemical transformation would be ideal.

1.4. Molecular recognition of methyllysines and methylarginines by synthetic host molecules

One key feature that a successful enrichment needs to have is pan-specificity. Pan-specificity is the ability to bind one PTM, without depending on the sequence of peptide (or, in general case, any other biopolymer) on which the target PTM is found (Figure 1.13). In molecular recognition terms, the neighboring groups should have minimal effect on binding strength. Without pan-specificity, an enrichment reagent brings significant selection bias to the experiment. Sequences that bind strongly to the enrichment reagent will appear to be overexpressed while low binding sequences would go undetected. True pan-specificity is difficult to achieve. IMAC-based phosphoproteomics enrichment reagents are likely the closest to achieving complete pan-specificity. As discussed in section 1.3.5, pan-specific antibodies for *N*-methyllysine or arginine are not available.

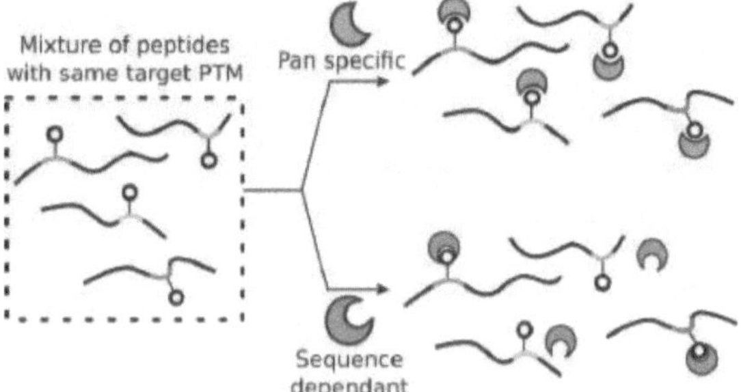

Figure 1.13. Difference between pan-specific and sequence dependent host molecules.

The underlying idea behind this book work is that small host molecules are better suited than proteins to be pan-specific binders for methylated peptides. This motivation stems from the known reason behind the poor performance of supposedly pan-specific antibodies. Their large binding surface engages amino acids neighboring the target sidechain (Figure 1.10) which leads to the binding and enrichment of peptides that don't have the desired PTM, but bind because of favorable interaction with the surrounding sequence (false positives), and loss of peptides that do have the desired PTM, but a disfavorable surrounding sequence suppresses their binding with the agent (false negatives). A small molecular host, designed to selectively bind one of the *N*-methylated sidechains, can be envisioned to have better size compatibility with its guest molecules. Because their binding surface is of comparable dimensions to methylated sidechains, it should intrinsically have limited contact with the surrounding sequence. This should, in principle, enable us to develop host molecules that bind *N*-methylated residues on different peptide sequences with similar strength—the very definition of pan-specificity. Additionally, small hosts that are synthesized and purified to homogeneity will have better batch-to-batch reproducibility, ease of scale up, and potential to tolerate harsh conditions without degrading.

Small molecular hosts for ammonium ions are abundant in the literature.[156] Thousands of molecules that range from macrocycles[157] to clip-shaped hosts[158] to metal complexes[159] have been reported and are too numerous to review here. In aqueous solution, the binding is largely dictated by hydrophobic effect because quaternary ammonium ions are hydrophobic in nature.[160] Significant contributions from cation-π interactions also promote binding. Methyl reader proteins found inside cells employ this strategy perfectly.[161] Their active site consists of aromatic amino acids, tryptophan, phenylalanine and tyrosine, preorganized to form an "aromatic cage" (Figure 1.14).[162] Side chains of methylated K/R are sequestered inside this hydrophobic cage which, at the same time, also engages in cation-pi interaction with the methylated side chain.[163] Taking a lesson from this, multiple aromatic cage containing molecules have been published and shown to discriminate between methylated lysines. This subset of the literature will be reviewed below.

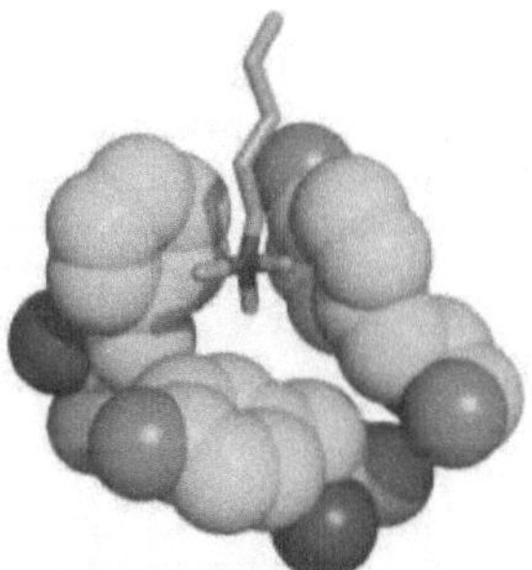

Figure 1.14. Recognition site of an *N*-methyl binding PHD domain binds Kme3 using an aromatic cage made up of two tyrosine and one phenyl alanine (PDB 3ASK).[164]

1.4.1. Acyclic hosts

All the hosts described below can be classified as macrocycles. Before we begin our discussion it is useful to ask if that is a strict criterion to design a successful host? Do we need macrocycles? The answer is yes and the concept behind it is called preorganization.[165] According to this, complexation between a host and guest would be stronger when the number of conformations that a host must sample in order to receive the guest is lower. Macrocycles with their restricted rotation are better suited (or otherwise called, "better preorganized") to receive the guest molecules. An example of an acyclic host (and its limitation) can be seen in works of Whiting et al.[166,167] Their series of trisindole based trianionic hosts (Figure 1.15) bind Kme3 very weakly ($K_d \sim$ 4000 μM). Only a highly hydrophobic guest like tetrabutyl ammonium chloride binds with moderate affinity ($K_d \sim$ 130 μM). The structure is highly flexible and require six different C-C bonds to rotate to a narrow range of angles to makes favorable contact with the guest molecule. Entropic penalties rule out any enthalpic compensation and the complex is weak.

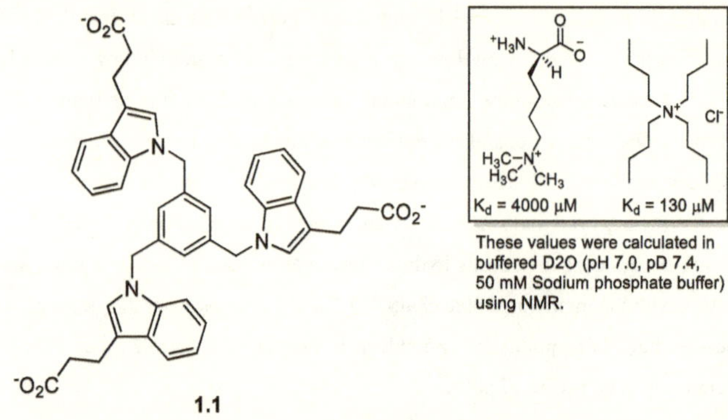

Figure 1.15. Whiting's trisindole receptor has little preorganization and shows weak binding to Kme3.[166] More hydrophobic guests (e.g. tetrabutyl ammonium) are better suited. K_d values reported by the authors for complex between 1.1 and these guests in shown in the box.

1.4.2. Calix[4]arenes

Calix[4]arenes (**1.2**) are phenolic metacyclophanes that are synthetically versatile.[168] Hydrogen bonding among the lower-rim phenols makes the "cone" conformation, where phenols face each other like walls of a room, more stable than all other conformations.[169] This generates a hydrophobic cavity, akin to "aromatic cage" of reader proteins, within the structure that has been used to sequester a plethora of guest molecules.[170–172] *p*-tetrasulfonatocalix[4]arene (**PSC**) is a water-soluble analogue which was the first molecule to be shown to bind trimethyllysine selectively over free lysine (70-fold).[173] Carboxycalixarene (**1.3**), another water-soluble counterpart, also shows affinity for Kme2/3 over free K (>10 fold).[174]

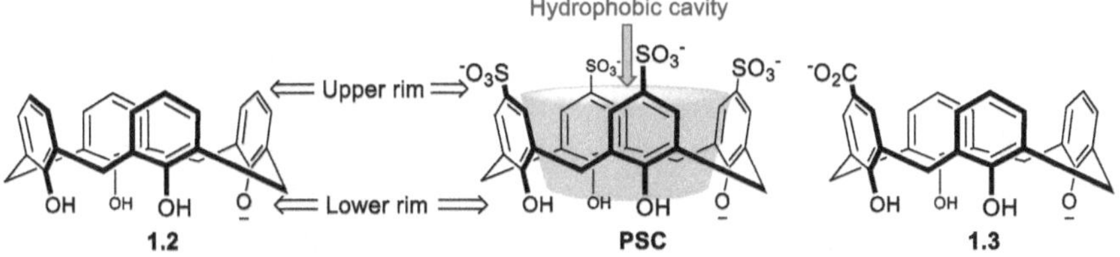

Figure 1.16. Parent calix[4]arene (1.2) and two simple water soluble counterparts (PSC and 1.3). The intramolecular hydrogen bond at the lower rim is responsible for generating the hydrophobic cavity (shown in blue for PSC) in all calix[4]arene derivatives.

Table 1.1. K_d for select *p*-sulfonatocalixarenes derivatives, against *N*-methylated guests, published in literature.

Compound	K_d for unmethylated guest (μM)		K_d for methylated guest (μM)		Selectivity	Method[c]	Ref
PSC	K	1900	Kme3	27	71	NMR	173
	ARTKQTAY	46	ARTKme3QTAY	5.0	9	ITC	176
	TARKSTGY	101	TARKme3STGY	7.2	14	ITC	176
	AARKSAPY	220	AARKme3SAPY	5.4	41	ITC	176
	GGVKKPHY	128	GGVKme3KPHY	9.1	14	ITC	176
1.3	K	>500	Kme3	60	>8	NMR	174
			ARKme3STGGK	50		NMR	174
			PATGGVKme3KPHRY	65		NMR	174
A	K	2300	Kme3	260	9	NMR	175
	AARKSAPY	13.3	AARKme3SAPY	0.70	19	Fl	178
AP0	K	2400	Kme3	16	150	NMR	175
	AARKSAPY	19	AARKme3SAPY	0.75	25	Fl	178
1.4	K	4800	Kme3	480	10	NMR	175
	AARKSAPY	11.3	AARKme3SAPY	0.88	13	Fl	178
1.5	AARKSAPY	>500	AARKme3SAPY		-	ITC	
				>500			176
1.6	AARKSAPY	>500	AARKme3SAPY	85	>6	ITC	176
1.7	H4[a]	17.8	H3K27me3[b]	8.9	-	SPR	179
1.8	H4	1.8	H3K27me3	0.39	-	SPR	179
1.9	H4	0.73	H3K27me3	0.39	-	SRP	179

[a] H-SGRGKGGKGLGKGGAKRHRKGGK(BIOTIN)-NH₂. [b] AcRKSTGGKAPRKQLA--TKAARKme3GGK(biotin)-NH₂. Reference 175, 176 and 178 contain binding data for multiple other substituted calix[4]arenes which are omitted for brevity.[c] NMR -Nuclear Magnetic Resonance, ITC -Iso Thermal Calorimetry, Fl -Fluorescence, SPR -Surface Plasmon Resonance

The synthetic versatility of calixarenes allowed Daze et al. to explore effects of upper[175] and lower[176] rim modifications of **PSC** separately. Both substitution sites proved to be highly impactful at changing methyllysine binding but only upper rim substitution with aromatic groups provided the desired boost to binding strength. Specifically, addition of an upper-rim phenyl group increased the selectivity for trimethyllysine over unmethylated lysine (as the free amino acids) from 70-fold to 150-fold (Table 1.1). NMR studies support total inclusion of N-methyls and the neighboring ε-CH$_2$ group inside the cavity. The extended aromatic arm likely provides favorable CH-π interactions to the guest. This is supported by the NMR shifts in lysine δ and γ protons, and with the observation that electron withdrawing groups on the phenyl substituent decrease binding strength. The picture is completely different for free lysine binding to **PSC** and its derivatives. Free lysine has a more side-on geometry of interaction with **PSC**.[173,177] The cationic nitrogen on the side chain of free lysine interacts with the sulfonates near the upper rim while δ, γ and ε CH$_2$ sag deeper into the cavity (Figure 1.17.b).

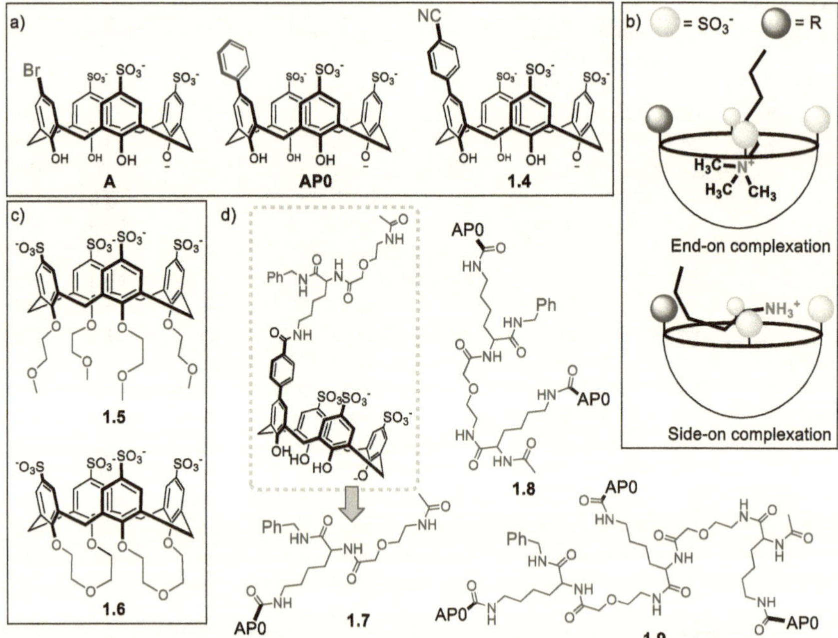

Figure 1.17.a) Selected upper rim functionalized calix[4]arenes from the work of Daze et al.[175,176] **b) NMR suggests that upper rim arylated calix[4]arenes bind K and Kme3 using two very different modes of binding. More hydrophobic Kme3 sinks deeper into the cavity while K is more hydrophilic and floats near upper rim. c) Lower rim functionalized calix[4]arenes by Daze et al. show loss of affinity towards methylated hosts. d) Calix[4]arene based multivalent host molecules by Kimura et al. K$_d$ for all these hosts in summarized in Table 1.1.**

While the selectivity observed for Kme3 as the free amino acid is encouraging, peptides pose a bigger challenge. Increasing the length of peptides comes with an increase in overall charge which can lead to non-specific binding through electrostatic interactions. Because there is no difference in the charge state of a methylated and unmethylated peptide, this increased electrostatic interaction would be same for all and would counter the driving force behind the observed Kme3 selectivity. Tabet et al. showed that upper rim substituted PSC analogs remain moderately selective for the methylated forms when tested against peptides with +2 charge.[178] When going from a free amino acid to doubly charged peptides, the selectivity drops from 150-fold to 25-fold for compound **AP0** while other calixarenes have similar losses. Interestingly, carboxycalixarene experience almost no loss in selectivity even when tested against peptides with +4 charge.[174] This is probably due to less negative charge on carboxycalixarene (−1) verses sulfonated calixarene (−4) which reduces the overall contribution from non-specific interactions. Another example, that supports this hypothesis, is multivalent hosts from Kimura et al. The authors used a trisulfonated calixarene from Daze et al. and conjugated it to a peptide backbone to make mono-, di- and tri-valent hosts (Figure 1.17.d).[179] Each step came with a severe reduction in selectivity, so much that the trivalent host, with −12 charge, was only 1.3-fold selective for methylated peptide over unmethylated peptide.

All calixarene based hosts have impressive binding strengths. Hosts like **AP0** and **1.4** bind Kme3 containing peptides with sub-micromolar dissociation constants, already stronger than that of naturally evolved methyl reader proteins.[163,180,181] In addition, different groups have used different sets of peptides to study calixarene-Kme3 complexation and have found calixarenes to be selective for methylated peptides over unmethylated ones to some degree over all examples (Table 1.1). This broad specificity has allowed researcher to developed calix[4]arene based assays to study both chemical and mechanical properties of nucleosomes and related proteins.[178,182–185]

1.4.3. Disulfide cyclophanes

More proof of the benefits of mimicking natural aromatic cage motifs came from the group of Marcey Waters at the University of North Carolina. Their approach involves the use of dynamic combinatorial chemistry (DCC) to generate a diverse family of macrocycles held together by disulfide bonds. The molecules in this library and building blocks are all in equilibrium with each other due to dynamic exchange of disulfides (Figure 1.18). Exchange of building blocks among members of the library generates a large combination of molecules with different shapes and sizes in just one experiment. The underlying idea of DCC is that in presence of suitable templating guest the equilibrium shifts towards the library member which forms the most stable complex (Figure 1.18). This host molecule can be easily identified by looking for amplification in concentration post guest addition (usually done by HPLC). The big advantage of DCC is quick diversification of molecules and a combined qualitative estimation of their binding strength in one experiment without investing lot of resources in synthesis and purification.

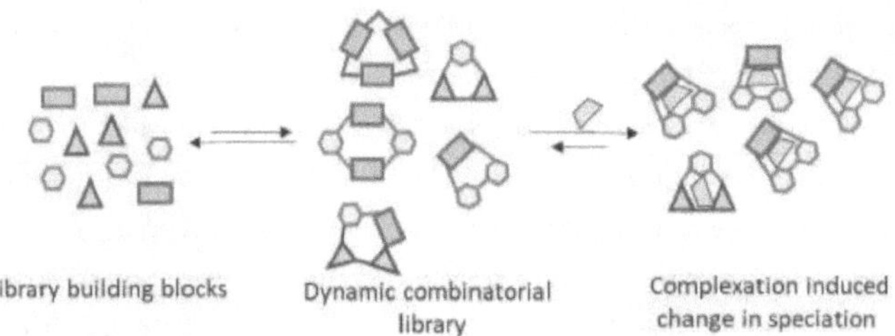

Figure 1.18. Working principle of Dynamic Combinatorial Chemistry (DCC).

Ingerman et al. showed the utility DCC by using a trimethylysine containing dipeptide as a guest to template the formation of dithiamacrocycle **A₂B** as a potential trimethyllysine host from a small set of eight equilibrating dithiol building blocks (Figure 1.19, Blue box).[186] After full characterization of **A₂B**, they found that it binds Kme3 containing histone peptides with low micromolar affinity ($K_d = 25$ µM) and ~50-fold selectivity. In later work, the same set of building blocks, when equilibrated with a dimethylated arginine as a guest molecule, led to identification of a different macrocycle **A₂D** which discriminated between isomeric dimethylarginines.[187] With 7-fold selectivity for Rme2a over Rme2s containing peptides and submicromolar affinity, this was the first molecule to bind the asymmetric arginine isomer selectively over the symmetric counterpart. However, **A₂D** is still overall Kme3 selective (1.3-fold over Rme2a).

Waters and coworkers have since improved upon both these results by introducing small changes in the original building blocks (Figure 1.19). Swapping **B** with **N**, led to identification of **A₂N**, another Kme3 selective host with much improved affinity ($K_d = 300$ nM).[188] Introducing a different building block, **I**, allowed them to synthesize the first dimethyllysine (Kme2) selective macrocycle, **A₂I**.[189] The selectivity achieved is low (1.1-1.3 fold) but it still is an impressive feat as sequestering Kme2 selectively is a formidable challenge (see Chapter 5). Compared to Kme3, the Kme2 sidechain is less hydrophobic. The size difference between the two is small and size alone does not give much basis for discrimination. At the time of writing of this book, it remains (3 years after publication) the only host demonstrated to bind more strongly to dimethyllysine than trimethyllysine. More recently, Mullins at al. overcame **A2D**'s inability to bind Rme2a over Kme3 by going back to the drawing board and starting with new building blocks.[190] Their rationale behind this was to promote a more box-shaped macrocycle in order to prevent the more spherical Kme3 from entering. They reformed the library with **E**, **G** and **N**, known to form box shaped receptors, and identified and isolated N_2G_2 (Figure 1.19), as the first Rme2a selective host that was selective over both Rme2s and Kme3 containing peptides (K_d =1.2 µM for Rme2a versus 13 µM for Kme3).

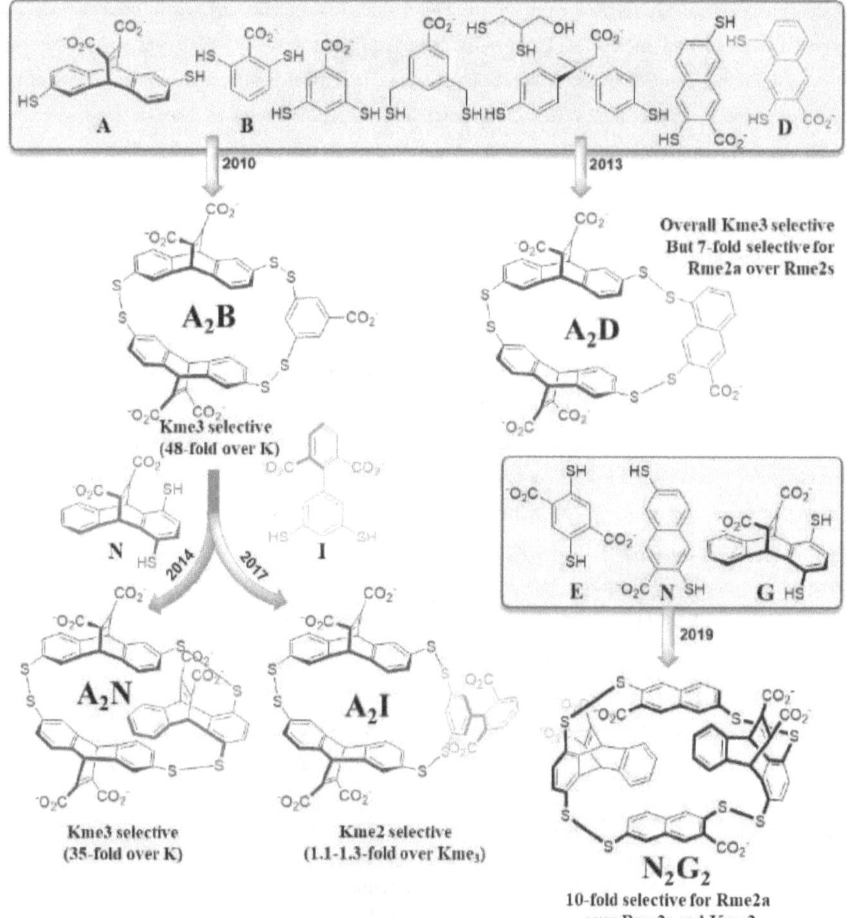

Figure 1.19. Diverse cast of dithiamacrocycles (and their building blocks) published by Waters and coworkers. The main features of host-guest binding behavior are explained for each molecule.

These dithiacyclophanes bind methylated peptides using very similar molecular recognition elements to those of calix[4]arenes.[191,192] Like calix[4]arenes, they contain a hydrophobic cavity lined with aromatic rings which can provide multiple cation-π interactions.[188] Both molecules depend on electrostatic interactions with peripheral anionic groups to orient the guest towards their cavity.[193] The significant advantage of these dithiacyclophanes over calix[4]arenes is the ease with which their shape can be predictably changed. **N_2G_2** is good example where the authors intentionally directed the chemistry towards more box-shaped host molecules. Comparatively, calix[4]arene's cavity shape and size is sensitive to substitution but almost impossible to predict or rationalize (which will become a major theme of this book). A major drawback of the Waters group hosts, however, is the lability of disulfide bonds which are highly sensitive to pH, soft electrophiles, and reducing environments. This poses a threat as these dithiacyclophanes can easily fall apart in biochemically relevant conditions. Despite these hurdles, Waters and coworkers were able to use them to develop a fluorescent sensor array for detection of combinatorial histone modifications.[194]

1.4.4. Resorcinarenes

Resorcinarenes are meta cyclophanes like calixarenes but derived from resorcinol instead of phenol. With similar shape, size and synthetic protocols, they can be expected to have similar binding behavior. Cyanoresorcinarenes are proven receptors for quaternary ammonium ions (K_d = ~1-2 μM).[195] Installation of cyano groups lowers the pKa of the phenol groups, which generates a negatively charged rim on the molecule. Analogous to the sulfonates of calixarenes and the carboxylates of dithiacyclophanes, these phenolates (red oxygens in Figure 1.20) can provide primary attractive interactions needed to encapsulate a positively charged guests. In line with this reasoning, Peacock et al. showed that a simple tetracyanoresorcinarene (**1.10**) binds Kme3 with 50-fold selectivity over free K (Figure 1.20).[196]

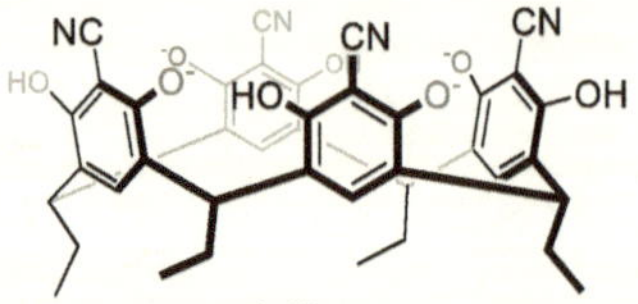

1.10

K_d (Kme3) = 21 μM
K_d (Kme2) = 68 μM
K_d (Kme1) = 476 μM
K_d (K) = >1000 μM

> 50 -fold selective for Kme3

Figure 1.20. Tetracyanoresorcinarene (1.10) from the work of Peacock et al. is a 50-fold selective host for Kme3.

Curiously however, bulk of published work on resorcinarenes as a host for *N*-methylated guests utilizes a strategy that goes completely against the norm. Dalcanale and coworker have focused on neutral analogues of resorcinarenes where the upper rim of the molecule is stapled together with alkyl phosphonates.[197] These resorcinarene scaffolds (called **Tiiii**, Figure 1.21.a) use their in-facing P=O bonds to provide attractive ion-dipole interactions when engaging an ammonium guest. One derivative of **Tiiii**, compound **1.11**, shows impressive affinity (K_d = 0.9 μM) for Kme1 in methanol (Figure 1.21.b).[198] Detailed binding studies with other methylated amino acids were not done which prohibits us from making any comments on selectivity. Also, as expected, binding in aqueous solution is drastically weaker (Figure 1.21.b). Dalcanale and coworkers are more motivated towards utilizing host-guest chemistry in conjugation with nanotechnology to develop sensors (Figure 1.21.c).[197,199,200] To this effect, despite reduced binding in aqueous solution, **Tiiii** based resorcinarenes have shown much promise. The methylenes of **Tiiii** provide an easy position for functionalization away from binding face. This position was utilized to anchor resorcinarenes on microstructures, like a dielectric resonantor and microcantilever, in order to develop sensors for *N*-methylated ammonium guests.[197,199]

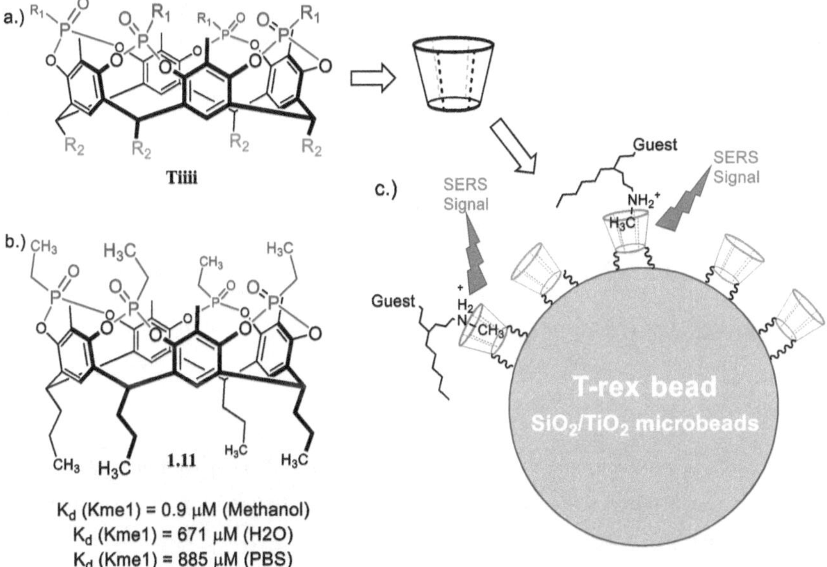

Figure 1.21.a) Dalcanale's tetraphosphonated resorcinares (Tiiii), b) A derivative of Tiiii, compound 1.11, shows impressive affinity for *N*-methyllysine in non-aqueous media, c) One of many published strategies by Dalcanale and coworkers to utilize Tiiii based molecules in development of fucntional sensors.[199]

While Dalcanale and coworkers have shown that neutral host molecules can be utilized to recognize *N*-methylated guests, Hooley and coworkers go one step further by synthesizing an array of resorcinarene molecules that contains a positive, a neutral and a negatively charged host (Figure 1.22, Compound **1.12-1.14**).[201,202] Their strategy involves elongation of the resorcinarene cavity by addition of benzimidazole walls.[203] Due to their elongated shape, these molecules show interesting aggregation behavior which the authors cleverly utilize to their own benefit. They introduce dyes whose photophysical properties change upon aggregation. Chemically modifying these dyes with an alkyl ammonium tail led to their coaggregation with resorcinarenes (Figure 1.22).[201–203] Introduction of guest causes a displacement of dye from the aggregate which can be tracked spectrophotometrically. A combinatorial array of different dyes and hosts led to a sensor array that can not only detect the presence or absence of *N*-methyl amino acids, but also relay information on the site of methylation within a given peptide.[202]

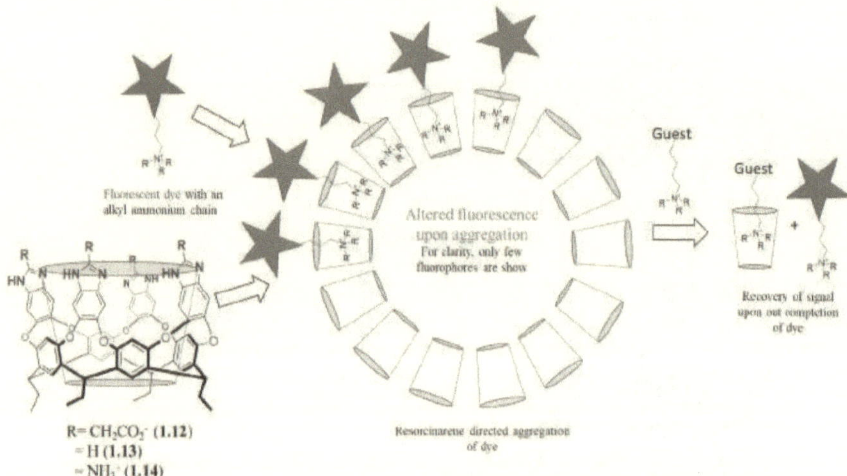

Figure 1.22. Hooley and coworker's resorcinarene-based deep cavitands have been used to develop a dye displacement-based sensor array for combinatorial detection of methylation marks on histone derived peptides.

1.4.5. Pillar[5]arenes

A black sheep among cyclophanes, a decacarboxylated pillar[5]arene (**1.15**) has been shown to bind unmethylated lysines stronger than methylated ones (K_d = 550 µM for K versus 770 µM for Kme3).[204] The authors cite some form of steric occlusion of the larger Kme3 guest as a mechanism for selectivity. But the overpowering effect of electrostatics cannot be ignored. The authors themselves point out that two negative portals of decacarboxylated pillar[5]arene (each with -5 charge) is highly compatible to dicationic guests that can thread through the macrocycle (Figure 1.23.b). Free lysine with its two primary ammonium ions perfectly fits this picture. Hydrogen bonding between lysine NH_3^+ groups and host COO^- groups must also play some role. Kme3, on the other hand can neither provide these hydrogen bonds nor thread through the cavity. The trimethylated sidechain of Kme3 is hydrophobic enough that it would prefer being sequestered rather than exposed to water at the charged rim of the host (Figure 1.23.c). This drives the selectivity towards non-methylated amino acids. This idea is reinforced when authors find a 10-fold reduction in binding affinity between free K and a tripeptide with K in the middle where the positive charge of α-amino group of K is neutralized.[204]

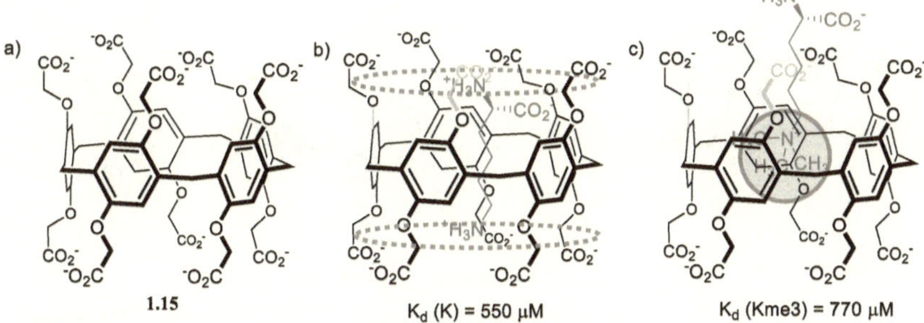

Figure 1.23.a) Decacarboxylated pillar[5]arene (1.15) is a water soluble host that prefers unmethylated K over Kme3. The difference arises from two different binding modes of K and Kme3, b) Lysine threads through the pillar[5]rene cavity making contacts with both hydrophilic portals, c) Kme3 is likely to be encapsulated within the cavity and only engage one side of the molecule.

1.4.6. Cucurbiturils

An antithesis to all biomimetic efforts containing "aromatic cages," cucurbit[7]uril (**CB7**) has no aromatic groups, but has the highest selectivity for the free amino acid Kme3 (K_d = 0.5 μM with ~3500 fold selectivity over K).[205] This selectivity is even higher than the any known reader protein, the record for which is held by ING2 which is 1500-fold selective for Kme3 over K.[180] Like other guests of cucurbituril, the binding strength of Kme3 can only be explained by the desolvation of its highly rigid hydrophobic cavity (Figure 1.24.b). An excellent example of the non-classical hydrophobic effect, cucurbituril's cavity contains high enthalpy water molecules which are released upon guest binding.[157] A tight fit between **CB7** and Kme3 results in thorough desolvation of the cavity and is the reason proposed for its strong binding. Ion-dipole interactions at the rim are assumed to provide secondary attraction. Strong match between size of host and guest can once again be seen to have an important role. **CB6**, comparatively smaller to **CB7**, binds only weakly to methylatedlysines (exclusion complex).[205]55 It is noteworthy that **CB7**, unlike every other molecule discussed here (except **1.11**), is neutral but still has the most impressive affinity and selectivity. This serves as a powerful example that while electrostatics most certainly help, there are other answers. Derivatization of **CB7** is difficult which makes it a challenge to translate this result into a useful assay. The only attempt in literature tries to utilize native **CB7** molecule to separate methylated peptide from unmethylated ones in a capillary electrophoresis (CE) setup.[206] However, **CB7** performs underwhelmingly compared to p-sulfonatocalix[4,6]arenes

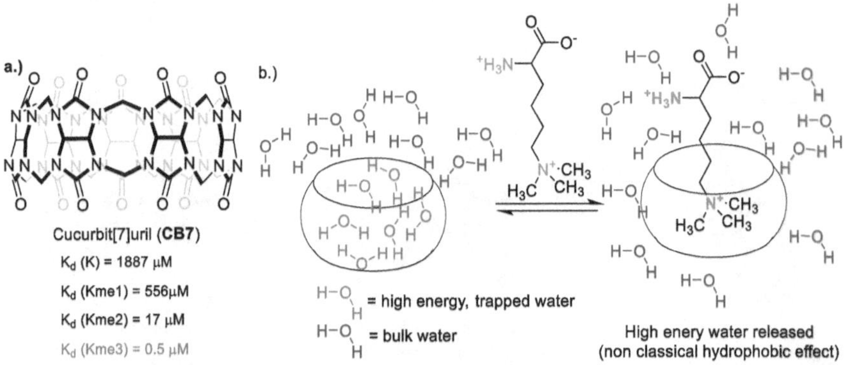

Figure 1.24.a) Cucrbit[7]uril and its binding profile against different methylated versions of lysine.[205] b) The high affinity is a result of thorough desolvation of high-enthalpy water molecules but a tightly fitting guest molecule, Kme3.

1.5. Supramolecular host as affinity reagents, an unexplored avenue.

While the molecules discussed above have impressive supramolecular properties, it is desirable to take chemistry from idealized host-guest binding studies into more complicated avenues. One more advanced goal would be to design new biochemical assays for detection of methylated peptides. Most groups whose work is represented here have taken steps towards it. The Hof lab at University of Victoria were first to publish a calix[4]arene based sensing arrays for methylated peptides derived from histones.[182] Other groups have demonstrated utility of their molecules in affinity labelling[194], enzyme assay development[201], SERS based detection[199] etc. The common thread for these studies is their focus on sensing or detection of methylated peptides, or changes in concentration of methylated peptides.

The long-term goal that motivates this book work is development of enrichment reagents for methylated peptides. Methyllysine enrichment poses interesting set of problems which cannot be completely answered by current techniques. Large scale discovery-based proteomics requires pan specific enrichment reagent which are unavailable.[126] Supramolecular hosts offer a potentially useful alternative and have not been actively studied in the context of proteomics. At the time of completion of this book two such works have been published. One example, by Lee et al., utilizes calixarenes and cucurbiturils to separate peptides based on methylation states using capillary electrophoresis (CE).[206] These hosts are used to induce a change in electrophoretic mobility of peptides. Methylated peptides when encapsulated by the host show large shift in retention time. However, CE is not a suitable front-end technique for proteomics. The instrument is run under sealed conditions and is strictly an analytical technique. Efforts to join CE with MS do exist in literature but are not as successful as LC-directed proteomics experiment.[207] The second published example was completed in the Hof group during this book research with my own contributions and is reported in Chapter 3.

This sets the stage for introducing the goals of this book. While this work falls in the realms of synthetic and supramolecular chemistry, it is done while keeping a long-term goal in mind which is to find host molecules that not only bind methylated peptides with some selectivity over unmethylated peptides, but can also be used to separate thousands of methylated peptides from highly complex biological mixtures.

1.6.   Summary and goals of this book

This Chapter summarizes the importance of N-methylation of lysine and arginine and identifies the problems associated in the field of methyl proteomics. A broad statement describing the required properties of a potential enrichment agent is stated above. This can be broken down further to smaller, achievable goals:

1. We need to find stronger-binding hosts ($K_d < 1\,\mu M$) for Kme3 peptides.
2. We need to find host molecules that are selective for lower methylation states (e.g. Kme2).
3. We need to find host molecules that can be immobilized on a solid phase and used as in affinity chromatography for methyl peptide enrichment.

To achieve these goals, I will present new synthetic approaches for the creation of calixarenes with different modifications that change their binding properties and/or that allow one to attach them to solid support. Chapter 2 will present a diversity-based approach, in which dozens of new upper-rim modified calixarenes are made and tested in parallel for binding to methylated peptides. Chapter 3 will present the results of early studies carried out on model proteomics samples with one host that emerged from this initial search. Chapter 4 will report new efforts to make lower-rim substitutions that allow attachment to solid support, with continued studies of methyl peptide binding. Chapter 5 will focus on one new calixarene analog, which shows Kme2 selectivity and can be attached to solid support. Chapter 5 also finishes with a small proof-of-concept study for methyl peptide proteomics using a real cell lysate. Chapter 6 provides the context and future directions for this book work.

Chapter 2. Upper rim functionalized *p*-sulfonatocalix[4]arenes. A diversity-driven approach to discover new hosts targeting methylated peptides

Contributions

The idea behind this approach was first conceived by Kevin Daze and FH. Synthetic methods towards each scaffold were developed by KD and AS. Synthesis of all compounds, including calix[4]arenes and peptides, and their binding studies were done jointly by KD, JL, Trevor Henderson and AS. AS performed all curve fitting and analysis of the binding data. AS also did all follow up studies on bigger scales.

2.1. Foreword

Chapter 1 establishes the need for enrichment agents in the field of methyl proteomics. It also establishes the potential advantage that small molecular hosts can have towards this goal. We choose to explore this avenue using functionalized *p*-sulfonatocalix[4]arenes, a highly studied water soluble host. This choice was made due to the depth of literature available towards functionalization of calix[4]arenes, and their abilities in general to bind methyllysines. This Chapter describes our work with upper rim functionalized *p*-sulfonatocalix[4]arenes. It describes, in detail, the synthetic methods established to regioselectively modify their upper rim. It describes the primary binding assay utilized throughout this book. We use these methods of develop a library of host molecules and study their binding properties towards methylated peptides in hopes to gain insights into their structure-activity relationship.

2.2. Motivation

The motivation for this Chapter comes from a previously published work from Hof group where Daze et al. showed that the introduction of a phenyl substituent on the upper rim of *p*-sulfonatocalix[4]arene increased the affinity and selectivity for Kme3 containing peptides (Figure 2.1).[175] The affinity of this new biphenyl containing trisulfonated calixarene **(AP0)** was found to be comparable to naturally found proteins albeit the data was reported only for the free amino acid Kme3. NMR studies showed that this new class of upper-rim aryl-substituted calixarenes directly engaged the alkyl side chain of amino acid guests as it projects out of the deepest part of the cavity (Figure 2.1). The side chain CH-α protons of amino acid guests experienced large upfield shifts when using aryl-substituted calixarenes. A follow up study by Tabet et al. found that the dissociation constants of these hosts decreased from low micromolar to submicromolar when determined for peptides containing Kme3 instead of free amino acids.[178] This can be reasonably attributed to increased role played by electrostatics of the anionic groups surrounding the binding pocket. Free Kme3 has a net positive charge of +1 while the peptides used by Tabet et al. had net +2 charge.

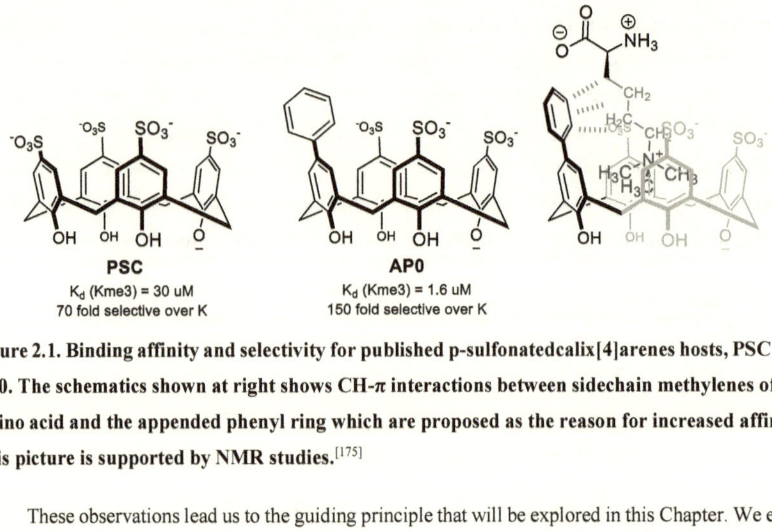

Figure 2.1. Binding affinity and selectivity for published p-sulfonatedcalix[4]arenes hosts, PSC and AP0. The schematics shown at right shows CH-π interactions between sidechain methylenes of the amino acid and the appended phenyl ring which are proposed as the reason for increased affinity. This picture is supported by NMR studies.[175]

These observations lead us to the guiding principle that will be explored in this Chapter. We expect that the substituting different aryl groups at the upper rim of calixarenes would change the binding behaviour of calixarenes (by changing the shape of its cavity) in ways that are beneficial for certain guests, but not easily predictable (Figure 2.2). We can predict with reasonable confidence that the deeper cavities of such calixarenes will be well suited for guests with cationic sites next to alkyl chains, such as lysine and its derivatives. In unpublished work before this Chapter was started, we tried to design hosts that are specific for lower methylation states (Kme2, Kme1) over Kme3 using simple predictions of hydrogen bonding patterns, but in all cases were unsuccessful. We suspect that this is because the effect of substitution on calixarene's conformation was completely left out of our designs.

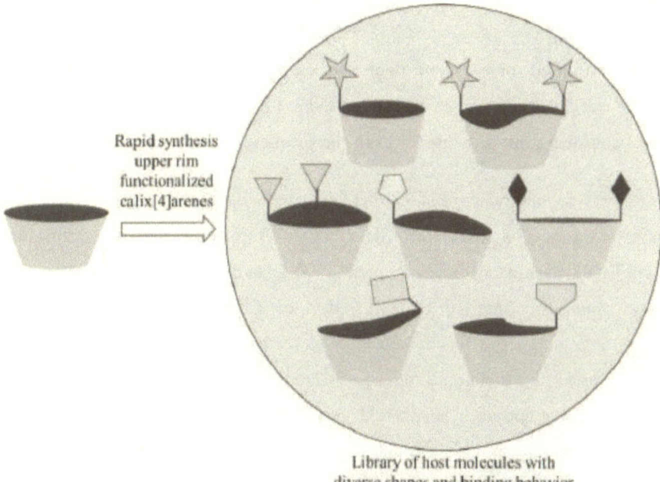

Figure 2.2. A schematic representation of the guiding principle of this project. Substitutions on the upper rim change the shape of calix[4]arenes in unpredictable manner which in turn changes their binding behavior.

It is beyond the scope of this book to thoroughly review all literature on calixarene conformations[169,208–211], but some relevant details are included here. Calix[4]arenes with unprotected lower-rim phenols often favour a C_{4v}-symmetric cone conformation (Figure 2.3), favoured by intraannular hydrogen bonds between the phenols[211]. Throughout this book, calixarene macrocycles will be represented in this conformation unless otherwise stated, because showing all possible conformations of every calix[4]arene in simple chemdraw images is an impossible task. Despite this convention, it must be appreciated that in aqueous conditions calix[4]arenes spend only a fraction of their time in cone shape. The intramolecular H-bonds between phenols that favour the cone conformation experience competition from water molecules, allowing the calix[4]arenes to explore a multitude of conformations rapidly. Substitutions on calixarenes changes the relative population of these conformations which results in altered host behavior[212].

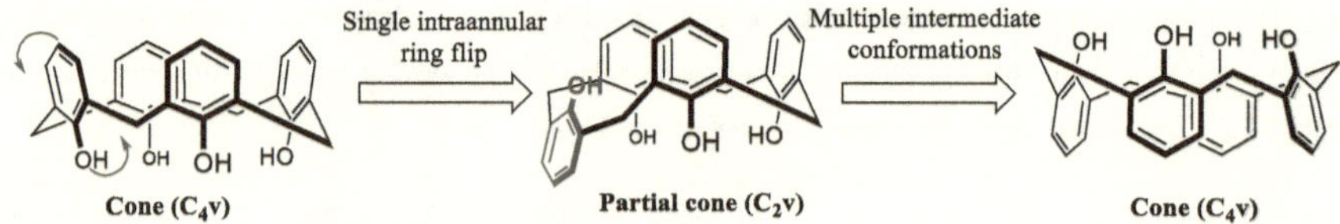

Figure 2.3. A simplified mechanism depicting calixarene's conformational dynamics. Each phenolic unit of calix[4]arene can do an intraannular flip independent of other phenols resulting in multitude of conformation possible for even the simplest calixarene.

Our attempts to rationalize and predict the conformational changes in calix[4]arenes upon substitution with aryl groups have been unsuccessful. Our efforts to design hosts with selectivity for any lysine analog but Kme3, the most hydrophobic methyl PTM, have similarly failed. And so, we turn towards a diversity-driven approach which will be discussed in this Chapter (Figure 2.2). The strategy involves rapid synthesis of large number of functionalized p-sulfonatocalix[4]arenes and their screening against a set of methylated peptides in a medium-throughput fashion.

Our goal with this approach is two-fold. First, by synthesizing and screening a library of host molecules with distinct modifications at the upper rim against a set of PTM containing peptides, we aim to find hosts with strong affinity and selectivity towards different methylation states. The substitutions which bring the biggest change in the binding behavior of calixarene will be scrutinized further. Second, we wish to use molecules from the library and establish ways to use them as enrichment agents (Chapter 3). This is a more collaborative effort between many members of Hof group whose contributions will be noted appropriately. We expect that such enrichment agents can be used to identify otherwise elusive methylation sites in proteomics experiments.

2.3. Design of the library

The library-based approach described in this work starts with synthesis of five different scaffolds (Figure 2.4). Scaffold **A-D** are upper rim (poly)brominated analogues of sulfonated calix[4]arenes which are designed to be used as substrates in Suzuki coupling. Coupling of different boronic acids with aryl bromide will provide an (poly)arylated calix[4]arene with bigger hydrophobic surface to engage guest molecules. These aryl groups will be attached to the calix[4]arene scaffold using a Csp2-Csp2 bond which is rigid but with some rotational degrees of freedom.

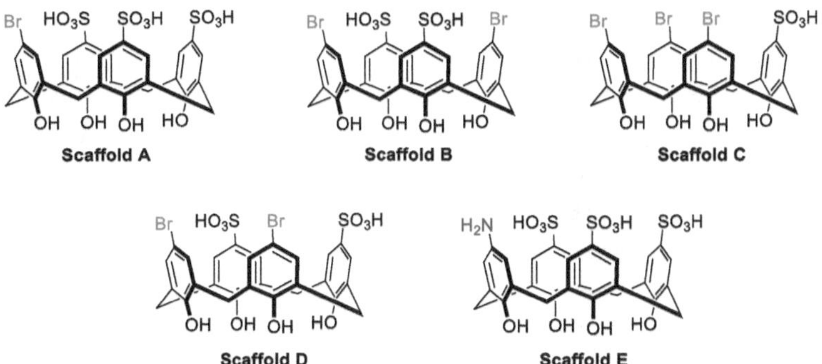

Figure 2.4. Structures of five different scaffolds (A-E) that were used in diversification of upper rim of *p*-sulfonatocalix[4]arenes. Reliable synthesis of these scaffolds is suitable scale was the first challenge towards goals of this Chapter.

Scaffold **A** is continuation from already established work of Kevin Daze[175,178]. Scaffold **B-D** were designed to test the effect of deeper and more encapsulating hydrophobic pockets on the affinity for guests like methylated K/R. Scaffold **B** and **D** are isomers but with different symmetry. Arylation of these scaffold is expected to provide a host with very different shape[212]. Scaffold **C** is expected to generate the most hydrophobic library members.

Scaffold **E** is a mono-functionalized calix[4]arene, similar to scaffold **A**, but with an amine group instead. It was included with an intention to use it in coupling reactions with aryl sulfonyl chloride, carbonyl chlorides and isothiocyanates. This will generate aryl functionalized calix[4]arene with very different connectivity. Compared to the Csp2-Csp2 bond generated by Suzuki coupling on scaffold **A-D**, the amide, sulfonamide and thiourea linkage generated using scaffold **E** are expected to be more flexible.

2.4. Synthesis of Scaffolds A-E

Most of upper rim derivatisation strategies for calix[4]arenes rely on the highly active para position of phenolic subunits of the macrocycle. Due of the electron donating nature of -OH group the para position is susceptible to electrophilic aromatic substitution (EAS). Direct reaction of calix[4]arene with an electrophile however leads to a mixture of all possible isomers and degrees of substitution (Scheme 2.1). In order to selectively synthesize one scaffold, the reactivity of phenols has to be modulated. Selectively converting phenols of calix[4]arene to a less activating –OR group would stop any electrophilic substitution on the corresponding para position. To the best of our knowledge, the first such example of guided reactivity on calix[4]arene's upper rim was demonstrated by Berthalon et al. who showed that partially esterified calix[4]arenes can be used to selectively *de-tert*-butylate the upper rim of calix[4]arene.[213] Esterification of phenols lowers the electron donating capabilities of oxygen significantly. Strategies for regioselective esterification at lower rim exist in literature which should help us to synthesize each scaffold[214–216]. Our strategy closely follows that of Berthalon et al. where esterification is used to direct the reaction at the upper rim and then removed via saponification. Except, we start with dealkylated calix[4]arenes and use the esters to direct electrophilic aromatic substitution instead of Friedel-Crafts reactions. The following paragraphs will describe synthesis of each individual scaffold in detail.

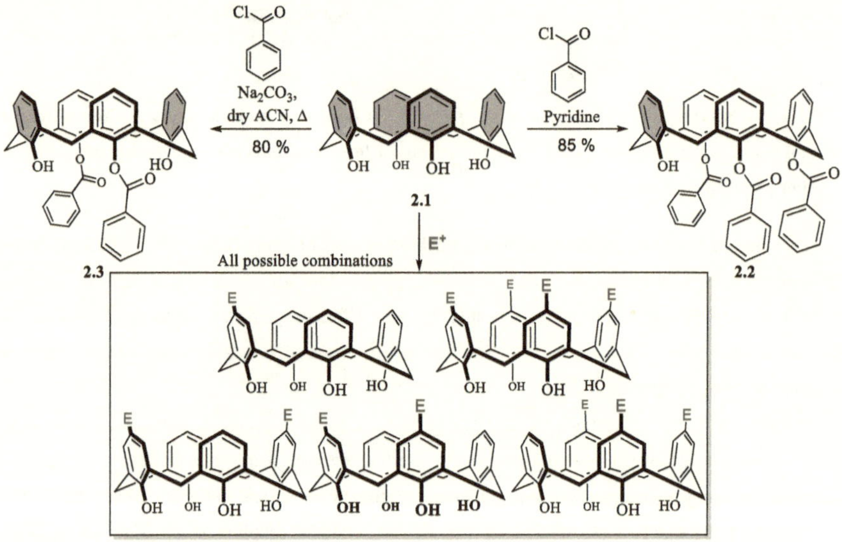

Scheme 2.1. Strategy to regioselectivity functionalize calixarene at the upper rim. Unsubstituted calixarene undergoes rapid EAS to generate mixtures of all possible products. Benzoylation at the lower rim modulates the reactivity and allows for a more controlled reaction on active phenols (shaded red) only.

2.4.1. Synthesis of scaffold A

Selective mono-functionalization of upper-rim positions was achieved through the tribenzoylated intermediate **2.2** (Scheme 2.2). Compound **2.2** is prepared by exhaustive benzoylation of the lower rim using excess benzoyl chloride in pyridine[214]. Tetrabenzoylated product is rarely formed due to steric crowding of benzoates at the lower rim. This gives us an advantage and high equivalents of benzoyl chlorides can be added to drive the reaction to higher yield without the risk of over benzoylation. The benzoylated phenols are electron poor and not active towards electrophilic aromatic substitution. The remaining phenol can be easily brominated (**2.4**) to near quantitative yields by reacting it with bromine in chloroform. The benzoate esters can then be readily cleaved (**2.5**) using methanolic NaOH which makes the remaining three phenols active. These can be now sulfonated using sulfuric to give scaffold **A**. The total number of steps involved (starting from *tert*-butylcalix[4]arene) is five. But these are high yielding steps that rarely require purification. The most damaging step to overall yield is in fact the last step, i.e. sulfonation, which is a reversible reaction and requires forcing condition to give only moderate yield with some tetra- and di- sulfonated products as impurities.

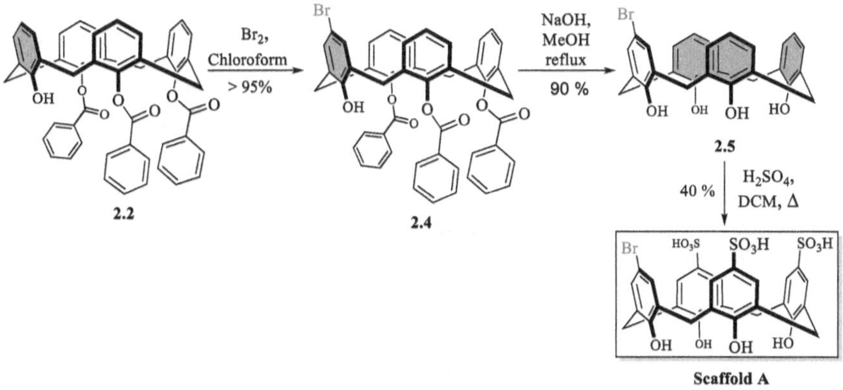

Scheme 2.2. Regioselective synthesis of scaffold A. Red coloured phenols are active towards electrophilic aromatic substitution.

2.4.2. Synthesis of scaffold *B*

Selective diametric bromination of calix[4]arene was achieved using dibenzoylated intermediate **2.3** (Scheme 2.3). It is well established in the literature that the reactivity at lower rim of calix[4]arene is affected by the choice of base. Both *O*-esterification[215] *and O*-alkylation[217] can be directed towards different outcomes by choosing appropriate base and counterion. Our choice of base was sodium carbonate in acetonitrile which directs benzoylation to happen exclusively on 25 and 27 position of calix[4]arene. Just like the synthesis of scaffold A, benzoylated phenols are deactivated and do not react with electrophile which allows us to install two bromines on distal phenol rings. Cleavage of esters and sulfonation of remaining phenols using the same procedures as above gives us scaffold **B** (Scheme 2.3).

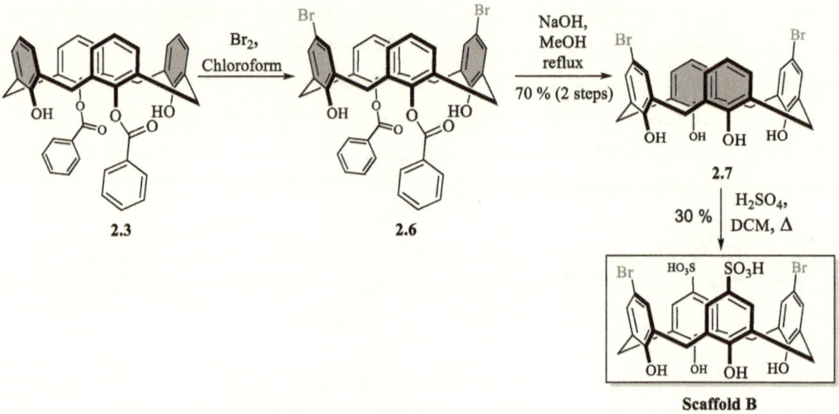

Scheme 2.3. Regioselective synthesis of scaffold B. Phenols that are shaded red are active towards electrophilic aromatic substitution.

2.4.3. Synthesis of scaffold C

Scaffold C was synthesized using similar procedure as scaffold A except bromination and sulfonation steps were exchanged (Scheme 2.4). This allowed us to install three bromine and one sulfonate on the upper rim of the calixarene, exactly opposite to scaffold A.

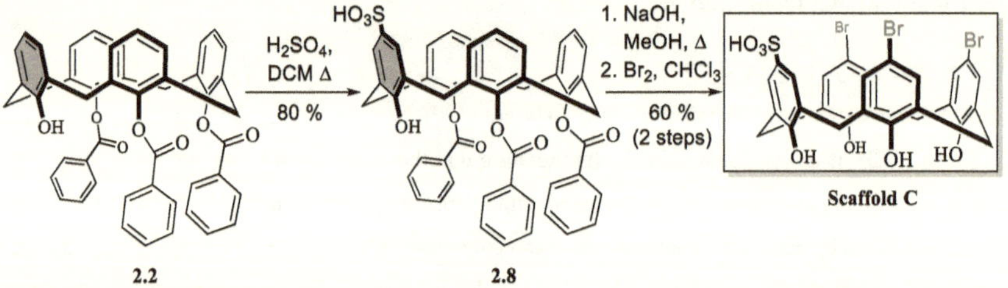

Scheme 2.4. Regioselective synthesis of scaffold C. Phenols that are shaded red are active towards electrophilic aromatic substitution.

2.4.4. Synthesis of scaffold D

Brominating adjacent phenols of calix[4]arene requires a different dibenzoylated isomer. Synthesis of this isomer has been investigated in detail by Lynch et al. and, we adapted and improved this protocol successfully in our lab[215]. In a transesterification step, the 25,27 dibenzoylated isomer (**2.3**) can be transformed to 25,26 dibenzoylated (**2.9**) isomer in presence of Cs_2CO_3 (Scheme 2.5). Cesium ion's template effect seems to be the driving factor behind this reaction. ^{13}C shifts of the **2.9** suggest a 1,2 alternate conformation while the starting reagent, **2.3**, shows a flattened cone conformation[218]. It is expected that a soft ion like cesium would coordinate with the soft pi electrons of the arene rings on calix[4]arene. This coordination likely drives the calix[4]arene away from the flattened cone towards a 1,2 alternate conformation which forces the phenolic -OH closer to ester carbonyl. The migration happens as a result of intramolecular nucleophilic attack from phenolate on the ester carbonyl in this conformation. Once the migrated calix[4]arene (**2.9**) is obtained, the rest of the scheme is a continuation from scaffold **A** and **B** synthesis. Unsubstituted phenols are easily brominated in presence of Br$_2$. The ester can then be cleaved and remaining two para positions sulfonated to give scaffold **D** (Scheme 2.5).

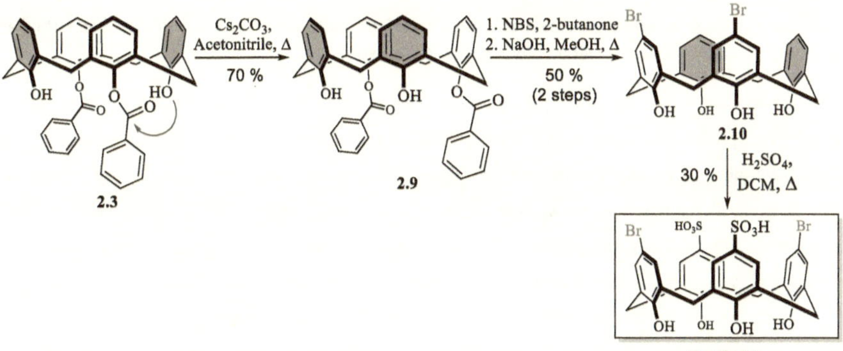

Scheme 2.5. Regioselective synthesis of scaffold C. Intramolecular attack by phenol on one of the benzoate ester is shown on 2.3. Red coloured phenols are active towards electrophilic aromatic substitution

2.4.5. Synthesis of nitro-calix[4]arene (P-NO2) and scaffold E

While the main strategy for synthesis of scaffold **E** is same as scaffold **A**, it was more challenging to reproducibly synthesize it in appreciable yield and purity. The key steps that required major investment were nitration of tribenzoylated intermediate **2.2** to **2.11** (Scheme 2.6) and reduction of nitro group of **P-NO2** to amino group of scaffold **E** (Scheme 2.7). The original nitration step involved reacting **2.2** to a mixture of HNO$_3$/H$_2$SO$_4$. This step hardly ever proceeded cleanly. Acid-catalyzed cleavage of esters would always lead to a mixture of nitrated products being generated which had to be purified using a time consuming chromatography. A cleaner nitration reaction was achieved by first using Menke conditions[219] (Cu(NO$_3$)$_2$ in acetic anhydride). However, the best results were obtained by using from Topchiev's nitration conditions (metal nitrates plus Lewis acid)[220]. A combination of metal nitrates with AlCl$_3$ and BF$_3$ were tested and the best conditions were found to be a combination of KNO$_3$ and BF$_3$.Et$_2$O (Scheme 2.6). Mono-nitrated calixarene (**2.11**) can then be debenzoylated and sulfonated in the same fashion as described above to give us its water-soluble analogue **P-NO2**.

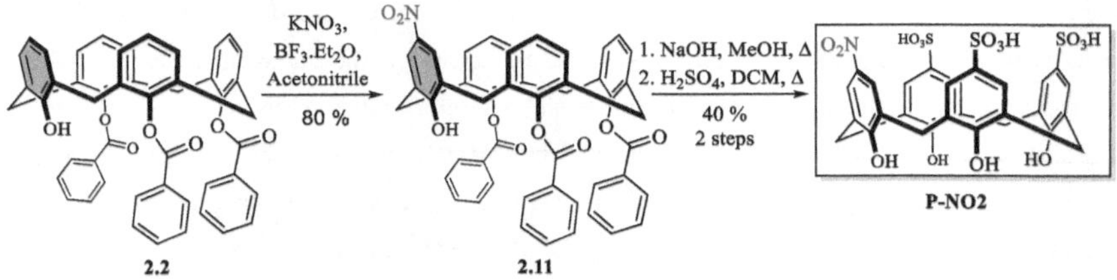

Scheme 2.6. Regioselective synthesis of mononitrotrisulfonato calixarene (Ap-NO2). Red coloured phenols are active towards electrophilic aromatic substitution.

Reduction of **P-NO2** to scaffold **E** at first seemed trivial. Both Raney Ni and Pd/C with H_2 were able to accomplish the transformation. However, it turned out to be very difficult to characterize the product. Mass spectrometry of crude reaction mixture suggested complete conversion to product however it was difficult to purify using reverse phase high pressure liquid chromatography (RP-HPLC), where the peaks appeared very broad. Even after purification, NMR peaks of the product were found to be too broad for any meaningful assignment. It was later realized that scaffold **E** was highly unstable under these conditions. The 4-aminophenol subunit of scaffold E can easily oxidize in air to iminoquinone then hydrolyze to quinone, sometimes reducing back to hydroquinone if put back in the presence of reducing agent (Scheme 2.7.a). Due to lack of NMR characterization, this oxidation route was only recognized by a detailed inspection of mass spectrometry data which itself was challenging since molecular weight of all the byproducts differ only by 1-2 units (Scheme 2.7.b). This resulted in them going undetected under the MS isotope pattern for a long time at the beginning of this project. The oxidation of scaffold **E** seem to be accelerated in basic conditions required by Raney Ni and Pd/C to the point where most of the product would be oxidized before the next step in the scheme. Initial attempts required reduction of **P-NO2** to scaffold **E** and then setting up the next reaction immediately without any purification in hope to get some final product.

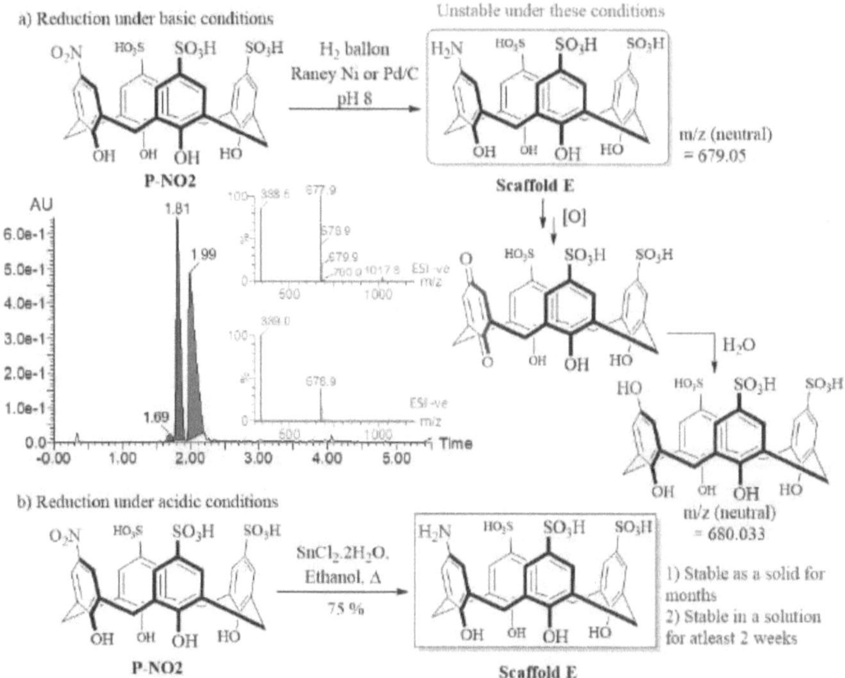

Scheme 2.7 a) The original attempts at reduction of nitrocalix[4]arene (P-NO2) to aminocalix[4]arene (Scaffold E). All attempts using this strategy led to poor yield and rapid degradation of product. UHPLC-MS chromatogram of a relatively cleaner sample shows Scaffold E along with its oxidized state (hydroquinone) after being left for few days in solution. b) A successful reduction scheme under acidic conditions using SnCl₂.2H₂O.

A few attempts to convert easily available scaffold **A** to scaffold **E** using copper(I) catalysed amination chemistry were also made but were unsuccessful. Success came by going back to reduction of **P-NO2** but using acidic conditions instead (Scheme 2.7.c). A straightforward reduction using SnCl₂.2H₂O in acidic ethanol was found in literature and successfully reproduced in the lab[221]. The reaction is clean and the scaffold **E** crashes out of solution which makes the purification very easy. The byproducts are stannous hydroxide and stannic acid which are both insoluble in water and easily removed.

2.5. Synthesis of Library

Once the synthetic route to scaffolds **A-E** was established, the library members were synthesized using either Suzuki coupling of boronic acids to brominated scaffolds A-D (Scheme 2.8, left box) or amide/sulfonamide/isothiocyanate coupling to scaffold **E** (Scheme 2.9). Due to poor solubility of these scaffolds in organic solvents, aqueous reaction conditions were needed. Leadbeater et al.'s microwave assisted ligand free Suzuki coupling was successfully adapted for this project.[222] Compound **AP2** from the library was further coupled to Rink amide resin preloaded with tyrosine or aspartic acid (Scheme 2.8, right box). Compound **AP7** and **AP9** were obtained upon cleavage using TFA. In addition, some of the compounds were isolated as *C*-terminal acids, likely due to acid-catalyzed amide hydrolysis of **AP7** and **AP9** and are included in the library (**AP6** and **AP8**).

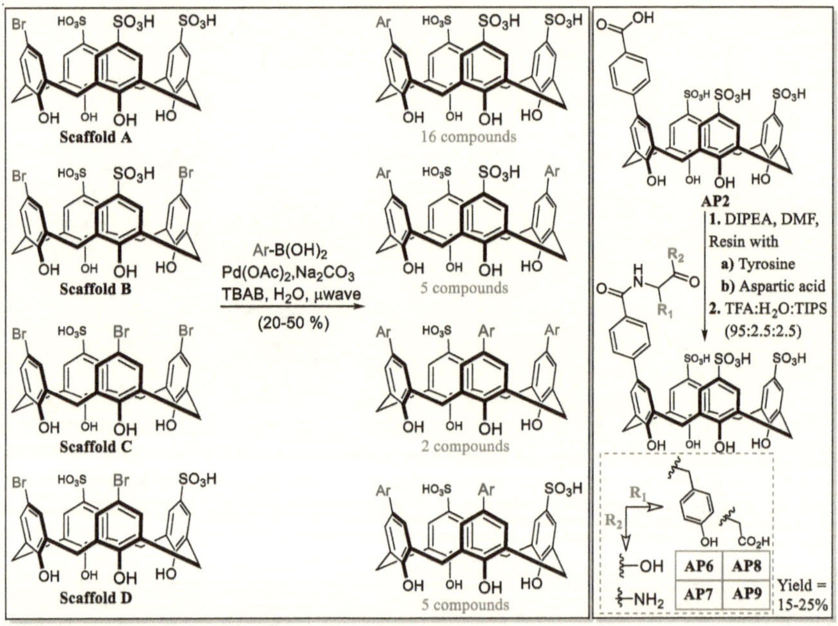

Scheme 2.8. (Left box) Scaffold A-D were used as substrates in Suzuki coupling to synthesis 28 new arylated calix[4]arene molecules. (Right box) One of these molecules, AP2, was further coupled on solid phase peptide synthesis resin preloaded with either tyrosine or aspartic acid. Cleavage with TFA led to both amide and acid containing product being produced. All 4 molecules (AP6-9) were also included in the library

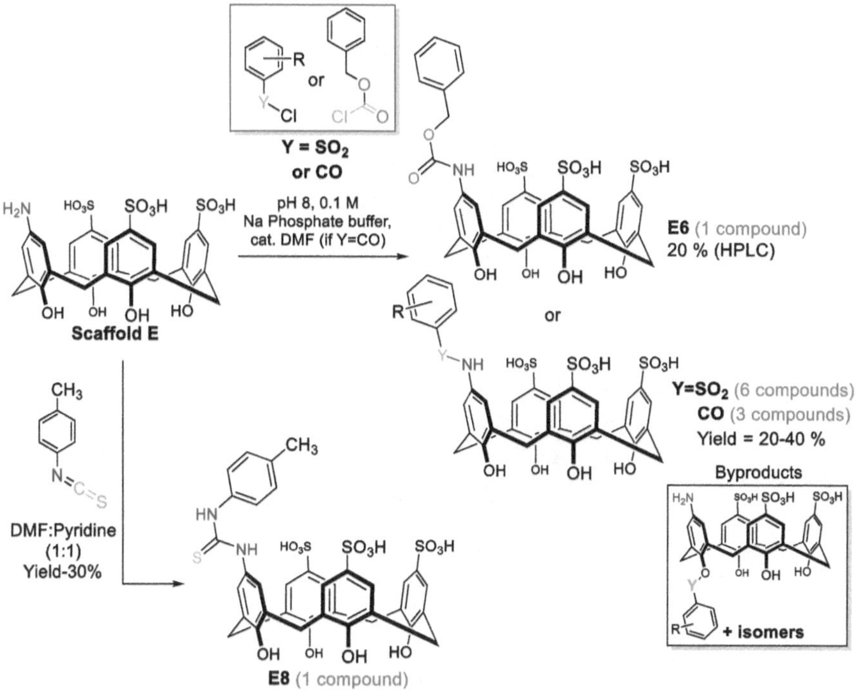

Scheme 2.9. Scaffold E was used in various coupling reactions which led to synthesis of 11 novel hosts with more flexible sulfonamide, amide, carbamate and thiourea linkers.

Similarly, coupling reaction with scaffold **E** (Scheme 2.9) needed to be performed in aqueous conditions. The challenge to this was supressing the concomitant hydrolysis of reactants, the answer to which lies is proper pH and ionic strength control of the reaction conditions.[223] Sulfonamide coupling between sulfonyl chloride and aromatic amines has been shown to proceed with high yield at pH 8 and was successfully adapted for this project.[224] Acyl chlorides were the most difficult to work with not only due to their rapid hydrolysis (multiple order of magnitude faster than others) but also due to their tendency to form esters at the lower rim instead of an amide at upper rim. The lower rim esterified byproducts were seen only minimally during the sulfonamide reactions but were major products during amide couplings. To mitigate the hydrolysis of acyl chlorides, they were first dissolved in small volume of DMF and added as an aliquot to a solution of scaffold **E** at pH 8. This lowered their hydrolysis rate enough for nucleophilic attack to occur. No solution was found to lower the esterification byproducts, but we found that post-reaction treatment with NaOH will hydrolyse all esters formed. This made the purification step easier, but the side reactions still significantly lowered the yield of the reaction. Isothiocyanates and chloroformates were much easier to work with because they reacted exclusively with the amine and not with the lower-rim phenols.

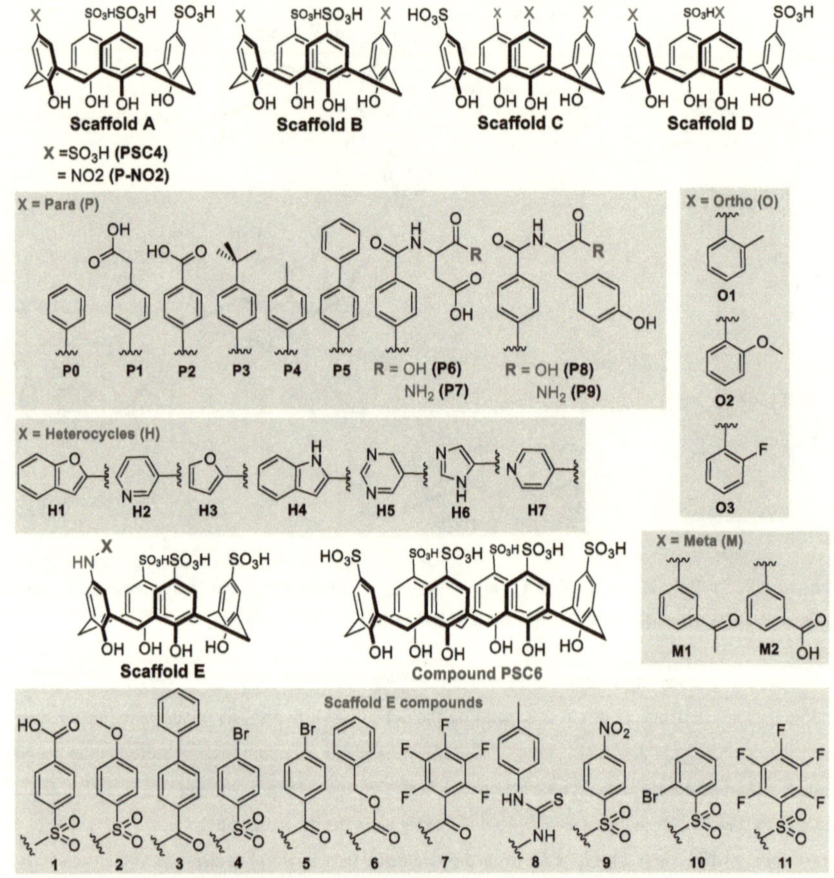

Figure 2.5. Enumerated list of all fragments in the library. Name of each library member discussed in this Chapter is amalgamation of the scaffold name and the fragment name.

There are total 46 molecules in the final version of the library that was used for binding studies (Figure **2.5**). Scaffold **A** and **E** are overrepresented. This is a result of unexpected low solubility of compounds generated from scaffold **B-D**. These scaffolds themselves have poor solubility in water, a problem that only gets exacerbated once additional aryl functionalities are added on. Compounds generated from these scaffolds also show strong dimerization which limits our ability to study their host guest properties in a high-throughput fashion.[225] Scaffold **A** and **E** also occasionally show dimerization however it is a relatively weaker and they can be assumed to be in monomer form at the concentration tested (5 μM and below). Some of these calix[4]arene dimers have been a subject of detailed study by other Hof group members, but are not a topic of this book.[226–228]

Figure 2.5 can also be used to generate the code used to refer compounds in the library. In this case, the name is simply an amalgamation of the scaffold and the fragment number. For example, compound **E2** is fragment **2** attached to scaffold **E** position X. Boronic acids used in this study are further divided into four groups namely para (**P**), ortho (**O**) and meta (**M**) substituted boronic acids and heterocyclic boronic acids (**H**). Using this code, compound **AO2** would be fragment **O2** attached to scaffold **A** at position X.

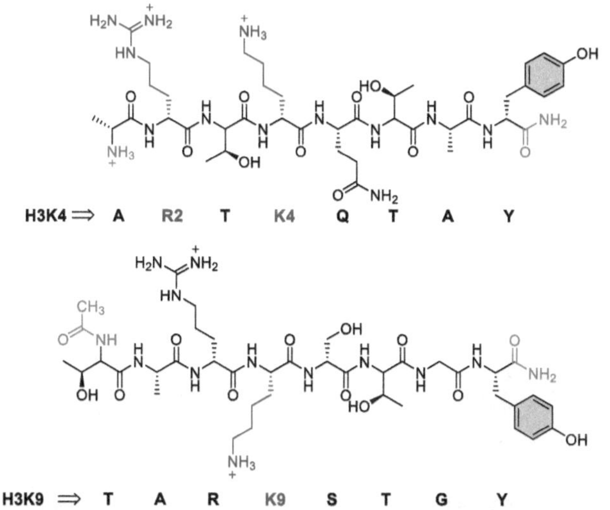

Figure 2.6. (Top) The H3K4 sequence derived from human H3 histone tail. A total of 7 peptides were derived from this sequence by substituting methylated amino acids at highlighted position (Blue), R2 and K4. (Bottom) Base sequence of H3K9 peptide derived from human H3 histone tail. Only one peptide was derived from this sequence with position K9 replaced by Kme3. Note that only first 7 amino acid in both sequences are derived from H3 and addition of Y at C-terminus was done as UV photoprobe to assist in purification. N- and C-terminus of peptides were acylated/amidated to represent their true charge state in vivo.

At this point, it is worthwhile to discuss the nomenclature of guests used in this study. All guests are peptides and are eight amino acids long. The base sequence of all peptides is derived from the *N*-terminal tail of histone 3 (H3) (Figure 2.6). The eighth position, however, is tyrosine in all cases, which was added as a strong UV absorbing molecule for UV-guided detection during HPLC purification of peptides. Each peptide has one specific post-translational modification, identity and position of which is mentioned in the name (Table 2.1). H3K4me2 for example is an 8-mer peptide which first seven amino acid from H3 tail sequence (H3(1-7)). It has a tyrosine (**Y**) at its last position and the fourth lysine is *N,N*-dimethylated (hence K4me2). All peptides had their *N*-terminus free and *C*-terminus acetylated. The exception is H3K9me3 peptide whose base sequence is H3(6-12) instead of H3(1-7) and has both its *N*- and *C*-terminus acetylated.

Table 2.1. Name and charge of all peptide guest with modification listed at appropriate places on the base sequences described in Figure 2.6

Base sequence	Modification	Guest Name	Charge
	None	H3K4	+3
	K4 = Kme1	H3K4me1	+3
	K4 = Kme2	H3K4me2	+3
H3K4	K4 = Kme3	H3K4me3	+3
	K4 = Kac	H3K4ac	+2
	R2 = R2me2a	H3R2me2a	+3
	R2 = R2me2s	H3R2me2s	+3
H3K9	K9 = Kme3	H3K9me3	+2

2.6. Binding assay

We choose to compare the strength on binding between calix[4]arenes and peptides by estimating their equilibrium dissociation constant (K_d). The mathematical definition of K_d for a typical 1:1 binding equilibrium between host (H) and guest (G) is shown in red box in Figure 2.7. Strength of binding has an inverse relation with K_d. Stronger binding between H and G will result in higher formation of HG and hence lower K_d. A significant portion of time in this project was spent trying to find an ideal assay to determine K_d for calix[4]arene-peptide complexes. The methods explored included nuclear magnetic resonance (NMR) titration, isothermal titration calorimetry (ITC), surface plasmon resonance (SPR), fluorescence polarization (FP) and indicator displacement assay (IDA). IDA and ITC ended up being the only techniques that gave any significant response. Due to low to sub micromolar K_d that was expected of our system, the ideal concentration of titrants had to be in the low micromolar region, which rendered NMR unsuitable.[229] FP was briefly tested as a candidate, but the signal was too low to extract any meaningful data. SPR, which was tested in more detail, also had the same flaw. The working principles for both SPR and FP require large relative changes in molecular size upon complexation. This is not the case in our system where the size of the host-guest complex is comparable to free host and guest molecules. Barely any signal was observed in FP. We were able to detect binding using SPR, but the overall signal was too low for confident estimation of binding constants. ITC is capable of generating binding curves at concentration closer to what we desire but was discarded because it would have required large amount of peptides to be synthesized (a full titration including scouting and duplicate runs usually consumes about 6-13 mg of peptide per calix[4]arenes).

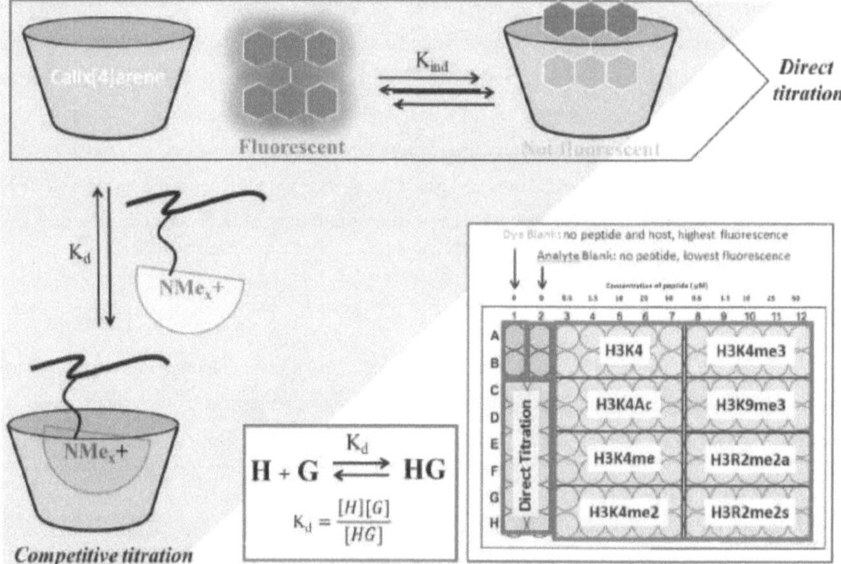

Figure 2.7. Scheme for indicator displacement assay used in this book. The fluorescent indicator of choice was lucigenin for entirety of this work. Lucigenin's fluorescence is quenched upon complexation with *p*-sulfonatocalix[4]arenes. The fluorescence is recovered upon titration of peptides because the equilibrium shifts more towards "free" indicator. All titrations were performed in a 96 well plate, the layout of which is shown in blue box at bottom right. A total of 9 duplicate titration (1 direct, 8 competitive) were performed for each compound per plate. These were used to determine K_d value, as defined in red box, for each calix[4]arene-peptide pair.

IDA was our choice of assay to study binding between peptides and calix[4]arenes. This technique has been popularized by the groups of Werner Nau and Eric Anslyn.[230–234] The working principle involves titrating a dye with the host first.[230,231] Macrocyclic encapsulation of dyes results in change in their fluorescence intensity.[235] The change is fluorescence upon titrating host molecule to a fixed concentration of dye (Direct titration) can be used to build a binding curve and extract indicator binding constant (referred as K_{ind} in Figure 2.7). Next, to a fixed concentration of host-dye complex, guest molecules are titrated.[230,231] This is an example of a competitive titration where guest and dye are competing for fixed concentration of host molecules. This competition shifts the equilibrium between host and dye away from their complex. This would result in an opposite shift in fluorescence intensity than direct titration (Figure 2.7).

The method has been adopted in our group by previous labmates, Sara Tabet (MSc) and Janessa Li (BSc Honors).[178] Lucigenin was our choice as a dye and shows decreased fluorescence upon binding with calix[4]arene almost universally.[235] The assays were designed to be performed in 96 well plate with host concentration fixed at 5 µM, lucigenin concentration fixed at 500 nM and peptide concentrations varied from 0-50 µM (details of procedure are explained in Section 2.9.6). All titrations were done in duplicate. The whole plate data would then be analysed using an in-house developed Python code which would extract 9 set of duplicate titration data and fit them using equations appropriate for each kind of titration.[236–238]

The K_d value ascribed to each host-guest pair is the average of two experiments and is tabulated in Table 2.3 (Section 2.9.6). Errors are reported as average standard error of fit parameters, calculated from covariance matrix of nonlinear regression.

2.7. Trends in binding

Figure 2.8 summarizes binding constants estimated for each library member in the form of a heat map. A clear diversity in binding behavior of library members can be easily seen with some compounds completely losing all affinity (**E10**) and some losing all selectivity (**DO3**). However, most compounds in the library are selective binders for H3K4me3. Some other trends and comparisons are noteworthy.

2.7.1. Scaffold A versus Scaffold E: mono substituted trisulfonated calix[4]arenes

Scaffold **A** seems to be the generally best performing scaffold. Every scaffold **A** member binds H3K4me3 with K_d < 1 µM. **AH6** has the lowest affinity for H3K4me3 (K_d = 1.0 µM) but is still stronger than reader proteins found *in vivo*.[163,180,181,239] The selectivity for H3K4me3 over H3K4 is also very uniform among scaffold **A** members with most compounds being 20-30 fold selective for H3K4me3. Notable exceptions are **AH2** (3 fold selective) and **AP1** (>100 fold selective). Scaffold **E** produced more diverse behavior. Some scaffold **E** members (**E2** and **E4**) seem to have completely lost the ability to bind methylated peptides while some still show outstanding binding and selectivity for H3K4me3. **E5** from this subset is particular interesting because it barely shows any binding to H3K4 and H3K4ac while maintaining strong binding to H3K4me3 (K_d = 0.4 µM, > 15,000 fold selective over H3K4). Within scaffold **E**, all sulfonamides (except **E1**) show heavily reduced binding to methylated peptides. To our surprise, compound **E2**, **E4**, and **E10** (sulfonamides) fail to bind any peptide. **E9** and **E11** show minimal binding but only when tested at 10-fold higher concentration than other host molecules. Other Scaffold **E** members (amide, thiourea, carbamate linkage) fare better but are much weaker than Scaffold **A** compounds with respect to affinity.

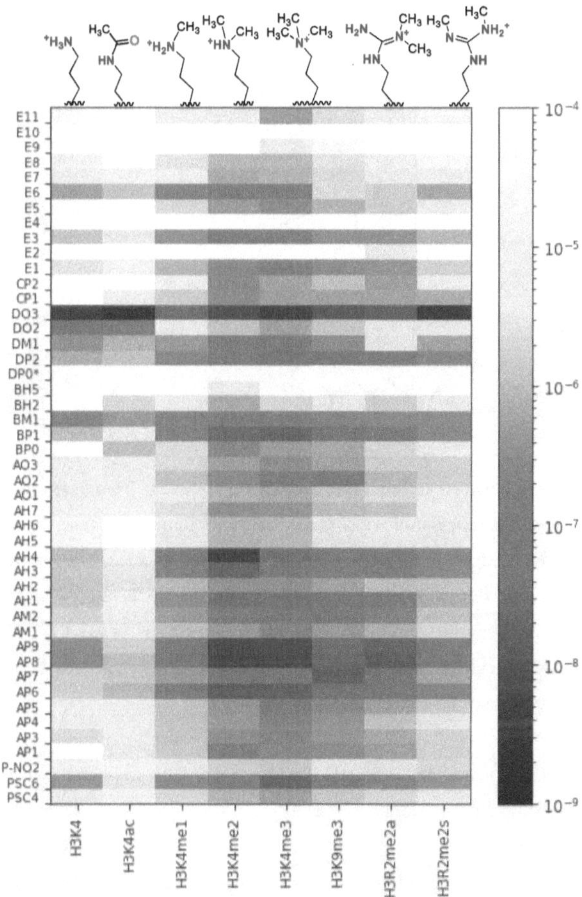

Figure 2.8. Complete summary of all binding constants determined in this Chapter. The constants are reported as dissociation constants, K_d, and so a lower value signifies stronger binding. This means that darker bands in this plot denote stronger binding. Compounds and peptides are labelled according to the nomenclature discussed earlier. The structure of the PTM installed on each peptide is shown at the top to aid the reader.

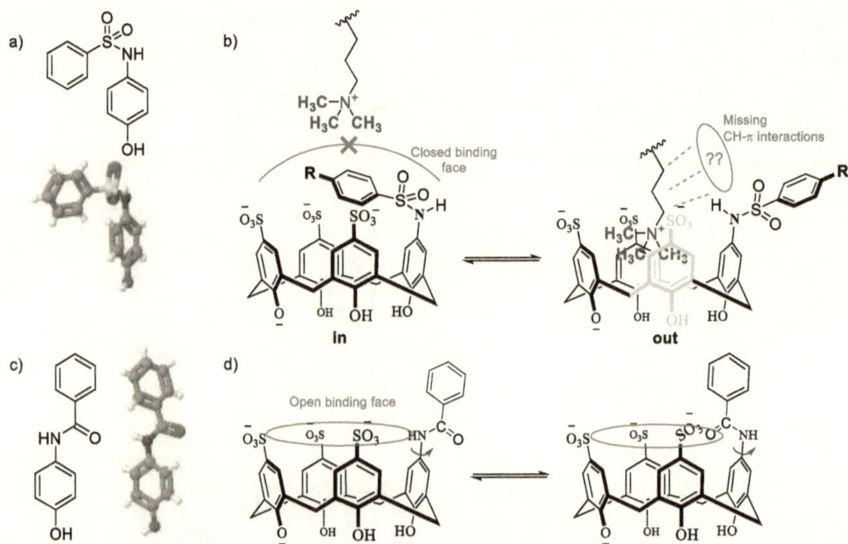

Figure 2.9. Difference between sulfonamide and amide members from scaffold E. a) Chemdraw and crystal structure of a representative *N*-phenylbenzenesulfonamide showing the bent structure (CCDC id NUPZEE).[240] b) Rotation around C(aryl)-N bond generates two major conformers of sulfonamide arm, in and out, will both have reduced affinity. c) Chemdraw and crystal structure of a representative *N*-phenylbenzamide showing the rod like structure (CCDC id VOFCAW).[241] d) Rotation around C(aryl)-N bond generates two conformers for amide, but both show open binding face unlike sulfonamides.

It is difficult to rationalize all the diverse behavior from scaffold **E** members with this limited data, but the behavior of sulfonamides definitely stands out. All except **E1** have much lower affinity to methylated peptide and, also, lucigenin. We attribute this behavior to the inherently bent shape of sulfonamide bonds. A quick search in Cambridge Crystallographic Data Center (CCDC) reveals the acute nature of C(aryl)-N-S bond angle in *N*-aryl sulfonamide bond (Figure 2.9.a). Comparatively, *N*-aryl amides are rod-shaped molecules (Figure 2.9.c). The bent shape of sulfonamide is likely to generate two major conformers, "in" and "out", based on the direction of tilt of *S*-aryl arm (Figure 2.9.b). The "in" conformer completely blocks the binding cavity. The "out" conformer, while not sterically imposing, removes any possibility of CH-π interaction with between aryl group and side chain methylenes which would also lower the affinity. **E1** is the only sulfonamide that produced good binding curves. We believe this is because of presence of a carboxylate at the *para* position of the *S*-aryl arm (R in Figure 2.9.b) of **E1**. The presence of a negatively charged group could shift the equilibrium towards the "out" conformer due to electrostatic repulsion with negatively charged sulfonates. The "out" conformer of **E1**, while not providing any CH-π interactions, is likely to engage neighbouring positively charged groups on our peptide (arginine and free amine at the *N*-terminus) once again through the carboxylate. These attributes set **E1** apart from other sulfonamides and it shows decent binding to all peptides tested. All other molecules derived from scaffold **E** also have flexible arms and likely exist in multiple equilibrating conformers but, as shown for amides (Figure 2.9.d), the binding cavity is only mildly perturbed and doesn't lead to massive reduction in affinity.

2.7.2. Scaffold A versus Scaffold B and D (mono- versus diarylated calix[4]arenes)

As discussed before, evaluation of compounds generated from scaffolds **B-D** ran into unexpected solubility issues. The scaffolds themselves are difficult to dissolve in water and become even more insoluble after arylation. Only 12 molecules from these three scaffolds were soluble enough to be evaluated. Due to limited data available it is difficult to rationalize any of the observed binding behavior. If anything, their chaotic behavior reinforces our belief in the fact the substitutions affect the binding behavior of calix[4]arene in ways that are difficult to predict. For example, **DP0** completely loses its ability to bind lucigenin (a first in our experience) while its isomer, **BP0**, binds lucigenin as expected. **AP0** (not included in Figure 2.8) also binds lucigenin and has been reported elsewhere to bind methylated peptides.[178] **DP0** is the only Suzuki-coupled product on any scaffold not to bind lucigenin and we were unable to calculate its binding constants against our peptides. This observation becomes even more outstanding when we note that **DO2** and **DO3** show the expected binding to lucigenin. These compounds differ from **DP0** only in terms of a single *ortho* substitution at each appended aryl ring (Figure 2.10). Completely opposite to the comparison of **DP0** and **BP0**, **BM1** and **DM1** have nearly identical binding profiles.

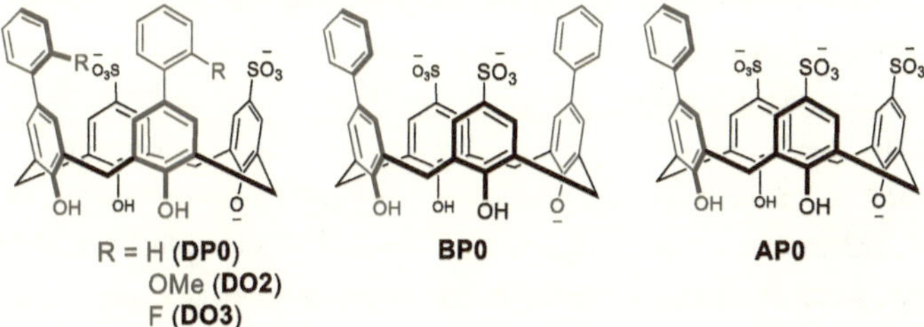

Figure 2.10. Structure of DP0, its isomer BP0 and closely related compounds DO2 and DO3. AP0 is also shown for comparison. Despite much similarity, DP0 doesn't bind to lucigenin while all others do.

DO2 and **DO3** also show increased affinity towards non-methylated peptides, opposite of **AO2** and **AO3** (Table 2.2). This is a stark departure from the norm. While there are few isolated examples in the library that show marginal selectivity for H3K4 over H3K9me3, without exception all molecules in the library are selective for H3K4me3 over H3K4, a combination of stronger electrostatics and hydrophobic affect. **DO2** is 16-40 fold selective for unmethylated peptides over all methylated peptides. **DO3** binds every peptide strongly (K_d = 3–35 nM), with small selectivity for H3K4 and H3K4ac.

Table 2.2. Dissociation constants (in nM) of ortho aryl functionalized Scaffold A and D compounds show reversed selectivity.

Host	Methylated peptides (charge)		Nonmethylated peptides (charge)	
	H3K4me3 (+3)	H3K9me3 (+2)	H3K4 (+3)	H3K4ac (+2)
DO2	610 ±70	1500±200	36 ±62	50 ±38
DO3	4.8 ±0.1	11 ± 1	3.8 ±0.3	1.9 ±0.3
AO2	110 ±20	80 ±20	4100 ±200	7000 ±200
AO3	330 ±30	660 ±60	4300 ±100	6200 ±3100

Average of two titrations. Errors are reported as average standard error.

We believe the driving force for this anomalous behavior lies in the restricted rotation of the embedded biphenyl arm of calixarenes (highlighted in blue in Figure 2.10). Planar configuration of biphenyls is more unstable than staggered configuration due to steric crowding of ortho substituents (Figure 2.11.a).[242] Depending on the bulkiness of substituents, rotation around C-C bond can either be mildly restricted or completely stopped (atropoisomerism).[243,244] The biphenyl arm embedded in our arylated calix[4]arenes must adopt planar configuration upon guest inclusion (Figure 2.11.b). **AO2** and **AO3** have methoxy and fluorine substituents respectively which have moderate steric bulk.[243] The energetic planar conformation (of biphenyl arm) is likely compensated by the favourable CH-π and electrostatic interactions gained upon encapsulation of Kme3 side chain. **DO2** and **DO3**, on the other hand, not only have two ortho substituted biphenyl embedded in their structure, they also have present less electrostatic attraction (-4 charge on **AO2/3** vs -3 charge on **DO2/3**). Additionally, proximity of ortho substituents in **DO2** and **DO3** is bound to increase steric clash significantly more than just twice of their scaffold **A** counterparts (Figure 2.11.c). We believe these factors combined raise the energetics of planar conformation of biphenyl arms to the point that they can not be easily compensated by guest encapsulation. Staggered conformation must also reduce the cavity volume, inhibiting bulky guests from entering. However, a simple size exclusion cannot explain the binding profile since H3K4me2/1 bind weaker than H3K4me3. A high-end computational study is required to find the precise conformation **DO2** and **DO3** adopt when molecules enter their cavity, which we believe might differ for each peptide.

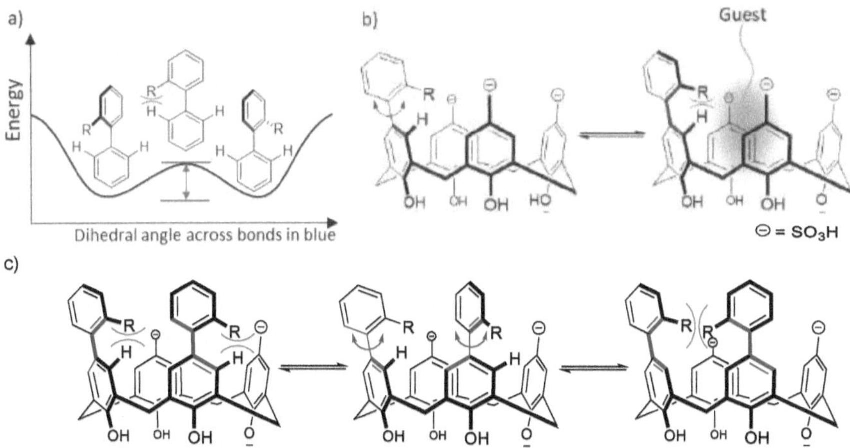

Figure 2.11.a) (Top Left) A typical energy profile of ortho substituted biphenyls shows the planar conformation with 0 dihedral angle (across bonds in blue) to be more energetic than staggered conformation. The energy barrier for rotation (shown in red) increases with the bulkiness or R group. b) (Top Right) Scaffold A derived molecules adopt planar configuration despite the steric clash due to favourable host-guest interactions. c) (Bottom) Scaffold D derived compounds can adopt two planar conformations both of which have higher steric clash (compared to scaffold A).

2.7.3. *Deeper pockets lead to increase general selectivity for methylated over unmethylated peptides*

Most compounds derived from Scaffold **C** had poor solubility and are not included in this study. The solubility improved enough to allow us to K_d determination when the coupled phenyl groups had charged groups on them. Fragment **P1** and **P2** are two such aryl groups each with a *para* carboxylic acid which, at the pH tested, has a negative charge. Coupling of these fragment does not reduce overall charge on calix[4]arene, unlike any other fragment in our library. **PSC, AP1, BP1, CP1** and **CP2** have the same charge (-5) but differ with respect to their hydrophobic surface area which incrementally increases with each aryl arm attached. We note that the overall selectivity towards methylated peptides increases concomitantly. If selectivity for H3K4me3 over H3K4 is taken as an example, the selectivity increases from 7-fold for **PSC**, to >200 for either scaffold **C** compound (Figure 2.12) Similar trend can be seen for selectivity for H3K9me3 over H3K4.

The reason for this increase is two-fold. First, the K_d for H3K4 increases when going from **PSC** to **CP1/2** (Figure 2.12). Interestingly, there is a big increase in K_d(H3K4) when going from **BP1** (K_d = 1.0

uM) to **CP1/2** (K_d = >100 μM). Second, opposite to the trend for H3K4, K_d for H3K4me3 decreases as more phenyl rings are added (Figure 2.12). Once again, the trend breaks at **CP1/2** where it slightly decreases. The net effect is an overall increase in selectivity (defined as K_d(H3K4)/K_d(H3K4me3)). Since the charge on this subset of host and guest molecules are same, this behavior can be easily reasoned based on hydrophobicity of incoming guest side chain. Kme3 has more hydrophobic sidechain and would prefer to be sequestered by the enveloping cavity of **CP1/2**. Side chain of K, on the other hand is hydrophilic and prefers a more side-on binding mode where the ammonium and methylene stay close to the negatively charged upper rim of calix[4]arene (Orange box, Figure 2.12). We reason that the carboxylates of **CP1/2** form a second negatively charged rim which prevents encapsulation of lysine side chain. The possibility to form charged hydrogen bonds with between ammonium and carboxylate must also plays its part in preventing it from entering the cavity.

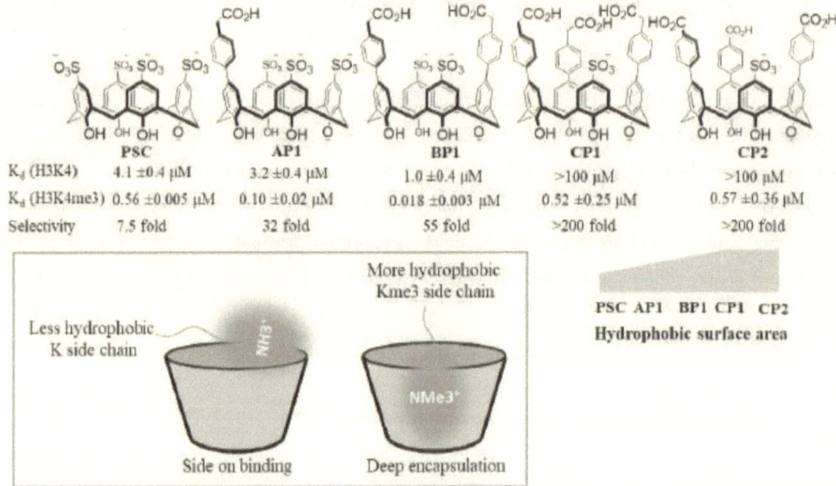

	PSC	AP1	BP1	CP1	CP2
K_d (H3K4)	4.1 ±0.4 μM	3.2 ±0.4 μM	1.0 ±0.4 μM	>100 μM	>100 μM
K_d (H3K4me3)	0.56 ±0.005 μM	0.10 ±0.02 μM	0.018 ±0.003 μM	0.52 ±0.25 μM	0.57 ±0.36 μM
Selectivity	7.5 fold	32 fold	55 fold	>200 fold	>200 fold

Figure 2.12. Subset of library member with same charge as PSC but deeper cavity show increasing selectivity for methylated peptides. CP1 and CP2 show most impressive selectivity (higher than cucurbiturils which hold the current record). Two competing modes of binding for unmethylated and methylated ammonium guests are shown in the orange box. The new negatively charged portal of CP1/2, formed by carboxylates, likely prevents H3K4 side chain from entering the cavity.

A caveat of increasing the hydrophobicity of calix[4]arene's cavity is that while the selectivity of scaffold **C** compounds for H3K4me3 over H3K4 increases, significantly higher than the record set by cucurbiturils, their preference among methylated peptides goes down. **CP1** specifically binds all methylated peptides with impressive strength (K_d = 300-1500 nM). **CP2** shows a more broad range of binding constants for our test peptides (K_d = 270-6200 nM) and, curiously, is 8-fold selective for H3R2me2a over H3R2me2s. Comparatively, **CP1** binds both methylated arginine containing peptides with identical strength. It is difficult to explain this effect with current set of data, but we can assert that this must be because of the difference between how alkyl and aromatic carboxylate interact with H3K4me1. **CP1** is, thus, an excellent candidate for pan-selective methylated peptides binders.

2.7.4 Electrostatics drive peptide-calix[4]arene interactions in complicated, shape dependant way

When we included **PSC6** in our library we assumed that because of its high negative charge (-7 compared to -5 for **PS4**), it would afford more non-specific interactions which would improve binding but decrease selectivity. We were right about binding but wrong about selectivity. **PSC6** has improved K_d for all peptides tested but shows improved selectivity for H3K4me3 over H3K4 (22 fold) than **PSC4** (7.5 fold). Comparison of **BP0** and **BP1** paints a completely opposite picture. They both have similar groups appended but **BP0** has only -3 charge compared to **BP1** (-5 charge). In a complete reversal from all example discussed above, **BP0** is much more selective to H3K4me3 (>180 fold) and H3K9me3 (>140 fold) over H3K4 to which it doesn't seem to interact in our assay. In a single library, we have competing evidence for role of electrostatics.

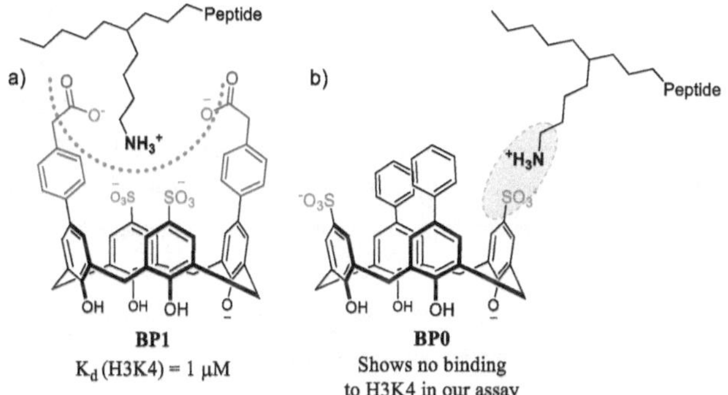

Figure 2.13. Proposed difference between binding modes of a) BP1, and b) BP0 to H3K4 peptide.

These examples cannot be explained by a simple encapsulation model. They once again stem from big differences in the mode of binding of unmethylated and methylated side chains (Orange box, Figure 2.12). Unmethylated lysine side chain prefers side on interaction with sulfonated calixarene where ammonium stays close to negatively charged sulfonates. Kme3 side is actually encapsulated within the cavity and the methylated ammonium is embedded deeper and away from sulfonates. This brings the entire peptide backbone closer and improves electrostatic contact. This is likely the reason behind improved selectivity of **PSC6**.

The case of **BP0** is more subtle. Because of way lysine interacts and the position of sulfonates on **BP0**, H3K4 is unlikely to be encapsulated and more likely to interact with the calix[4]arene's "walls" (Figure 2.13.b). Installing carboxylates on the phenyl arms, as in **BP1**, gives it an alternate position to adhere to (Figure 2.13.a). It must be noted that our choice of assay, IDA, has an inherent bias because it wouldn't show any signal for guest molecules that do not interact with the cavity of calixarenes. Such is the case for **BP0** and H3K4 while **BP1** does interact with H3K4 through its upper rim and hence shows signal in our assay. H3K4me3, on the other hand, is easily encapsulated inside the cavity of **BP0** due to combined electrostatic interactions and hydrophobic effect of the Kme3 side chain. Comparatively, **BP1** also encapsulates Kme3 easily and is in fact a stronger binder than **BP0** (this can be safely attributed to difference in electrostatics). The impressive selectivity of **BP0** stems from the completely nonexistent encapsulation of H3K4 within its cavity while still binding H3K4me3s.

2.7.5. *Dimethyllysine selective calix[4]arenes from the library, failed hits but valuable lessons*

While the big underlying goal of this work was to find strong binder (pan-selective) for methylated peptides, we were also very keen of finding calix[4]arene based hosts with strong selectivity towards lower methylation states of lysine. To date, waters group's disulfide containing cyclophane **A₂D** remains the only example of a compound that shows selectivity for Kme2 containing peptides over Kme3. Because of a difference of only one methyl group between the two guests, this has been formidable goal to achieve and, as discussed is section 2.2, all our designed efforts have failed. We were pleased to see multiple compounds from our library show marginal selectivity for H3K4me2 (1.5-2 fold). Compounds like **BP0, E3**, **E7**, **CP1** and **CP2** are few examples that stand out. However, we were also worried that such small difference may be due to biases within our assay. We instead, unfortunately, gravitated towards compounds like **AH3**, **AH4**, **BH2** and **BH5** which show 3-, 6-, 3- and 32-fold selectivity towards H3K4me2 respectively in our IDA. **BH2** and **BH5** were detected to be false positives very early on. Direct titrations of both were difficult to reproduce and we thought that additional equilibria might be responsible for this behavior. **AH3** and **AH4** showed expected curves and a significant time of this project was dedicated to replicating the IDA results in hopes to uncover the much sought after dimethyllysine selectivity. Both these compounds proved to be false positives and a big drain of time and efforts.

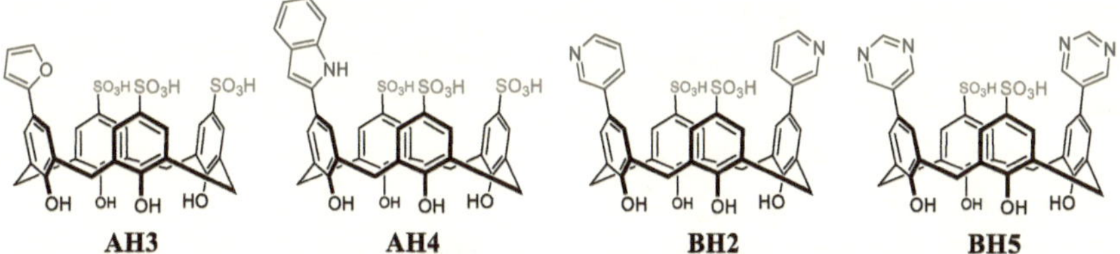

Figure 2.14. Heterocycle containing calix[4]arenes that appear Kme2 selective in our assays but are false positives.

We had failed to appreciate that all our dimethyllysine selective hits were heterocycle substituted calixarenes. In fact, almost all heterocycle substituted calixarenes in the library show marginal dimethyllysine selectivity. These heterocycles were difficult to synthesize to begin with. The Suzuki coupling gave the worst yields with heterocyclic boronic acids. In addition, multiple other chemical pathways becomes possible when a highly electron donating group like phenol is conjugated to an electron accepting heterocycle.[245–247] Notably, lot of these heterocyclic calixarenes are inherently fluorescent. Their fluorescence is also likely highly susceptible to hydrogen bonding and presence of metals. The upper rim of our calixarenes, where these heterocycles were positioned, is a negatively charged portal and likely proved to be an aid in this regard. Additionally, electron density from phenols flows into heterocycles making the heteroatom very nucleophilic. Presence of acidic protons in proximity is ripe combination of multiple ring opening reaction that heterocycles are known to take part in.[248] This proton can be delivered intramolecularly by the sulfonates or intermolecularly by the phenols. These conditions made both **AH3** and **AH4** to rapidly degrade into coloured byproducts. We failed to characterize these byproducts because in most cases the NMR peaks were broad and unassignable. **AH3** in particular was a subject of long investigation by us. It degrades into a green coloured compound even in solid state. Multiple attempts to curb this rapid degradation by modifying reaction conditions also failed. In the end, after multiple false leads, we completely abandoned the heterocycle substituted calixarenes and while they are included in this Chapter for completion, we draw no conclusions from them other than caution for future works.

2.8. Conclusion

The library of calix[4]arene based hosts that are central to this Chapter are all novel host that have never been synthesized outside our lab. There diverse binding profile prove our underlying idea, and motivation for this Chapter, that the shape and size of calix[4]arene cavity is heavily influenced by the substitution at the upper rim which in turn would alter their host-guest properties. And while we were not able to identify a dimethyllysine selective calix[4]arene, we were able to synthesize a subset of highly selective host for methylated peptides. These hosts (**CP1**, **BP0** etc.) have stronger affinity and selectivity than a) any published hosts in literature, b) known proteins that detect methylated peptides inside the cells. We can also now draw a meaningful structure activity relationships from this project that can help us in future works:

a. Sulfonamides at the upper rim block calix[4]arene's cavity unless it has a negative charged group at *para* position.

b. *ortho* substituted phenyl rings, when installed at the upper rim, cause a big change in population of conformations of calix[4]arenes.

c. If two *ortho* substituted phenyl rings are installed at proximal phenols of calix[4]arenes (Scaffold **D**), the host molecule is forced to adopt conformations where the biphenyl arm are in staggered conformation to avoid steric clash. This prevents the π face of aryl ring from interacting with the host.

d. Unless hydrophobic and/or methylated, ammonium guest prefer side on interaction with the calix[4]arene. This behavior can be exploited by changing position sulfonates such that the unmethylated lysine/arginine chain do not interact with the cavity.

e. Deeper hydrophobic cavity of arylated calixarenes promotes selectivity towards methylated peptides. Installation of additional negative charged groups enhances binding strength at the expense of selectivity.

f. Poly-arylated calixarenes have reduced solubility even when sulfonated. A minimum of 3 negative charge is needed to maintain decent solubility, but -4 makes it much easier to work with.

g. Aggregation behavior of calix[4]arenes was completely unaccounted for in this work but becomes more and more important as more hydrophobic groups are added on to the calix[4]arene. While the concentration utilized in this study (5 µM or less) will keep such equilibria to a minimum, it might be the reason behind few exceptional behaviors in our library.

h. Electronically conjugating heterocycles to the phenols at the upper rim leads to rapidly degrading and/or fluorescent calix[4]arenes which should be avoided. A non-fluorescent assay must be used to study their host-guest properties if needed.

Due to low toxicity of our hosts, calix[4]arenes with exceptional binding profiles identified in this Chapter can be promising tools in chemical biology. As discussed before (Section 2.2), we envision them to be useful in enrichment protocols for proteomics studies. At last we identify few key candidates from this library that can be used towards this goal. These host molecules and their utility will be demonstrated in Chapter 3.

2.9 Supporting information

2.9.1. General considerations

Proton NMR (^{1}H NMR) were recorded on a either Bruker Avance 300 MHz or 500 MHz spectrometer at 23 °C and processes with Topspin 3.6.1. Proton chemical shifts are reported in parts per million (ppm) downfield from tetramethylsilane and are referenced to residual proton in the NMR solvent. Data are represented as follows: chemical shift, multiplicity (s = singlet, d = doublet, t = triplet, q = quartet, m = multiplet, br = broad), integration, coupling constant in Hertz. Carbon NMR chemical shifts (^{13}C$\{^1$H$\}$ NMR) were recorded at 125 Hz at 23 °C and processes with Topspin 3.6.1. Carbon chemical shifts are reported in parts per million (ppm) downfield from tetramethylsilane and are referenced to the carbon resonances of the solvent except if the solvent is D_2O, in which case a 5% MeOH solution in D2O was used as an eternal reference. Infrared (IR) spectra were obtained using either a Perkin Elmer Spectrum-2 (ATR-FTIR) or a Perkin Elmer 1000 FT-IR spectrometer. Data are represented as follows: frequency of absorbption (cm^{-1}), intensity of absorption (s = strong, m = medium, w = weak, br = broad). Melting points were collected on a Gallendkamp Melting Point apparatus. Microwave reactions were performed on a Biotage Initiator microwave in a heavy glass microwave vial.

All chemicals were supplied by Sigma-Aldrich and used as received, unless otherwise indicated. **PSC4** and **PSC6** was purchased from TCI America. Fmoc-Lys(Me3)-OH was purchased from GL Biochem. All other amino acids purchased from Chemimpex.

2.9.2. HPLC purification

All compounds and peptides were purified using RP-HPLC on Phenomenex Luna C18(2) column (21.2 mm x 250 mm) with 5 μm particle and 100 Å pore size. The preparative RP-HPLC purifications were performed on either a Shimadzu Prominence HPLC or a Thermo-Dionex HPLC/MS system using a gradient of acetonitrile (ACN) in water as mobile phase. Both mobile phase eluents were spiked with 0.1% TFA. The compounds were detected using a DAD set to 280 nm. A typical gradient started at 90:10 mixture of H_2O and ACN and went down to 10:90 H2O:ACN in over 20 min.

Analytical LC-MS characterization was done on either a Waters Acquity-H UPLC-MS, using a BEC C18 column with 1.7 μm particle size, or a Thermo-Dionex HPLC/MS using a Jupiter 4u Proteo 90-A column with 4 μm particle size (4.6 mm x 250 mm, Phenomenex). A gradient identical to preparative HPLC was used as mobile phase except with small differences on Waters system where the solvents were spiked with 0.4 % formic acid and the gradient time was only 4 minutes. The compounds were detected using a DAD set to 280 nm and an inline mass spectrometer.

2.9.3. Peptide Synthesis

All peptides were made on an automated solid phase peptide synthesizer LIBERTY1 (CEM). The synthesis was done on Rink Amide resin (Chem Impex) using Fmoc chemistry and microwave assisted coupling methods programmed in the machine. After final coupling, the N-terminus of H3K9me3 was acetylated on resin using a 30:20:50 pyridine/acetic anhydride/DCM mixture for 2 hours. All other peptides were synthesized with free N-terminus. After synthesis, the peptides were cleaved from the resin as C-terminal amide using a 95:2.5:2.5 mixture of TFA, H_2O and triisopropylsilane. The reaction continued for 4 hours under gentle bubbling of N_2. After that, the resin was filtered and washed with TFA (3 x 5 mL). The combined TFA solution was concentrated on a rotavap. Ice cold diethyl ether was then used to precipitate crude peptides which were then air dried and then purified as described in Section 2.9.3. The gradient used for purification started from 5% acetonitrile in water (with 0.1% TFA), increased to 20% acetonitrile in 15 minutes and then up to 90% in next 15 minutes. The peptides were characterized on Waters UPLC (as described in Section 2.9.2). The analytical traces for all peptides are shown in Appendix B.

2.9.4. Synthesis of scaffolds

Compound 2.2,[249] 2.3,[215] 2.5[250] and 2.9[215] were synthesized made according to literature and their spectral properties matched published results.

Scaffold **A**. Compound **2.5** (200 mg, 0.38 mmol, 1 eq.) was dissolved in a minimal amount of DCM in a round bottom flask (RBF) with an attached reflux condenser. Concentrated H_2SO_4 (0.5 mL, 20 eq.) was added and the reaction mixture was stirred at 60 °C for 3 hours. The product is observed to be precipitating out of the reaction mixture during this time. After 3 hours, the DCM supernatant can be decanted off and checked for left over starting material using TLC. If significant amount of starting material is still present then reaction can be restarted in a fresh reaction vessel. The combined precipitate is then suspended in a minimal amount of EtOAc and poured into centrifuge tubes (50 mL) and topped off with Et_2O. After two rounds of decanting, resuspending in Et_2O and centrifugation, the precipitate is air dried to provide crude product as a grey powder. This crude can be used in further reactions without purification. Analytical samples can be obtained by further purifying it using the HPLC. Isolated yield after HPLC: 110 mg (40 %). Mp: >250 °C (dec). **IR (ATR, cm^{-1}):** 3198s br, 1454s, 1280s, 1160s, 1037s, 883w, 847w, 808w, 786w,

626m, 567m, 542w, 408w. **^{1}H NMR (500 MHz, 10% D2O in MeOD-d4):** δ 3.94 (br s, 4H), 4.02 (br s, 4H), 7.25 (s, 2H), 7.60 (s, 2H), 7.61 (s, 2H), 7.64 (s, 2H) **^{13}C{^{1}H} NMR (125 MHz, D2O):** δ 30.1, 31.2, 112.7, 127.1, 127.4, 127.8, 128.8, 129.0, 129.4, 129.5, 131.8, 137.1, 137.3, s148.3, 151.4, 152.0. **HR-ESI-MS:** Calculated for $C_{28}H_{23}BrO_{13}S_3Na$ $[M+Na]^{+1}$ = 766.9376, Found 766.9355.

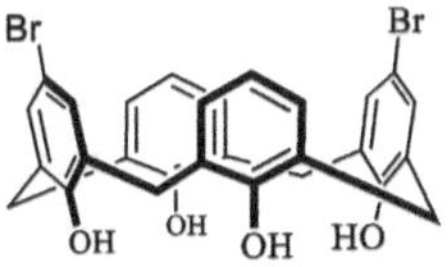

Compound **2.7**. Compound **2.3** (400 mg, 0.54 mmol) was suspended in (30 mL) chloroform in a RBF and bromine (0.22 mL, 4.32 mmol, 8 eq.) was added with stirring. The reaction was stirred for 4 hours at room temperature in open atmosphere. The reaction was quenched by adding 100 mL of 5 % (w/v) aqueous solution of $Na_2S_2O_3$. This results in formation of large quantities of yellowish precipitate. The entire solution was taken into a seperatory funnel and organic layer was separated. The organic layer was then washed several times with the same $Na_2S_2O_3$ solution until the aqueous layer remained clear. The organic layer was then washed with brine and the dried over Na_2SO_4 before concentrating *in vacuo* to give yellowish solid as crude debenzoylated material which was used further without purification. First, the material obtained was resuspended in methanol in a RBF. The RBF was equipped with a condenser and NaOH (2.1 gm, 54 mmol, 100 eq.) is added to the suspension which is then set to reflux overnight. After this, the reaction mixture is quenched by adding 150 mL 1 M HCl. More HCL can be added to bring the pH below 1. The resulting precipitate is filtered and washed with 200 mL hot hexane to remove benzoic acid. The precipitate can be further purified on silica gel chromatography using DCM as eluent to obtain analytical samples. Isolated yield – 215 mg (70 %). Mp: >250 °C (dec). **IR (KBr pellet, cm^{-1}):** 3141 m br, 2950 w, 1605 w, 1593 w, 1464 m, 1469 s, 1263 m, 1210 s, 878 m, 860 m, 829 m, 750 s. 665 m, 583 m. **^{1}H NMR (300 MHz, CDCl$_3$):** δ 3.51 (br s, 4H), 4.19 (br s, 4H), 6.80 (t, 2H, J = 7.5 Hz), 7.08 (d, 2H, J = 7.6 Hz), 7.15 (s, 4H), 10.05 (s, 4H, exchangeable). **^{13}C{^{1}H} NMR (125 MHz, CDCl$_3$):** 31.6, 114.2, 122.7, 127.6, 129.5, 130.3, 131.7, 148.0, 148.9. **HR-ESI-MS:** Calculated for $C_{28}H_{23}O_4Br_2$ $[M+H]^{-1}$ = 580.9958, Found 580.9779.

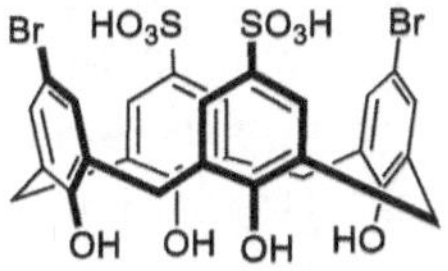

Scaffold **B**. Compound **2.7** (100 mg, 0.19 mmol, 1 eq.) was dissolved in a minimal amount of DCM in a RBF with an attached reflux condenser. Concentrated H_2SO_4 (0.25 mL, 20 eq.) was added and the reaction mixture was stirred at 60 °C for 3 hours. The product is observed to be precipitating out of the reaction mixture during this time. After 3 hours, the supernatant can be decanted off and checked for left over starting material using TLC. If significant amount of starting material is still present then reaction can be restarted in a fresh reaction vessel. The combined precipitate is then suspended in a minimal amount of EtOAc and poured into centrifuge tubes (50 mL) and topped off with Et_2O. After two rounds of decanting, resuspending in Et_2O and centrifugation, the precipitate is air dried to provide crude product as a grey powder. This crude can be used in further reactions without purification. Analytical samples can be obtained by further purifying it using the HPLC. Isolated yield after HPLC: 43 mg (31 %). **^{1}H NMR (500 MHz, 10% D2O in MeOD-d4):** δ 3.93 (br s, 2H), 7.21(s, 4H), 7.63 (s, 4H) **^{13}C{^{1}H} NMR (125 MHz, 1:1 MeOD-d4:D$_2$O):** 31.2, 114.4, 128.0, 129.1, 130.6, 132.6, 138.0, 148.6, 152.8. **HR-ESI-MS:** Calculated for $C_{28}H_{21}Br_2O_{10}S_2$ $[M-H]^{-1} = 738.8948$, Found 738.8941.

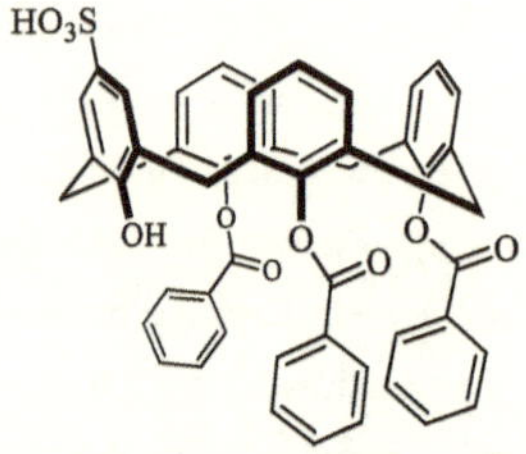

Compound **2.8**. Compound **2.2** (1.5 g, 2.07 mmol) is dissolved in 35 mL DCM in a RBF with stirring. H_2SO_4 (0.221 mL, 4.14 mmol, 2 equiv.) is added slowly and the reaction is fitted with a reflux condenser, placed under argon and refluxed at 80 °C overnight. Over time the solution adopts a pink colour and a pink/purple precipitate is observed. The reaction is allowed to reach room temperature and washed with 1 M HCl (with 5% MeOH). The organic layer is washed with 1:1 1 M HCl:Brine and then dried over Na_2SO_4, filtered and the solvent evaporated to dryness. The crude solid is recrystallized using 1:1 DCM:MeOH to yield 1.402 g of off-white powder was isolated in 83% yield. Mp: >250 °C (dec.). **IR (ATR, cm^{-1}):** 3524 w br, 3062 w, 2927 w, 1725 s, 1601 m, 1585 m, 1467 w, 1451 s, 1259 s, 1165 s, 1093 m, 1022 s, 756 m, 705 s, 657 m, 551 m. **^{1}H NMR (DMSO-$d6$, 300 MHz):** δ 3.41-3.82 (m, 8 H), 6.25 (s, 1 H), 6.48 (d, 2 H, J = 7.1 Hz), 6.56 (t, 2 H, J = 6.7, 14.8 Hz), 6.64 (t, 1 H, J = 8.5, 15.0 Hz), 6.82 (d, 2 H, J = 7.7 Hz), 6.87-6.91 (m, 2 H), 7.27 (d, 2 H, J = 7.4 Hz), 7.43 (t, 2 H, J = 8.1, 15.4 Hz), 7.48 (s, 2 H), 7.63 (t, 5 H, J = 8.0, 15.7 Hz), 7.86 (t, 2 H, J = 7.5, 16.0 Hz), 7.93-7.96 (m, 1 H), 8.08-8.10 (m, 3 H). **^{13}C{^{1}H} NMR (DMSO-$d6$, 75 MHz):** δ 167.2, 164.2, 163.4, 152.5, 147.9, 146.2, 140.1, 134.0, 133.7, 132.7, 132.7, 132.4, 131.9, 130.7, 130.5, 130.5, 130.3, 129.5, 129.4, 129.1, 128.7, 128.5, 128.4, 128.1, 127.8, 127.6, 126.1, 124.9, 124.3, 36.6, 30.3. **HR-ESI-MS:** Calculated for $C_{49}H_{35}O_{10}S$ $[M-H]^{-1} = 815.1956$, Found 815.1937.

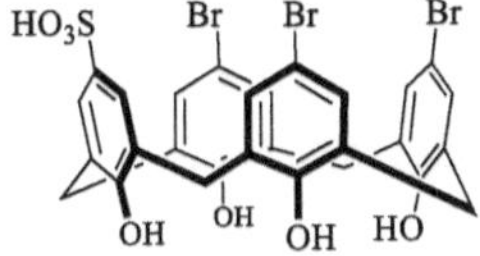

Scaffold C. Compound **2.8** (0.52 g, 0.633 mmol) is dissolved in 25 mL MeOH, in a RBF with stirring. NaOH (2.5 g, 63.3 mmol, 100 equiv.) is added and a reflux condenser is attached and the RBF is placed under argon. The solution is refluxed at 85 °C for 3 hours. The reaction is allowed to cool to room temperature, and solvent is concentrated under reduced pressure. The MeOH slurry is poured into 100 mL ice cold 1 M HCl (brings pH <1 by pH paper) and a white precipitate is observed. The precipitate is filtered and washed with hexanes (5 x 25 mL) to remove benzoic acid. The solid is air dried to yield 0.239 g of an off white powder (debenzoylated product) which is reacted further without purification. This material is dissolved in 20 mL chloroform, in a RBF with stirring. Bromine (0.35 mL, 6.7 mmol, 10.6 eq. with respect to compound **2.8**) is added, dropwise. The solution is allowed to stir under argon for 3 hours, during which a precipitate is observed to form. The solution is transferred to a centrifuge tube and centrifuged, and the chloroform is decanted and fresh chloroform is added, centrifugation and washing is repeated until the chloroform remains clear. The solid is air-dried, washed with 1 M HCl and centrifuged and the HCl is decanted. The solid is slurried in MeOH, transferred to a RBF and the solvent is removed under reduced pressure to yield 167 mg of an off white powder in a combined 60 % yield. Mp: >250 °C (dec.) **IR (KBr pellet, cm^{-1}):** 3164 br, 1458 s, 1206 s, 1041 s, 921 m, 863 s, 810 m, 582 m, 530 w. **^{1}H NMR (DMSO-d6, 300 MHz):** δ 3.87 (br, 8 H), 7.27 (s, 2 H), 7.40-7.44 (m, 6 H). **^{13}C{^{1}H} NMR (DMSO-d6, 75 MHz):** δ 150.6, 149.6, 149.1, 131.0, 130.9, 130.9, 130.5, 130.4, 127.0, 126.4, 111.9, 111.8, 30.2, 29.7. **HR-ESI-MS:** Calculated for $C_{28}H_{21}Br_3O_7S$ [M-H]$^{-1}$ = 736.8485, Found 736.8478.

Compound **2.10.** Compound **2.9** (100 mg, 0.135 mmol) was dissolved in 2-butanone (10 mL) in a RBF and *N*-bromosuccinimide (56 mg, 0.32 mmol, 2 eq.) was added with stirring. The reaction was placed under argon and stirred overnight at room temperature. The reaction is evaporated under reduced pressure to dryness, and crudely purified by column chromatography (SiO2, DCM), to yield 93 mg (78%, crude) of an off white powder. This material was then dissolved in 10 mL MeOH with NaOH (470 mg, 11 mmol, 100 equiv.), fitted with a reflux condensor, placed under argon and refluxed at 80 °C for 3 hours. After cooling to room temperature, the solvent was concentrated under reduced pressure. The MeOH solution is poured into 100 mL ice cold 1 M HCl to make the solution pH <1 (to pH paper) and a white precipitate was observed. This precipitate is filtered and washed with 1 M HCl and allowed to air dry. Once dry the solid is washed with numerous amounts of hexanes (5 x 25 mL), to remove benzoic acid. This was then subjected to column chromatography (SiO2, DCM) which supplied 54 mg as an off white powder in 68% yield (53% over 2 steps). **IR (ATR, cm^{-1}):** 3152 s br, 2947 w, 1595 w, 1555 w, 1467 m, 1450 s, 1402 w, 1249 m, 1212 m, 1203 s, 879 w, 861 w, 832 m, 791 s, 748 s, 668 m, 572 m, 520 m. **^{1}H NMR (CDCl3, 300 MHz):** δ 3.52 (br, 4 H), 4.21 (br, 4 H), 6.77 (t, 2 H, J = 8.0, 15 Hz), 7.07 (m, 4 H), 7.15 (t, 2 H, J = 2.2, 4.0 Hz), 7.20 (d, 2 H, J = 2.2 Hz), 10.05 (s, 3 H), 10.12 (s, 1 H). **^{13}C{^{1}H} NMR (CDCl$_3$, 125 MHz):** 31.5, 31.6, 31.7, 114.0, 122.7, 127.4, 128.3, 129.2, 129.5, 129.6, 130.6, 131.6, 131.9, 148.2, 148.7s. **HR-ESI-MS:** Calculated for $C_{28}H_{21}Br_2O_4$ [M-H]$^{-1}$= 578.9812, Found. 578.9808.

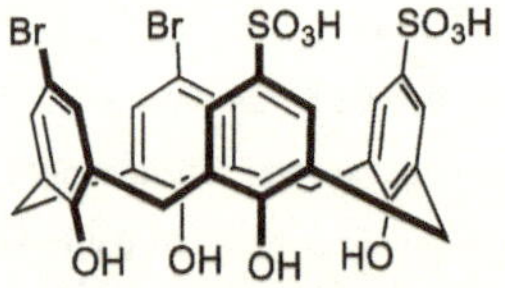

Scaffold D

Scaffold **D**. Compound **2.10** (100 mg, 0.19 mmol, 1 eq.) was dissolved in a minimal amount of DCM in a RBF with an attached reflux condenser. Concentrated H_2SO_4 (0.25 mL, 20 eq.) was added and the reaction mixture was stirred at 60 °C for 3 hours. The product is observed to be precipitating out of the reaction mixture during this time. After 3 hours, the supernatant can be decanted off and checked for left over starting material using TLC. If significant amount of starting material is still present then reaction can be restarted in a fresh reaction vessel. The combined precipitate is then suspended in a minimal amount of EtOAc and poured into centrifuge tubes (50 mL) and topped off with Et$_2$O. After two rounds of decanting, resuspending in Et$_2$O and centrifugation, the precipitate is air dried to provide crude product as a grey powder. This crude can be used in further reactions without purification. Analytical samples can be obtained by further purifying it using the HPLC. Isolated yield after HPLC: 37 mg (28 %). **^{1}H NMR (500 MHz, 10% D$_2$O in MeOD-d4):** δ 3.90 (3 br s overlapping, 8H), 7.00 (s, 2H), 7.17 (s, 2H), 7.68 (s, 2H), 7.74 (s, 2H). **^{13}C{^{1}H} NMR (125 MHz, 10% D$_2$O in MeOD-d4):** 30.8, 31.8, 32.2, 114.4, 128.2, 128.3, 129.0, 129.1, 130.7, 131.2, 132.6, 132.8, 139.3, 149.6, 152.4. **HR-ESI-MS:** Calculated for $C_{28}H_{21}Br_2O_{10}S_2$ [M-H]$^{-1}$ = 738.8948, Found 738.8941.

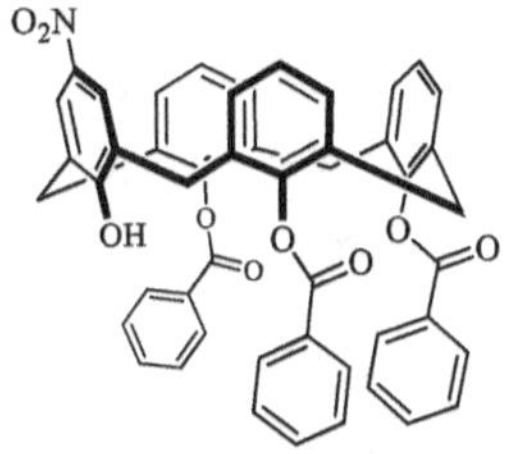

Compound **2.11**. Compound **2.2** (650 mg, 0.88 mmol, 1 eq.) and potassium nitrate (177 mg, 1.76 mmol, 2 eq.) were added to an oven dried RBF equipped with a magnetic stir bar. The RBF was then filled with N_2 using a schlenk line and sealed with a septa. Dry ACN (30 mL) and $BF_3.Et_2O$ (0.22 mL, 1.76 mmol, 2 eq.) were added, in that order, using a syringe and the reaction was left to stir overnight. The reaction was quenched by adding 200 mL 1 M HCl which resulted in a yellow precipitate. The precipitate was first filtered, then redissolved in DCM and washed with brine. The organic layer was drived over Na_2SO_4 and concentrated *in vaccuo* to give 700 mg of crude product. The crude can be purified by silica gel chromatography using 25 % ethyl acetate in hexane as the mobile phase (product $R_f = 0.2$) to give 550 mg product (80 % yield) as a yellowish solid. Mp: >250 °C (dec). **IR (KBr pellet, cm^{-1}):** 3524 m, 3061 w, 3026 w, 2936 w, 2249 w, 1727 s, 1720 s, 1595 m, 1510 m, 1450 m, 1337 s, 1258 s, 1159 s, 1059 s, 1093 s, 1059 s, 1028 m, 707 s. **^{1}H NMR (500 MHz, CDCl$_3$):** δ 3.61 (d, 2H, J = 14.8 Hz), 3.71 (d, 2H, J = 15.5 Hz), 3.81 (d, 2H, J = 15.5 Hz), 3.90 (d, 2H, J = 14.8 Hz), 6.31 (br s, 1H, exchangeable), 6.64 (m, 4H), 6.73 (t, 1H, J = 7.4 Hz), 6.90 (d, 2H, J = 7.4 Hz), 7.03 (dd, 2H, J = 3.7, 5.6 Hz), 7.27 (m, 4H), 7.52 (t, 4H, J = 7.6 Hz), 7.58 (m, 1H), 7.75 (t, 2H, J = 7.6 Hz), 7.96 (s, 2H), 8.01 (d, 2H, J = 7.6 Hz). **^{13}C{^{1}H} NMR (125 MHz, CDCl$_3$):** 32.7, 37.5, 125.1, 125.4, 126.5, 128.0, 128.1, 128.4, 128.5, 129.0 (2 overlapping carbons), 129.5, 130.7 (2 overlapping carbons), 131.0, 131.1, 131.3, 131.4, 133.4, 133.6, 133.9, 134.3, 140.5, 147.0, 148.4, 158.7, 164.0, 164.3. **HR-ESI-MS:** Calculated for $C_{28}H_{22}NO_{15}S_3$ [M-H]$^{-1}$ = 708.0157, Found 707.9931.

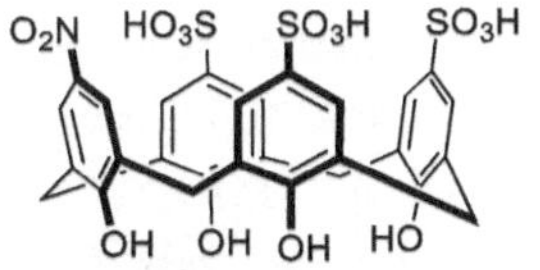

Compound **P-NO2**. Compound **2.11** (300 mg, 0.38 mmol, 1 eq.) is suspended in methanol (20 mL) in a round bottom flask equipped with a magnetic stir bar and a condenser. Sodium hydroxide (1500 mg, 38 mmol, 100 eq.) is added to the solution which is then set to reflux overnight. The reaction mixture is quenched by adding 150 mL 1 M HCl. More HCL can be added to bring the pH below 1. The resulting precipitate is filtered and washed with 200 mL hot hexane to remove benzoic acid. This debenzoylated product is used further without purification. First, the precipitate is dissolved in a minimal amount of DCM in a RBF with an attached reflux condenser. Concentrated H_2SO_4 (0.5 mL, 20 eq. with respect to **2.11**) was added and the reaction mixture was stirred at 60 °C for 3 hours. The product is observed to be precipitating out of the reaction mixture during this time. After 3 hours, the supernatant can be decanted off and checked for left over starting material using TLC. If significant amount of starting material is still present then reaction can be restarted in a fresh reaction vessel. The combined precipitate is then suspended in a minimal amount of EtOAc and poured into centrifuge tubes (50 mL) and topped off with Et_2O. After two rounds of decanting, resuspending in Et_2O and centrifugation, the precipitate is air dried to provide crude product as a grey powder. This crude can be used in further reactions without purification. Analytical samples can be obtained by further purifying it using the HPLC. Isolated yield after HPLC: 90 mg (40 %). Mp: >250 °C (dec). **IR (KBr pellet, cm^{-1}):** 3316s br, 1594w, 1521w, 1454w, 1342m, 1211s, 1155s, 1116s, 1040s, 895w, 808w, 786w, 746w, 665w, 651w, 626m, 559m. **^{1}H NMR (500 MHz, D_2O):** δ 3.89 (br s, 8H), 7.44 (s, 2H), 7.48 (s, 4H), 7.88 (s, 2H). **^{13}C{^{1}H} NMR (125 MHz, D_2O):** 31.0, 31.1, 125.7, 127.1, 127.2, 127.3, 128.1, 128.6, 128.7, 128.9, 136.4, 136.4, 141.7, 152.1, 152.3, 156.3. **HR-ESI-MS:** Calculated for $C_{28}H_{22}NO_{15}S_3$ [M-H]$^{-1}$ = 708.0157, Found 707.9931.

Scaffold **E**. Compound **P-NO2** (180 mg, 0.25 mmol, 1 eq.) and $SnCl_2.2H_2O$ (290 mg, 1.26 mmol, 5 eq.) was suspended in 20 mL ethanol in a round bottom flask. 1 M HCl was added dropwise with stirring till the solution became almost clear (1-1.5 mL). The round bottom flask was equipped with a condenser and flushed once with N_2 before setting it to reflux over night. During this time, lot of white precipitate crashed out of the reaction. The entire reaction mixture was cooled to room temperature and poured into centrifuge tubes (50 mL) and topped off with Et_2O. After centrifugation, the supernatant is discarded. The product is extracted from the precipitate by resuspending into H_2O and centrifugation (3 rounds). The combined aqueous fractions are lyophilized to give 130 mg of product as a white powder (Yield = 75 %). Mp: 180 °C (dec). **IR (KBr pellet, cm^{-1}):** 3445s br, 1471m, 1434m, 1183s, 1113s, 1046s, 892w, 794w, 741w, 671w, 654w, 629m, 548w. **^{1}H NMR (500 MHz, 15 % D_2O in MeOD-d4):** δ 4.00 (br s, 8H), 7.13 (s, 2H), 7.55 (s, 2H), 7.64 (m, 4H). **^{13}C{^{1}H} NMR (125 MHz, 15 % D_2O in MeOD-d4):** δ 31.5, 31.8, 124.6, 124.9, 127.7, 127.9, 128.0, 128.8, 128.8, 128.9, 131.0, 138.3, 138.9, 150.4, 151.8, 153.3. **HR-ESI-MS:** $C_{28}H_{25}NO_{13}S_3$ [M-2H]$^{2-}$ = 338.5171, Found 338.5169.

2.9.5. General synthetic procedure for the library compounds

2.9.5.1. General procedure for Suzuki coupling

Scaffold **A, B, C** or **D** (50 mg, 0.06 mmol, 1 eq.), R-B(OH)$_2$ or R-B(pin) (1.2 eq. with respect to number of bromines on the Scaffold used), tetrabutylammonium bromide (TBAB, 9.5 mg, 0.5 equiv., 0.003 mmol), Pd(OAc)$_2$ (2.8 mg, 20 mol%) and sodium carbonate (23 mg, 3.8 equiv., 0.218 mmol) are dissolved in 5 mL of deionized H$_2$O inside a microwave vial and irradiated in Biotage microwave reactor for 5 min at 150 °C. After the reaction vessel cools down to room temperature, 1 mL of 1 M thiourea is added and the mixture is stirred at 70 °C for 1 hour. This leads to precipitation of most of the palladium as a thiourea complex which is separated by filtration. The filtrate can be concentrated *in vacuo*, if required, before purification. The products are purified using an HPLC as described in Section 2.9.2. The fractions are scanned for desired m/z in MS and matching fractions are pooled and lyophilized. All products are characterized using either HPLC-MS or UPLC-MS as described in Section 2.9.2.

2.9.5.2. General procedure for sulfonamide coupling

Scaffold **E** (50 mg, 0.07 mmol, 1 eq.), and sulfonyl chloride (1.1 eq, 0.162 mmol) are dissolved in 6 mL of 1M Na$_2$HPO$_4$/NaH$_2$PO$_4$ buffer (pH 8) and stirred overnight at room temperature. The reaction mixture is taken up in centrifuge tubes and few drops of saturated FeCl$_3$ solution is added. This results in precipitation of most of the phosphate which can be then removed by centrifugation. The supernatant can be concentrated *in vacuo*, if required, before purification. The products are purified using an HPLC as described in Section 2.9.2. The fractions are scanned for desired m/z in MS and matching fractions are pooled and lyophilized. All products are characterized using either HPLC-MS or UPLC-MS as described in Section 2.9.2.

2.9.5.3. General procedure for amide coupling

Scaffold **E** (50 mg, 0.07 mmol, 1 eq.), is dissolved in 6 mL of 1M Na$_2$HPO$_4$/NaH$_2$PO$_4$ buffer (pH 8). Acyl chloride (1.2 eq.) is dissolved in 200 uL DMF and then added to the mixture which is then stirred overnight at room temperature. The reaction mixture is taken up in centrifuge tubes and few drops of saturated FeCl$_3$ solution is added. This results in precipitation of most of the phosphate which can be then removed by centrifugation. The supernatant can be concentrated *in vacuo*, if required, before purification. The products are purified using an HPLC as described in Section 2.9.2. The fractions are scanned for desired m/z in MS and matching fractions are pooled and lyophilized. All products are characterized using either HPLC-MS or UPLC-MS as described in Section 2.9.2.

2.9.5.4. General procedure for isothiocyanate coupling.

Scaffold **E** (75 mg, 0.11 mmol) and R-NCS (1.1 equiv.) were dissolved in 3 mL 1:1 DMF:pyridine while stirring in a RBF under argon. The reaction is stirred overnight at room temperature. The solution is diluted with 10 mL dH_2O and extracted with DCM (2 x 15 mL) and EtOAc (2 x 15 mL) and the aqueous layer is evaporated to dryness under reduced pressure. The crude solid is subjected to RP-HPLC purification as outlined in Section 2.9.2. Fractions containing product are pooled and lyophilized to afford the product as white powder in 30 % yield.

2.9.5.4. General procedure for amino acid coupled calix[4]arene (**AP6-AP9**)

2-chlorotrityl resin (24 mg, 0.012 mmol) was swelled in DMF for one hour. The resin was then filtered and suspended in a fresh batch of DMF. Fmoc-Tyrosine or Fmoc-Aspartic acid (0.071 mmol, 6 eq.) and diisopropylethylamine (DIPEA, 25 µL) are added to the resin (in DMF) and reacted under gentle bubbling of nitrogen (no stir bar) for 3 hours. The resin was washed multiple times with DMF before a second round of reaction (deprotection) was started with 20 % (v/v) piperidine in DMF. As before, the reaction mixture was agitated only by bubbling nitrogen. The deprotection was allowed to proceed for 1 hours after which the resin was washed with multiple rounds of DMF. Finally, calixarene **AP2** (30 mg, 0.031 mmol, 3 eq.) was added along with 25 µL of DIPEA as a solution of DMF. The reaction was allowed to continue in the same fashion as above. Post coupling, the resin was cleaved by reacting it to a 95:2.5:2.5 mixture of TFA, H_2O and triisopropylsilane for 4 hours, once again under gentle bubbling of N_2. The resin was filtered and washed with TFA (3 x 5 mL). The combined TFA solution was concentrated on a rotavap. Ice cold diethyl ether was then used to precipitate crude peptides which were then air dried and then purified as described in Section 2.9.3. Along with **AP7** and **AP9**, unintended products **AP6** and **AP8** are also obtained as a result of *N*-terminal hydrolysis.

2.9.6. Indicator Displacement Assay (IDA)

IDA was performed in NUNC 96 black-well plated with optically clear bottoms. A total of 9 titrations were performed in a single plate which included the direct titration of calixarene into lucigenin, and 8 competitive titrations of individual peptides into the calixarene-lucigenin complex. The plate layout is shown in Figure 2.15. Concentration of stock solutions of peptides and lucigenin were determined spectrophotometrically using A_{280} for peptides (ε=1280 $M^{-1}cm^{-1}$) and A_{410} for lucigenin (ε=8900 $M^{-1}cm^{-1}$).

Concentration of lucigenin was kept constant at 250 nM in all wells while the buffer concentration was kept constant at 10 mM sodium phosphate buffer at pH 7.4. Direct titration wells had varying concentration of calixarene and no peptides. All competitive titration wells had a fixed concentration of calixarene but varying concentration of peptides. To achieve consistency all reagents were added from one stock solution each with 10X concentration than required. 20 µL of each solution was added to all wells. All the remaining volume was made up using dH₂O such that total volume in each well was 200 µL.

Fluorescence of lucigenin was read across the plate using SpectraMax® M5/M5e microplate reader in top read mode. The excitation wavelength used was 369 nm while the emission was recorded at 485 nm. The plates were briefly centrifuged immediately prior to reading in a Beckman Coulter Allegra™ X-12R centrifuge. The raw data was then fed into an in-house written python code which split the data into 9 different concentration response which were plotted as dF_{obs} vs [calixarene], for direct titration, and vs [peptide], for competitive titrations. The program used established numpy and scipy libraries to fit the direct titration data using 1:1 binding isotherm (*vide infra)* to obtain K_{ind}. This K_{ind} was then used further to fit the remaining 8 data sets with competitive binding isotherms (*vide infra*) and extract K_d for each peptide and calixarene.

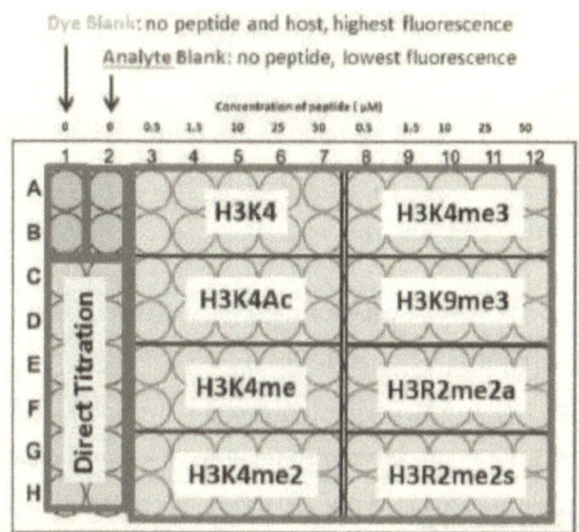

Figure 2.15. Plate layout for Indicator Displacement Assay (IDA) experiment.

All IDA data were fit as explained below:

1. For all the equations shown below:
   - $[H]^T$ is the total host concentration (host = calixarene, $[H]^T = [H]^{free} + [H]^{bound}$)
   - $[I]^T$ is the total indicator concentration (indicator = lucigenin, $[I]^T = [I]^{free} + [I]^{bound}$)
   - $[Pep]^T$ is the total peptide concentration ($[Pep]^T = [Pep]^{free} + [Pep]^{bound}$)
   - [HG] refers to host-guest complex. This would be calixarene-lucigenin complex in direct titration and calixarene-peptide complex in competitive titration.
   - $[H]^{free}$ is written as [H] for simplicity.

- dF_{obs} is change in fluorescence observed upon binding (negative for direct titration, positive for competitive titration). This is the raw data from the fluorometer.
- ΔF_{max} is the maximum change in fluorescence observed i.e. point where the curve plateaus. This is an iterable variable for all titration and was optimized.
- K_{ind} is dissociation constant of host-indicator (calixarene-lucigenin) complex. This is an iterable variable in direct titration and was optimized. In all competitive titrations, this is kept constant at a value determined by the direct titration
- K_d is dissociation constant of host-guest (calixarene-peptide) complex. This is an iterable variable in and was optimized

2. *Direct titration:* The general equation describing direct titration is a quadratic equation of the form[229]: $[HG]^2 + b[HG] - c = 0$, where $b = K_{ind} + [H]^T + [I]^T$ and $c = ([H]^T[I]^T)$. $[I]^T$ is kept constant while $[H]^T$ is varied. The general solution to quadratic equation can be used to determine $[HG]$ at any given $[H]^T$.

[HG] is used to fit the data using the equation shown below.[229] Both ΔF_{max} and K_{ind} are optimized.

$$dF_{obs} = \frac{\Delta F_{max}[HG]}{[I]^T}$$

3. *Competitive titration:* The general equation describing competitive titration is a cubic equation of the form[236]: $[H]^3 + a[H]^2 + b[H] + c = 0$, where $a = (K_d + K_{ind} + [H]^T + [I]^T + [Pep]^T)$, $b = (K_d*K_{ind}) + (K_{ind}*([Pep]^T - [H]^T)) + (K_d*([I]^T - [H]^T))$ and $c = -(K_d*K_{ind}* [H]^T)$. $[H]^T$ and $[I]^T$ are kept constant while $[Pep]^T$ is varied. A exact solution to this cubic equation has been published in literature[236] and can be used to determine $[H]$ at any given $[Pep]^T$.

[H] is used to fit the data using the equation shown below. Both ΔF_{max} and K_d are optimized.

$$dF_{obs} = \Delta F_{max}\left(1 - \frac{[H]}{K_{ind} + [H]}\right)$$

K_d values obtained for every host-guest pair are tabulated in Table 2.3. All binding isotherms are shown in Appendix C.

Table 2.3. Binding constants (K_d) for all compounds in the library against peptides derived from histone 3 tail as estimated by IDA.

Host	Binding constants against each peptide (K_d in µM)							
	H3K4	H3K4ac	H3K4me1	H3K4me2	H3K4me3	H3K9me3	H3R2me2a	H3R2me2s
PSC4	4.1 ±0.4	17 ±4	2.2 ±0.2	0.83 ±0.13	0.56 ±0.05	1.4 ±0.1	2.3 ±0.2	5.7 ±0.3
PSC6	0.38 ±0.04	2.9 ±0.6	0.26 ±0.04	0.058±0.015	0.017±0.007	0.084±0.015	0.22 ±0.04	0.37 ±0.04

P-NO2	17 ±11	16.2 ±50	7.6 ±2	1.8 ±0.6	1.4 ±0.5	6.6 ±3.5	3.0 ±2.7	2.1 ±1.3
AP1	3.2 ±0.4	8.9 ±1.3	1.0 ±0.2	0.27 ±0.03	0.10 ±0.02	0.28 ±0.03	0.44 ±0.05	1.5 ±0.1
AP3	2.1 ±1.0	6.0 ±1.6	1.2 ±0.2	0.31 ±0.3	0.19 ±0.02	0.42 ±0.03	0.65 ±0.6	1.8 ±0.2
AP4	3.8 ±0.7	8.0 ±1.7	1.2 ±0.1	0.46 ±0.07	0.19 ±0.03	0.41 ±0.05	1.2 ±0.1	3.2 ±0.9
AP5	4.3 ±0.3	7.7 ±0.9	1.3 ±0.7	0.44 ±0.08	0.14 ±0.01	0.31 ±0.07	0.60 ±0.04	1.6 ±0.2
AP6	1.6 ±2.0	0.88 ±0.40	0.12 ±0.01	0.013±0.001	0.015±0.001	0.039±0.001	0.049±0.001	0.23 ±0.02
AP7	1.8 ±0.5	1.4 ±0.6	0.56 ±0.07	0.051±0.015	0.073±0.020	0.018±0.04	0.13 ±0.04	0.71 ±0.10
AP8	0.27 ±0.03	1.0 ±0.3	0.046 ±0.001	0.0056 ±0.001	0.0066 ±0.0001	0.023 ±0.001	0.015 ±0.001	0.074 ±0.003
AP9	0.31 ±0.03	1.7 ±1.3	0.077 ±0.002	0.007 ±0.001	0.008 ±0.001	0.030±0.001	0.020±0.001	0.12 ±0.01
AM1	3.4 ±0.7	6.9 ±2.2	1.2 ±0.2	0.50 ±0.04	0.26 ±0.03	0.53 ±0.04	0.90 ±0.24	2.5 ±0.4
AM2	1.9 ±0.6	7.8 ±1.5	0.53 ±0.03	0.16 ±0.02	0.06 ±0.02	0.14 ±0.03	0.25 ±0.02	1.2 ±0.4
AH1	1.3 ±0.3	12 ±0.7	0.36 ±0.01	0.031±0.002	0.035±0.002	0.093±0.004	0.16 ±0.01	0.76 ±0.12
AH2	1.2 ±0.3	2.6 ±0.5	1.0 ± 0.4	0.40 ±0.09	0.38 ±0.09	0.64 ±0.17	1.2 ±0.4	1.3 ±0.6
AH3	1.3 ±0.3	3.0 ±1.0	0.21 ±0.03	0.019±0.003	0.055±0.008	0.19 ± 0.03	0.17 ±0.03	0.29 ±0.07
AH4	0.89 ±0.28	4.1 ±0.5	0.11 ±0.01	0.005±0.001	0.03 ±0.002	0.11 ±0.01	0.097±0.014	0.24 ±0.07
AH5	23 ±14	>100	8.3 ±4.2	0.90 ±0.3	1.0 ±0.3	3.4 ±1.2	3.8 ±1.3	9.0 ±2.1
AH6	23 ±14	>100	8.3 ±4.2	0.90 ±0.3	1.0 ±0.3	3.4 ±1.2	3.8 ±1.3	9.0 ±2.1
AH7	11 ±2	11 ±3	2.5 ±0.3	0.34 ±0.12	0.39 ±0.10	1.0 ±0.1	1.2 ±0.1	4.4 ±0.4
AO1	7.9 ±1.8	9.3 ±10	5.4 ±1.8	0.87 ±0.67	0.73 ±0.09	1.4 ±0.1	2.3 ±.02	4.5 ±1.1
AO2	4.1 ±0.2	7.0 ±0.2	1.1 ±0.7	0.29 ±0.03	0.11 ±0.02	0.08 ±0.02	0.92 ±0.15	1.6 ±0.4
AO3	4.3 ±0.1	6.2 ±3.1	2.2 ±0.4	0.57 ±0.08	0.33 ±0.03	0.60 ±0.06	1.6 ±0.2	1.8 ±0.6
BP0	>100	1.0 ±0.9	2.5 ±0.7	0.27 ±0.1	0.54 ±0.2	0.69 ±0.18	2.2 ±04	4.9 ±2.8
BP1	1.0 ±0.4	6.5 ±3.0	0.17 ±.0.05	0.024±0.003	0.018±0.003	0.042±0.003	0.054±0.004	0.27 ±0.01
BM1	0.16 ±0.11	0.41 ±0.13	0.18 ±0.09	0.11 ±0.05	0.10 ±0.06	0.21 ±0.06	0.27 ±0.10	0.29 ±0.13
BH2	>100	2.0 ±2.9	4.0 ±2.1	0.68 ±0.38	2.1 ±0.6	4.0 ±3.2	1.8 ±0.7	4.1 ±1.5
BH5	>100	n.d.**	>100	14 ±25	>100	>100	>100	>100
DP0*	n.a.	n.a.	n.a.	n.a.	n.a.	n.a.	n.a.	n.a.
DP2	0.70 ±0.06	1.1 ±0.03	0.25 ±0.06	0.10 ±0.02	0.039±0.01	0.071±0.015	0.088±0.017	0.26 ±0.05
DM1	0.36 ±0.12	0.77 ±1.8	0.44 ±0.60	0.12 ±0.03	0.083±0.015	0.32 ±0.08	3.5 ±3.2	0.59 ±0.02
DO2	0.036±0.062	0.050±0.038	4.4 ±1.4	1.3 ±0.4	0.61 ±0.07	1.5 ±0.2	4.1 ±0.7	4.7 ±2.5
DO3	0.0038 ±0.0003	0.0019 ±0.0003	0.023 ±0.001	0.0081 ±0.0001	0.0048 ±0.0001	0.011 ±0.001	0.016 ±0.006	0.0038 ±0.0002

CP1	>100	3.8 ±0.8	1.5 ±0.2	0.31 ±0.15	0.52 ±0.25	0.59 ±0.07	0.62 ±0.17	0.65 ±0.25
CP2	>100	n.d.**	6.2 ±7	0.27 ±0.13	0.57 ±0.36	1.2 ±0.5	0.55 ±0.28	4.4 ±3.5
E1	2.4 ±0.5	4.9 ±1.5	0.79 ±0.15	0.25 ±0.04	0.11 ±0.01	0.29 ±0.04	0.38 ±0.07	1.6 ±0.2
E2	n.d.**	n.d.**	>100	>100	>100	>100	9.5 ±10.2	>100
E3	1.1 ±0.2	2.7 ±1.1	0.35 ±0.07	0.050±0.006	0.083 ±0.015	0.32 ±0.07	0.36 ±0.08	0.79 ±0.08
E4	n.d.**	n.d.**	n.d.**	>100	>100	n.d.**	>100	>100
E5	57 ±420	>100	2.1 ±0.6	0.47 ±0.1	0.31 ±0.1	0.81 ±0.25	1.5 ±0.6	4.7 ±2.5
E6	0.51 ±0.15	1.1 ±3.5	0.17 ±0.08	0.057±0.065	0.083±0.045	2.2 ±1.5	1.5 ±2.6	0.60 ±1.6
E7	7.9 ±3.5	12 ±5	4.2 ±0.7	0.66 ±0.07	0.78 ±0.10	2.0 ±0.2	3.3 ±0.6	8.9 ±1.5
E8	15 ±7	>100	2.6 ±0.7	0.88 ±0.35	0.86 ±0.30	2.6 ±0.8	3.6 ±0.6	23 ±20
E9	>100	n.d.**	>100	>100	9.5 ±7.0	33 ±35	85 ±150	>100
E10	n.d.**	n.d.**	n.d.**	n.d.**	n.d.**	n.d.**	n.d.**	n.d.**
E11	39 ±30	>100	5.9 ±2	3.2 ±0.3	0.46 ±0.2	2.1 ±0.6	5.9 ±0.6	18 ±8

All values are average of two runs. Errors are reported as average standard error (of two titrations) of fit parameters, calculated from covariance matrix of nonlinear regression. *DP0 did not bind lucigenin and so we were unable to estimate any of its binding constants. ** n.d. applies to titration where response was so low that after blank correction ΔF values became negative. Our python based fitting program could not fit any such data set.

Chapter 3. A Calix-affinity method for enrichment of methylated peptide

Publications

This Chapter is adapted from a previously published paper.

Supramolecular affinity chromatography for methylation-targeted proteomics, Graham A. E. Garnett, Melissa J. Starke, Alok Shaurya, Janessa Li, Fraser Hof, *Anal. Chem.,* **2016**, 88, 7, 3697-3703.

Contributions

The idea behind immobilization of calixarenes on solid support and their use in enrichment studies was first conceived and performed by GG. MS and AS did follow up studies in preparation of the above-mentioned publication. All peptides and calix[4]arene for these enrichment studies were synthesized jointly by AS and GG. All solution-phase binding constants were determined by AS. GG performed the enrichment studies on the commercial histone sample in preparation of the proteomics study. Analysis of proteomics data was done by GG, MS and AS.

3.1. Foreword

Chapter 2 establishes a basic structure activity relationship for calix[4]arenes and methylated peptide binding. At the end of it, we are left with multiple host molecules with desirable properties that can claim to be potentially useful in chemical biology. For the goals of this book, one such property is the presence of additional functional groups that can be used to attach the molecules to different solid supports. The idea behind this is that, once immobilized, calix[4]arene modified solid support should be able to enrich analytes similar to antibodies in immunoaffinity purification (IAP). We begin this Chapter by identifying a subset of suitable molecules from the library in Chapter 2. Detailed investigations on each candidate's potential as an enrichment agent is the subject of GG's Master's book and would not be discussed in detail here. Only the late stage developments in the project that directly involved my own work will be presented and discussed in detail. This short Chapter has the goal of linking the results of Chapter 2 with the subsequent efforts that are reported in this book.

3.2. Motivation

Small molecule-based enrichment procedure does exist in literature but they are used for a different goal.[251–253] In a routine procedure, drug like small molecules are attached to a solid support and used as a affinity reagent to find their intracellular targets. This wide-spread methodology, called chemical proteomics, is typically done as a pull-down experiment, not chromatography. To the best of our knowledge, there is only one report that uses immobilized small molecules as a stationary phase for chromatography. In this report, Alves et al. purify antibodies by immobilizing nucleotides to the solid support and taking advantage of its affinity to antibodies nucleotide binding site to purify.[254]

No report describing use of immobilized small molecule in global discovery-based proteomics has been published. But we take note that a.) flat drug like molecules used in chemical proteomics can enrich their targets very efficiently, and b.) the enrichment efficiency of different analytes translates very well to solution based binding affinities. Compared to them, calix[4]arenes have a defined binding cavity which prompts us to hypothesize that their solution phase affinities should reflect neatly on their enrichment behavior if they are immobilized.

3.3. Molecules suitable as enrichment agents and their fate

Figure 3.1. Subset of molecules from the library that contain a carboxylic acid, and thus can be immobilized to an amine containing solid phase.

Chapter 2 leaves us with an array of hosts with intriguing properties. However, not all molecules in the library are suitable to be used as enrichment agents. Figure 3.1 shows a subset of molecules that we selected from the library in order to forward with this project. The primary criteria for selection of such molecules from the library was the presence of an additional chemical handle that can be used in later chemical steps. For enrichment agents, this step typically refers to their attachment to a solid phase, usually a permeable resin that is compatible with aqueous solutions. Amide coupling is ubiquitous in such applications. A quick look through our library showed that we had multiple compounds with a pendant carboxylic acid that can be used in coupling chemistry. We settled on these compounds and choose an amine containing cross-linked agarose resin (AffiGel-102, Biorad) as the solid support (Figure 3.2).

Compound **AP6** and **AP8** were discarded during the very initial stages due to the low yield of their synthesis. A typical coupling to resin step requires 7-10 mg of calix[4]arene while these calix[4]arenes were hard to make in more than 2-3 mg scale. We also realized early on that any calix[4]arene with more than one carboxylic acid led to severe cross coupling within the resin (Figure 3.2.b). This changed the physical properties of resin to the point that it would become non-malleable and impermeable to water. **BP1**, **CP1**, **CP2** and **DP2** were discarded for this reason. The two compounds that did show appreciable immobilization on resin without changing the physical properties were thus **AP1** and **E1** which were moved forward in this project.

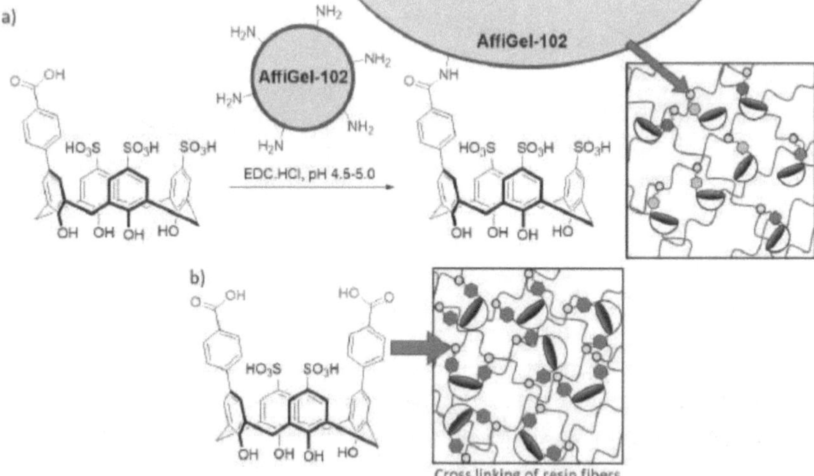

Figure 3.2. a) Scheme describing EDC based coupling of calix[4]arenes to AffiGel-102 containing single carboxylic acid, b) Cross coupling of resin when calix[4]arenes containing more than one carboxylic acid are coupled.

3.4. AP1 versus E1: small difference in connectivity, big change is efficacy

The coupling of **AP1** and **E1** to resin proceeded identically and smoothly. In a typical procedure, an Eppendorf tube with 1 mL of Affigel-102 resin was washed with water multiple times (using centrifugation, and discarding the supernatant). Precalculated amounts of **AP1** or **E1** were added as solutions in water and the pH of the reaction mixture was adjusted to 4.5-5.0 using 1 M HCl. After incubating for 10 min, ~10 mg of EDC was added to the tube and the reaction was followed by monitoring the concentration of calix[4]arene in the supernatant using HPLC. Throughout the reaction, the pH was monitored and maintained between 4.5 and 5.0. When the amount of calixarene remaining in the supernatant was low and remaining constant over time, we declared the coupling reaction done. The resin was washed several times with 2 M NH_4Cl and deionized H_2O.

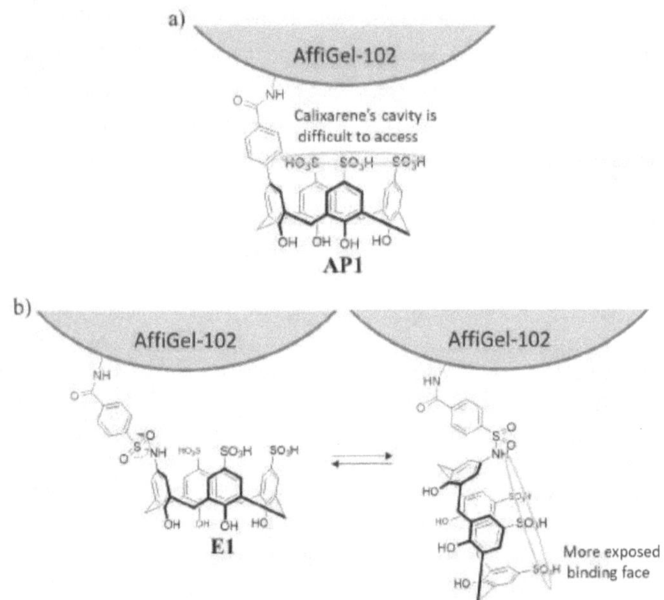

Figure 3.3. Difference between AP1 and E1. Upon immobilization, AP1's binding face is heavily obstructed (a), while E1's sulfonamide arm can rotate and become accessible to analytes (b).

A detailed comparison between performance of immobilized **AP1** and **E1** is reported elsewhere.[255] These resins were tested both in pull-down and chromatography-based enrichment methods. While enrichment results using a centrifugation-driven pull-down method were poor for both compounds, **E1** functionalized AffiGel-102 (hereafter called **E1.r**) performed exceptionally when used as the stationary phase in an affinity chromatography setup. We believe this to be the result of flexibility of sulfonamide arm. The fact that we are coupling a massive polymer at the binding face of calix[4]arene is not lost on us. **AP1's** binding face is presumable obfuscated by the resin while **E1's** sulfonamide bond allows it to be exposed more to the analytes (Figure 3.3).

We then explored the performance of **E1** in different chromatography form factors. **E1.r** was packed in to a 35 × 6.2 mm column and a 1 × 600 mm capillary column. Each trial column was sealed at both ends using 10 μm frits and then attached to a modified fast protein liquid chromatography (FPLC) system. A typical experiment consisted of first equilibrating the column with a 50 mM phosphate buffer at pH 7.5, called running buffer (RB), before injecting our sample (pure peptide or a mixture). The peptides were then eluted first by using RB, until 5-10 column volumes (CV) has been eluted, to generate a fraction of "unretained" peptides. Introducing an elution buffer (EB) with a high ionic strength (2 M NH_4Cl or NaCl) then generated a fraction of "retained" peptides. Peptides were detected as they eluted using a diode array detector set to detect the absorbance at 280 nm (Figure 3.4).

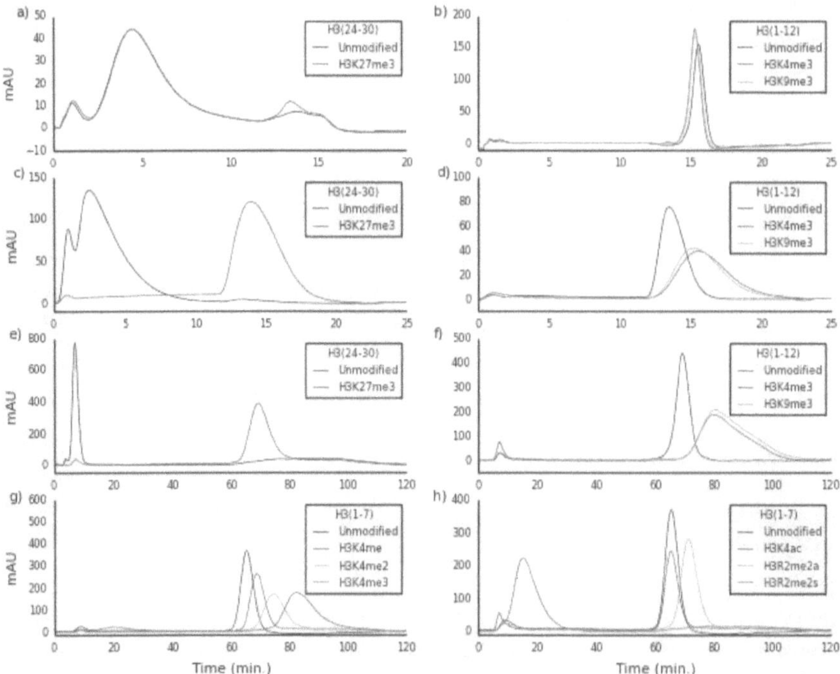

Figure 3.4: Chromatograms obtained from enrichment experiment using a commercially purchased SCX column (a-b), E1.r packed in a column (c-d) and E1.r packed in a capillary (e-h). All peptides were synthesized in lab and are based on histone 3 (H3) tail sequence. Figure reproduced with permission.

E1.r separated peptides based on their methylation states in both column and capillary chromatography form factors (Figure 3.4 c-h). In general, the difference in retention time between methylated and unmethylated peptides decreased as the peptide's charge increased. In the shorter, wider column form factor, this decrease in separation is severely limiting. The difference in retention time decreases from 13 min, for peptides with +2 charge, to ~ 1 min for peptides with +5 charge (Figure 3.4.c-d, Table 3.1). Significant overlap between higher charged methylated and unmethylated peptides limits its further use (Figure 3.4.d). Capillary chromatography, on the other hand, is much more satisfactory. Peptides with +2 charge are completely unretained if unmethylated (Unmodified in Figure 3.4.e, H3K4ac in Figure 3.4.h). There is significantly less overlap between methylated and unmethylated peptides of +5 charge when compared to **E1.r** packed in a column (Figure 3.4.f). Peptides of +3 charge are completely baseline separated however lower methylation states do overlap with unmethylated peptide (Figure 3.4.g). It is impossible to remove the effect of charge completely from this form of chromatography. All sulfonates are charged (negatively) at the pH of RB and retention due to a classical ion exchange chromatography mechanism between negatively charged sulfonates and positively charge amino groups is unavoidable. But ion exchange by itself cannot explain the significantly higher retention of methylated peptides on the column. To confirm this, we attempted pure ion exchange separation with these peptides using commercially available SCX columns (SPXL, GE Healthcare) but to no avail (Figure 3.4.a-b). The stationary phase of an SCX column is functionalized with sulfonates, allowing for a direct comparison to the sulfonated calixarenes.[256] The inability of SCX to differentiate between methylated and unmethylated peptides is a good evidence that the mechanism of separation in our columns is due to the sulfonated calix[4]arene's methyllysine-binding capabilities and not purely electrostatics.

3.5. Enrichment of a model proteomics samples

E1.r successfully split a model proteomics sample (calf thymus histone sample, proteolyzed using ArgC) into an unretained and a retained fraction (Figure 3.5.a). Analysis of all fractions, including the input fraction itself, using tandem MS showed that the retained fraction contained all trimethylated lysine containing peptides (Figure 3.4.c). Most dimethylated and monomethylated peptides also showed up in retained fraction. Methylarginine isn't well enriched as we see more of it in the input fraction. Most importantly, a.) none of the methylated peptides seen in the retained fraction are seen in the input fraction (Figure 3.5.d) and, b.) the percentage of peptides observed that do not contain a PTM decreases significantly from input fraction to retained fraction (Figure 3.5.b). This proves that **E1.r** is efficiently separating peptides based on the presence of PTMs, and that by reducing high levels of unmethylated peptides it allows us to detect methylated peptides that otherwise would have gone undetected.

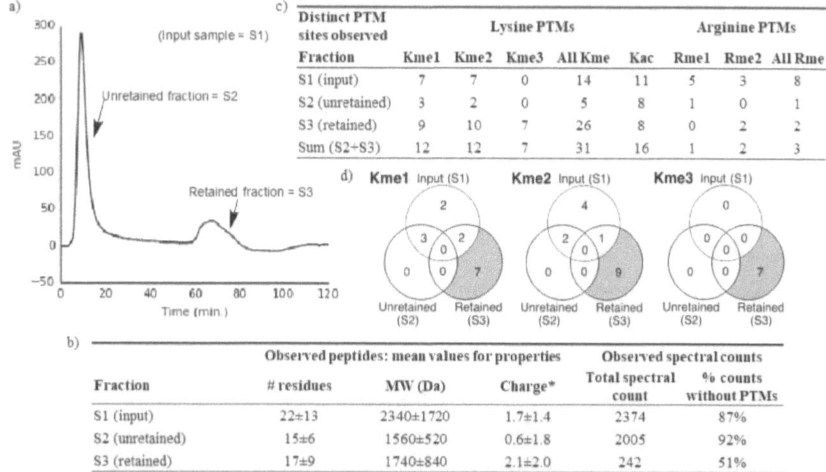

c)

Distinct PTM sites observed	Lysine PTMs					Arginine PTMs		
Fraction	Kme1	Kme2	Kme3	All Kme	Kac	Rme1	Rme2	All Rme
S1 (input)	7	7	0	14	11	5	3	8
S2 (unretained)	3	2	0	5	8	1	0	1
S3 (retained)	9	10	7	26	8	0	2	2
Sum (S2+S3)	12	12	7	31	16	1	2	3

b)

Fraction	Observed peptides: mean values for properties			Observed spectral counts	
	# residues	MW (Da)	Charge*	Total spectral count	% counts without PTMs
S1 (input)	22±13	2340±1720	1.7±1.4	2374	87%
S2 (unretained)	15±6	1560±520	0.6±1.8	2005	92%
S3 (retained)	17±9	1740±840	2.1±2.0	242	51%

Figure 3.5. Enrichment of ArgC-proteolyzed calf thymus histone sample using E1.r packed in a capillary. a) Chromatographic trace (UV detection at 280 nm) shows an unretained peak and a single broad retained peak, which were each collected and subjected to LC-MS proteomics analysis along with an input control sample. The raw list of peptides and PTMs identified by proteomics analysis are provided in the Experimental section. b) Analysis of peptide sequences identified from each fraction (input, unretained, and retained) suggests that short but highly charged peptides are retained on column. More PTMs are identified in retained fraction. * Mean solution-phase charges of the peptides were calculated by subtracting total anionic sidechains from total cationic sidechains in order to give an estimate of the charge of peptide at neutral pH. All uncertainties reported as standard deviations. c) Number of distinct methylated residues that were observed in each fraction. d) Venn diagram showing the distributions of peptides with PTMs uniquely identified in each fraction.

3.6. Shortcomings

Our attempts to use this **E1.r** with a real-world proteomics sample (cell lysates) did not work. Unlike calf thymus histone sample (purchased commercially), cell lysates have a plethora of non-peptidic biomolecules that can interfere with binding to solid phase affinity media. Our most common result for studies using samples derived from cell lysates that were passed over an **E1.r** column was the complete lack of retention (data not shown). Also, the percentage of non-methylated but positively charged peptides increases multi-fold when going from a purified histone sample to a much more complex cell lysate. We believe these factors add to significant competition for binding to the calix[4]arene cavity which leads to poor retention and methyl-peptide enrichment.

Another big shortcoming was that the overall yield of **E1** (over 8 steps) was low (~2-3%). Combined with the instability of scaffold **E** (Chapter 2*)*, we were never able to synthesize **E1** on more that 25 mg scale. Immobilization experiments usually required 10 mg **E1** per iteration and thus we would frequently have to go back and reinvest in synthesis and purification of **E1**. Despite this, the success with calf thymus histone sample (Section 3.5) is encouraging are proves the merit of pursuing such a strategy. We concluded from these studies that we needed an easily accessible calix[4]arene with a more open binding face (upon immobilization).

3.7. Conclusion

The findings in this Chapter prove that a supramolecular host can be utilized in affinity chromatography for enrichment of proteomics samples. However, we learn much more from our failures here. **AP1**'s failure in enrichment studies despite superior supramolecular properties suggests that we need to find a solution to immobilize calix[4]arenes without restricting their binding face first. Our current approach sets us up for failure at later stages for reasons that we could not have predicted. The fate of **BP1**, **CP1**, **CP2** and **DP2** also supports the same conclusion. Their selectivity towards methylated peptides in solution is excellent, but their structural features prevent their use in proteomics enrichment applications.

3.8. Supporting information

3.8.1. Synthesis of E1

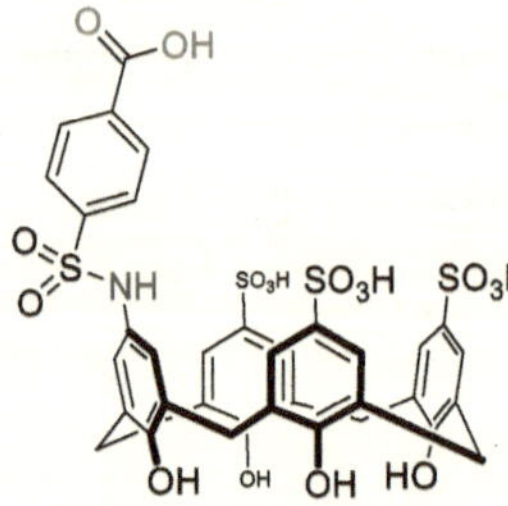

Scaffold **E** (50 mg, 1 eq, 0.073 mmol) and 4-chlorosulfonyl benzoic acid (18 mg, 1.1 eq, 0.08 mmol) are dissolved in 3 mL of 1 M Na_2HPO4/NaH_2PO4 buffer (pH 8) and stirred overnight at room temperature. The reaction mixture was purified on a Shimadzu Prominence HPLC system on a 9.4 mm x 250 mm semi-preparative Agilent Eclispse XDB-C18 5 µm column using a gradient that started at 90:10 mixture of H_2O:ACN and went up to 10:90 H_2O:ACN in 25 min (both solvents were spiked with 0.1%

TFA). After evaporation of solvents *in vacuo* 21 mg of product is obtained as white powder in 34% yield. **Mp:** 204 °C (dec). **IR (KBr pellet, cm^{-1}):** 3210s br, 1714s, 1474s, 1454s, 1401w, 1160s, 1110s, 1040s, 886w, 786w, 690w, 651m, 623m, 559w. **^{1}H NMR (300 MHz, D2O):** δ 7.52 (s, 2 H), 7.47 (d, 2 H, J = 1.8 Hz), 7.38 (d, 2 H, J = 1.8 Hz), 7.32 (s, 4 H), 6.70 (s, 2 H), 3.77-3.64 (br, 8 H). **^{13}C{^{1}H} NMR (75 MHz, D2O):** δ 166.7, 150.8, 150.6, 145.5, 139.7, 136.1, 135.6, 133.1, 130.1, 129.8, 128.7, 128.2, 127.8, 127.7, 127.1, 126.8, 126.6, 126.1, 122.4, 30.5, 30.4. **HR-ESI-MS:** Calculated for $C_{35}H_{29}NO_{17}S_4Na^+$ [M+Na]$^+$ = 886.0216, Found 886.0216.

3.8.2. Immobilization of E1 on Affi-Gel resin (E1.r) and preparation of enrichment columns

Calixarene **E1** was coupled to Affi-Gel-102 (primary amine functionalized, cross-linked agarose resin; Bio-Rad) using EDC as the coupling agent. Reaction pH was kept close to 5.0 and the extent of coupling was judged by monitoring changes in concentration of **E1** in the supernatant using HPLC. Once satisfactory amount of immobilization of **E1** was achieved on the resin, it was packed as a slurry in water into clear PFA capillary (IDEX 600 mm length, 1/16" x 1/8", I.D. x O.D.) or empty 1 mL columns (35 x 62 mm, Agarose Bead Technologies) using a syringe pump. Both ends of the capillaries were closed with fittings (IDEX, part. no. XP-335-CP), ferrules (IDEX, part. no. P-300NX-CP), frits (VICI-Jour part. no JR-1150-10P-5) and union (IDEX, part. no. P-703-01) before attaching it to an Agilent 1200 Series HPLC adapted for use as a Fast Protein Liquid Chromatograph (FPLC). The columns were then washed with 50 mM sodium phosphate + 2 M NH$_4$Cl at flow rates that kept the overall pressure <1 bar until the resin was evenly packed (visual inspection) and no material could be seen coming off the column (judged by low and stable UV detector readings). Care was taken to avoid formation of air bubbles, gaps and/or headspace in the column.

The 1 mL HiTrap SPXL (7 x 25 mm) ion exchange column (GE Healthcare) was used as purchased.

3.8.3. Chromatography method

Running buffer (RB) for all enrichments was 50 mM Na$_2$HPO$_4$ (pH 7.5). Elution buffer (EB) for all **E1.r** columns was RB + 2 M NH$_4$Cl. Elution buffer (EB) for SPXL column was RB + 2 M NaCl. Detection of eluted peptides was done using the UV detector set to observe A$_{222}$ and A$_{280}$. To fully equilibrate the columns to the current running conditions, 3 or more blank injection runs were performed before starting a batch of samples and 1 blank run was performed prior to new experimental runs each time the solvent/salt system was changed. The baseline shows consistent, minor variations arising from changes in salt concentrations. Chromatograms were baseline subtracted from a blank injection running the same method. The following method time-programs were used during all analyses.

3.8.3.1. 1 mL columns (**E1.r** or SPXL)

Flow rate = 1 mL/min. Total run time = 30 minutes. The method started with flushing the column for 10 minutes using pure RB followed by transition from RB to EB in a gradient fashion during next 5 minutes. Next, the column was flushed with full EB for 5 minutes before transitioning back to RB in a gradient fashion in 5 minutes. The column was flushed with full RB for final 5 minutes to condition it for the next sample.

3.8.3.2. Capillary column (**E1.r**)

Flow rate = 0.1 mL/min. Total run time = 120 minutes. The method started with flushing the column for 40 minutes using pure RB followed by transition from RB to EB in a gradient fashion during next 20 minutes. Next, the column was flushed with full EB for 20 minutes before transitioning back to RB in a gradient fashion in 20 minutes. The column was flushed with full RB for final 20 minutes to condition it for the next sample.

3.8.4. Proteolysis of histone sample

Calf thymus histone (Worthington Chemicals) was incubates with 100:1 (w/w) ArgC protease (Sigma-Aldrich) at 37°C for 18 hours in 100 mM NH_4HCO_3. After incubation, digested samples were frozen at -4oC and thawed immediately before use.

3.8.5. Proteomics protocols and data LC-MS/MS Analysis[257]

All proteomics samples were submitted for proteomics analysis at the UVic/Genome BC Proteomics Centre. All details below are as provided by the analytical service. The lists of identified peptides and PTMs were then manually collected and tabulated (Tables 3.1-3.3).

The collected fractions were first acidified by adding 10 μL formic acid and then desalted using a Thermo Scientific C18Stage Tips SP301 (200 μL). Following binding and washing, samples were eluted with 40 μL (80% v/v Acetonitrile, 0.1% v/v Formic acid), speed vacuumed to near dryness and rehydrated with 2% acetonitrile, 0.1% formic acid, water.

The samples were separated by on-line reverse phase chromatography using a Thermo Scientific EASY-nLC 1000 system with a reversed-phase pre-column Magic C18-AQ (100μm I.D., 2 cm length, 5μm, 100Å) and an in-house prepared reverse phase nano-analytical column Magic C-18AQ (75μm I.D., 15 cm length, 5μm, 100Å, Michrom BioResources Inc, Auburn, CA), at a flow rate of 300 nL/min. Solvents were A: 2% Acetonitrile, 0.1% Formic acid; B: 90% Acetonitrile, 0.1% Formic acid. After a 249 bar (~8μL) pre-column equilibration and 249 bar (~10μL) nanocolumn equilibration, samples were separated by a 55 minute gradient (0 min: 5%B; 45 min: 30%B; 2 min: 100%B; hold 8min: 100%B).

The chromatography system was coupled on-line with an Orbitrap Fusion Tribrid mass spectrometer (Thermo Fisher Scientific, San Jose, CA) equipped with a Nanospray Flex NG source (Thermo Fisher Scientific). Instrument parameters used for iontrap (IT-MS/MS) with HCD fragmentation are as follows -Nano-electrospray ion source: spray voltage 2.4kV, capillary temperature 275 °C; Survey MS1 scan: m/z range 400-2000, profile mode, resolution 120,000 FWHM@200m/z, number of microscan = 1, maximum inject time 50 ms. Internal calibration was done using siloxane mass (445.120024) as a reference. Data-dependent acquisition of Orbitrap survey spectra were scheduled at least every 3 seconds, with the software determining "Top-speed" number of MS/MS acquisitions during this period. The automatic gain control (AGC) target values for FTMS and MSn were 200,000 and 10,000 respectively. The most intense ions charge state 2-7 exceeding 50,000 counts were selected for CID ion trap MSMS fragmentation with detection in centroid mode. Monoisotopic Precursor Selection (MIPS) was enabled and Dynamic exclusion settings were: repeat count = 2; repeat duration = 15 seconds; exclusion duration = 60 seconds with a 10 ppm mass window. The ddMS2 IT CID scan used a quadrupole isolation window of 1.6 Da;IonTrap rapid scan rate, centroid detection, first mass: 100m/z, 1 microscan, 50ms maximum injection time and stepped collision energy 30% ± 5.

The samples were also analyzed by orbitrap (OT-MS/MS) with HCD fragmentation. Parameters used are as follows -Survey MS1 scan: m/z range 400-2000, profile mode, resolution FWHM@200m/z, number of microscan = 1, maximum inject time 50 ms. Internal calibration was done using siloxane mass (445.120024) as a reference. Data-dependent acquisition Orbitrap survey spectra were scheduled at least every 3 seconds, with the software determining "Top-speed" number of MS/MS acquisitions during this period. The automatic gain control (AGC) target values for FTMS and MSn were 400,000 and 10,000 respectively. The most intense ions charge state 2-7 exceeding 50,000 counts were selected for HCD MSMS fragmentation in the ion routing multipole. Monoisotopic Precursor Selection (MIPS) was enabled and Dynamic exclusion settings were: repeat count= 2; repeat duration= 10 seconds; exclusion duration = 10 seconds with a 10ppm mass window. The ddMS2 OT HCD scan used a quadrupole isolation window of 1.6 Da; 30,000 resolution Orbitrap scan, first mass: 100m/z, centroid detection, 1 microscan, 60ms maximum injection time and stepped collision energy 30% ± 5.

Data Analysis Parameters were as follows. Raw files were created by XCalibur 3.0.63 (Thermo Scientific) software and analysed with Proteome Discoverer 1.4.0.228 software suite (Thermo Scientific). Parameters for the Spectrum Selection to generate peak lists of the CID spectra (activation type: CID; s/n cut-off: 1.5; total intensity threshold: 0; minimum peak count: 1; precursor mass: 350-5000 Da) The peak lists were submitted to an in-house Mascot 2.4.1 server against UP_cow (24,461 sequences; 13,017,573 residues) database search as follows: precursor tolerance 5 ppm; MS/MS tolerance 0.8 Da;

Trypsin (or Arg-C) enzyme 2 missed cleavages; FT-ICR instrument type; fixed modification: none; variable modifications: acetyl(K), methyl (K,R), dimethyl(K,R), trimethyl (K). Percolator settings: Max delta Cn 0.05; Target FDR strict 0.01, Target FDR relaxed 0.05 with validation based on q-Value. The data collected by FT-HCD MS/MS used the following search parameters: precursor tolerance 5 ppm and MS/MS tolerance 15 mmu. The Mascot data was further analyzed for PTM content using Scaffold. Proteins with less than 50% probability of being present were omitted from the data analysis. All proteins identified in each fraction, and every residue with a confirmed PTM from the list Kme1, Kme2, Kme3, Kac, Rme1, and/or Rme2, are tabulated in Section 5.9.6.

3.8.6. Identified proteins and methylated sites in each fraction

Table 3.1. All identified proteins and methylation sites in non-enriched input sample.

S.No.	Protein name	Uniprot Accession Code	Probablity	Coverage	PTM residue	Kme1	Kme2	Kme3	Kac	Rme1	Rme2
1	uncharacterized	G3MWV5	100%	32%	K34	1					
2	uncharacterized	G3MWV5	100%	32%	K46				1		
3	uncharacterized	G3MWV5	100%	32%	K52				1		
4	Histone H1.2	H12	100%	22%	R33					1	
5	Histone H1.2	H12	100%	22%	K34	1					
6	Histone H1.2	H12	100%	22%	K46				1		
7	Histone H1.2	H12	100%	22%	K52				1		
8	Histone H2B	G5E6I9	100%	16%	None						
9	Histone H2B	F2Z4F9	100%	69%	None						
10	Histone H2B	E1B8G9	100%	9%	None						
11	uncharacterized	G3MWH4	99%	13%	None						
12	Histone H2B	F1MVX6	94%	19%	K140				1		
13	Histone H2B	G3N080	89%	0%	None						
14	Histone H3.1	H31	100%	51%	K19				1		
15	Histone H3.1	H31	100%	51%	K24				1		
16	Histone H3.1	H31	100%	51%	K28	1	1				
17	Histone H3.1	H31	100%	51%	K80	1	1				
18	Histone H3.3C	H3C	60%	17%	K80	1	1				
19	Histone H4	G3X807	100%	78%	R15					1	
20	Histone H4	G3X807	100%	78%	K16		1				
21	Histone H4	G3X807	100%	78%	R19					1	1
22	Histone H4	G3N2B8	80%	67%	K6				1		
23	Histone H4	G3N2B8	80%	67%	K9				1		
24	Histone H4	G3N2B8	80%	67%	K13				1		
25	Histone H4	G3N2B8	80%	67%	K17				1		
26	Histone H4	G3N2B8	80%	67%	R20					1	
27	Histone H4	G3N2B8	80%	67%	K21		1				
28	Histone H4	G3N2B8	80%	67%	R24					1	1
29	Histone H2A	F2Z4G5	100%	81%	R89						1
30	Histone H2A	F2Z4G5	100%	81%	K96	1	1				
31	Histone H2A	F2Z4G5	100%	81%	K100	1	1				
32	Histone H2A	A4IFUS	100%	81%	None						

33	Histone H2A	H2A2C	100%	81%	None
34	Histone H2A	E1BH22	100%	72%	None
35	Histone H2A	F1MLQ1	100%	48%	None
36	Histone H2A.J	H2AJ	56%	44%	None
37	uncharacterized	F1MMU4	100%	10%	None
38	Histone H2A.V	H2AV	90%	45%	None
39	Histone H2A.Z	H2AZ	98%	45%	None
40	High mobility group AT - hook 1	Q0VC27	100%	18%	None
41	Histone H2A	Q2HJ65	100%	14%	None
42	Heterochromatin protein 1-binding protein 3	HP1B3	100%	4%	None
43	uncharacterized	E1B8K6	100%	7%	None
44	KRT5 protein	A5D7M6	100%	9%	None
45	uncharacterized	E1B8N6	100%	5%	None
46	uncharacterized	F1N7I5	99%	1%	None
47	60S ribosomal protein	F1MK30	100%	12%	None
48	Vimentin	VIME	99%	3%	None
49	Histone H1.0	H10	85%	0%	None
50	Actin, aortic smooth muscle	ACTA	99%	9%	None
51	60S ribosomal protein	RL6	75%	0%	None
52	Histone H2A	Q1LZ92	100%	2%	None
53	uncharacterized	F6S1Q0	100%	4%	None
54	SWISS-PROT:P04264	P04264	99%	3%	None

Table 3.2. All identified proteins and methylation sites in Unretained fraction post enrichment.

S.no.	Protein name	Uniprot accession code	Probablity	Coverage	PTM residue	Kme1	Kme2	Kme3	Kac	Rme1	Rme2
1	uncharacterized	G3MWV5	100%	21%	K34	1			1		
2	uncharacterized	G3MWV5	100%	21%	K46				1		
3	Histone H2B	E1B8G9	100%	9%	R93					1	
4	uncharacterized	G3MWH4	100%	13%	None						
5	Histone H3.1	H31	100%	41%	K19				1		
6	Histone H3.1	H31	100%	41%	K24				1		
7	Histone H3.1	H31	100%	41%	K80	1	1				
8	Histone H3.3C	H3C	85%	17%	K80	1	1				
9	Histone H4	G3X807	100%	59%	None						
10	Histone H4	G3N2B8	76%	50%	K6				1		
11	Histone H4	G3N2B8	76%	50%	K9				1		
12	Histone H4	G3N2B8	76%	50%	K13				1		
13	Histone H4	G3N2B8	76%	50%	K17				1		
14	Histone H2A	A4IFUS	70%	20%	None						
15	Histone H2A	H2A2C	75%	12%	None						
16	Histone H2A	E1BH22	94%	11%	None						
17	Histone H2A	F1MLQ1	65%	20%	None						

S.no.	Protein name	Uniprot Accession code	Probability	Coverage	PTM residue
18	High mobility group AT - hook 1	Q0VC27	100%	24%	None
19	Histone H2A	Q2HJ65	100%	3%	None
20	Heterochromatin protein 1 - binding protein 3	HP1B3	96%	6%	None
21	uncharacterized	E1B8K6	100%	2%	None
22	KRT5 protein	A5D7M6	100%	5%	None
23	uncharacterized	E1B8N6	100%	5%	None
24	uncharacterized	F1N7I5	100%	2%	None
25	60S ribosomal protein	F1MK30	99%	3%	None
26	Vimentin	VIME	100%	6%	None
27	SWISS-PROT:P35527	P35527	100%	7%	None
28	Actin, aortic smooth muscle	ACTA	100%	3%	None
29	60S ribosomal protein	RL6	99%	8%	None
30	Histone H2A	Q1LZ92	100%	2%	None
31	uncharacterized	F6S1Q0	100%	9%	None

Table 3.3. All identified proteins and methylation sites in Retained fraction post enrichment.

S.no.	Protein name	Uniprot Accession code	Probability	Coverage	PTM residue	Kme1	Kme2	Kme3	Kac	Rme1	Rme2
1	Histone H1.3	H13	97%	24%	K17			1	1		
2	Histone H1.3	H13	97%	24%	K35	1					
3	Histone H1.2	H12	100%	24%	K17				1		
4	Histone H1.2	H12	100%	24%	K34	1					
5	uncharacterized	G3MWH4	99%	13%	None						
6	Histone H3.1	H31	100%	25%	K10		1	1			
7	Histone H3.1	H31	100%	25%	K15				1		
8	Histone H3.1	H31	100%	25%	K28	1	1		1		
9	Histone H3.1	H31	100%	25%	K37	1	1				
10	Histone H3.1	H31	100%	25%	K38	1	1				
11	Histone H4	G3X807	100%	28%	R19						1
12	Histone H4	G3N2B8	92%	40%	K6				1		
13	Histone H4	G3N2B8	92%	40%	K9				1		
14	Histone H4	G3N2B8	92%	40%	K13				1		
15	Histone H4	G3N2B8	92%	40%	K17				1		
16	Histone H4	G3N2B8	92%	40%	R24						1
17	Histone H2A	A4IFUS	73%	42%	None						
18	Histone H2A	H2A2C	71%	7%	None						
19	Histone H2A	E1BH22	63%	32%	None						
20	Histone H2A	F1MLQ1	63%	15%	None						
21	uncharacterized	F1MMU4	100%	20%	None						
22	Histone H2A.V	H2AV	100%	22%	None						

#	Protein	ID			Site			
23	Histone H2A.Z	H2AZ	96%	7%	None			
24	High mobility group AT-hook 1	Q0VC27	98%	18%	None			
25	Histone H2A	Q2HJ65	100%	3%	None			
26	Heterochromatin protein 1 - binding protein 3	HP1B3	100%	6%	K535		1	1
27	Heterochromatin protein 1 - binding protein 3	HP1B3	100%	6%	K538	1	1	1
28	Heterochromatin protein 1 - binding protein 3	HP1B3	100%	6%	K542	1	1	1
29	Heterochromatin protein 1 - binding protein 3	HP1B3	100%	6%	K544		1	1
30	Heterochromatin protein 1 - binding protein 3	HP1B3	100%	6%	K548	1	1	
31	Heterochromatin protein 1 - binding protein 3	HP1B3	100%	6%	K549	1	1	1

Chapter 4. Combining upper- and lower-rim substitutions on sulfonated calix[4]arenes

Contributions

The idea to study the effect of a single carboxymethyl substituent at the lower rim in combination with upper-rim substitutions was conceived of by me. All synthesis, NMR and binding studies were performed by me.

4.1. Foreword

Chapter 2 and 3 described our experience with upper rim functionalized calix[4]arenes. Our conclusion point towards both beneficial and detrimental aspects of directing all efforts towards the upper rim. While supramolecular properties can be easily tuned using this strategy, immobilization via the upper rim and use as an enrichment agent was encouraging but not satisfactory. One major disappointment was the inability of upper-rim functionalized calix[4]arenes to enrich real world samples (cell lysates) despite promising results on model proteomics samples (pre-isolated histones). This prompted us to look for alternate positions on the calix[4]arene skeleton for immobilization. We identified the lower rim as an attractive alternative (despite early disappointments) after a literature survey. This Chapter presents our attempts to synthesize calix[4]arenes containing a single pendant carboxylic acid at the lower rim, in combination with upper-rim substitutions that we had established during Chapter 2.

4.2. Motivation

The library of upper rim functionalized calix[4]arenes led to strong hosts for Kme3 containing peptides. However, two big problems were encountered when attempts were made to take the host molecules from the library and convert them to useful enrichment agents. First, not all members of the library had additional chemical handles (carboxylic acids, amines) which could be used to functionalize a solid support. This meant that only a fraction of molecules in the library could be tested as enrichment agents in their as-synthesized form. Second, the subset of molecules that did have an additional handle to attach to solid support did not have consistent performance after being immobilized. As discussed in Chapter 3, compound **AP1** performed poorly as an enrichment agent while **E1** was excellent, even though **AP1** was a stronger binder in IDA experiments (Figure 4.1). While our data prevents us from making a conclusive statement about the reason behind this disparity, we reasoned that this might be due to the fact that the solid support was attached to the binding face of the calix[4]arene (upper rim). This might block the ability of the host molecule to successfully encapsulate any guest peptide. **E1** proved better because it has extra rotational degrees of freedom around the sulfonamide bond which allows it to present its binding pocket to an incoming guest.

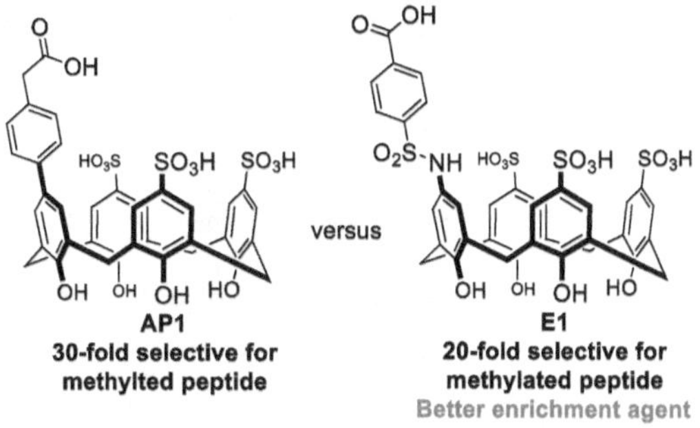

Figure 4.1. Solution based selectivity of upper rim functionalized calix[4]arenes could not predict their performance as enrichment agents.

We reasoned that calix[4]arene hosts which have additional functional handles at a position less obstructing to the binding face might be more suitable to investigate. This would allow us to study their host-guest properties in solution, easily immobilize them on a solid support, and expect some consistency in their host-guest behavior in solution and on solid support. One such position for substitution on the calix[4]arene scaffold is the lower rim phenols. Hundreds of reports where authors attach designed substituents at the phenolic position using an electrophilic linker have been published.[258–262] A plethora of applications have been proposed including electrochemically sensing anions,[263] extraction of rare earth metals,[264] and as a gene[265] and drug[266] delivery systems, to name a few.[267]

Lower rim substitution has a drastic effect on the conformation of calix[4]arenes that is somewhat more predictable than the upper rim substitution. Two competing effects are expected. First, substituting phenolic H with an R group diminishes its ability to form intramolecular phenol-phenol hydrogen bonds, which may lead to collapse of the cone-shaped calix[4]arene cavity.[176,268] Second, if the R group is bulky enough then the intra-annular movement of phenols is sterically hindered which restricts the calixarene into cone/partial-cone/pinched-cone conformations, or a dynamic combination of these conformations.[211,212] These phenomena were evident in Hof group's only, much earlier attempt towards lower rim functionalization of *p*-sulfonatocalix[4]arene which led to a total loss binding.[176] Converting all four phenols of **PSC** to ethers led to the complete collapse of the aromatic cage due to loss of hydrogen bonding among the lower rim phenols. Strapping the ethers together to form a more rigid and restricting substitution at the lower rim led to regeneration of the binding cavity in the preferred conformation. This allowed the calix[4]arene to regain some of its binding affinity towards methylated peptides.

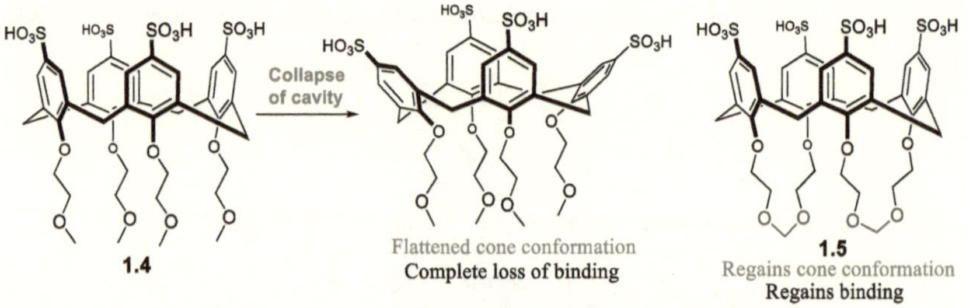

Figure 4.2. Previous attempts lower rim substitution demonstrate influence over the binding pocket.[176]

While results of our previous attempts on substitution at the lower rim at first deterred us from visiting the option again, we became aware of published works from the groups of Seiji Shinkai and Rocco Ungaro on water soluble lower-rim carboxymethylated calix[4]arenes (Figure 4.3) that retained a cone conformation in aqueous solutions.[269–271] This is a result unlike the collapse seen for glycol ether-type substituents (Figure 4.2). Studies on lower-rim carboxylmethylated calixarenes with either upper-rim *p*-tertbutyl groups or upper-rim *p*-sulfonato groups have been published (Figure 4.3). Shinkai et al. have commented on uranophilicity[269] of these calix[4]arenes while Ungaro et al. have studied the thermodynamics of ammonium ion encapsulation by these hosts.[270]

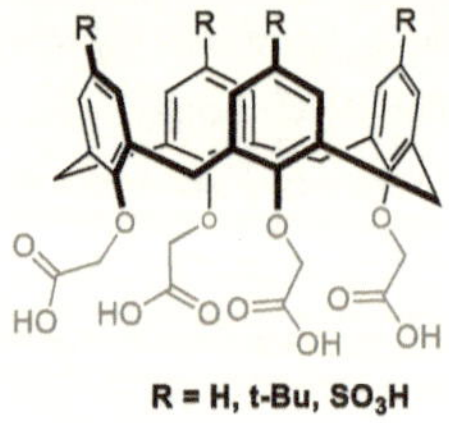

Figure 4.3. Published *O*-carboxymethyl calix[4]arenes that maintain cone conformations in aqueous solutions.[269–271]

Buoyed by these reports, we sought to adapt these modifications on to our scaffolds. We envisioned that remaking select library members on calix[4]arenes with carboxymethyl substituent at the lower rim would allow us to immobilize calix[4]arenes, which did not posses any additional chemical handle, on a solid support easily while also having less obstructing effect on the binding face of the calix[4]arene, therefore giving a more consistent correlation in host-guest properties in solution and after attachment to solid support.

The goal of this Chapter is to present our efforts towards the development of highly functionalized calix[4]arenes with both upper and lower rim modifications. We hope that this dually functionalized calix[4]arene will allow us to both modify its host-guest properties (by changing substituents at the upper rim) and expand the utility of these host via later attachments at the lower rim.

4.3. Design and synthesis

Shinkai and Ungaro's work focused on tetra O-alkylated calix[4]arenes which we immediately recognized to be not desirable for our cause. First, we recognized that tetra O-carboxymethyl p-sulfonato calix[4]arene will have a net -8 charge which will generate a huge amount non-specific electrostatic interactions. We reasoned that this would lower our chances to observe any selectivity for methylated peptides, all of which are positively charged. Second, we had already observed that calix[4]arenes with more than one carboxylic acid tend to cause crosslinking of solid support upon immobilization which would make it less permeable and not suitable as an enrichment agent.

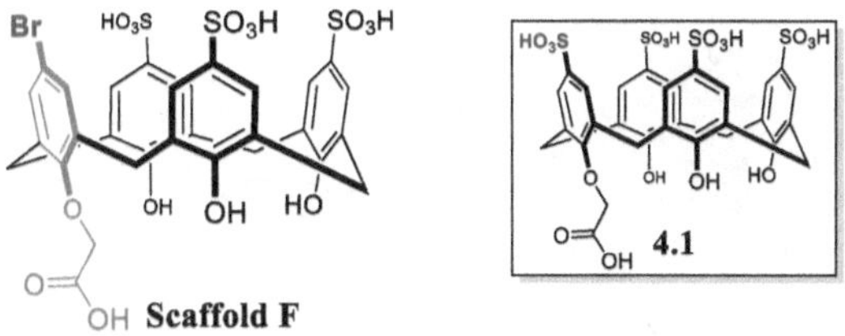

Figure 4.4. Our design for incorporating O-carboxymethyl on library members from Chapter 2 led us to Scaffold F. Unknown to us, a similar compound (4.1) has been synthesized and studied.[272]

We instead choose to test the effect of single carboxymethyl substitution on the lower rim of Scaffold **A**, our best performing scaffold from Chapter 2 (Figure 4.4). This decision was based on a reductionist approach. We hoped that by installing a single substitution at the lower rim, the changes in conformational properties of Scaffold **A** would be minimal. If this hypothesis were found to be correct, we would be able to pick and choose substitutions at the upper rim from the library in Chapter 2 and hope to replicate their supramolecular properties on our new scaffold with reasonable certainty. This would give us the ability to tune supramolecular properties using the upper rim substitutions and convert these hosts to a useful enrichment agent through the lower rim. During the preparation of this Chapter, we became aware of one such effort where the effect of single carboxymethyl substitution on amino acid complexation of p-sulfonated calix[4]arene was studied. Silva et al. synthesized **4.1** and curiously found it to bind negatively charged amino acid (Aspartic acid) over lysine and arginine.[272] We can't rationalize binding of these anionic guests with anything that we have observed, but at least the retention of a guest binding property is promising. While not exactly the same as our new Scaffold **F**, they conclude that mono O-carboxymethylation of calix[4]arene maintains its cone conformation which aligns well with our ideas. Their binding trends, however, must be replicated before commenting further.

The idea to place carboxymethyl at the same phenol as the one also presenting a bromine at the upper rim was not random. We were concerned about increasing synthetic steps to our already longer than ideal synthesis. However, triester calix[4]arene **2.4** presented a nice opportunity to integrate regioselective lower rim functionalization by adding just one extra step (Scheme 4.1). The optimized route to scaffold **F** starting from **2.4** is shown in Scheme 4.1. The only new step, *O*-alkylation with ethyl bromoacetate, was first attempted using K_2CO_3 and base and DMF as solvent but these conditions never recorded yields higher than 45 %. Addition of dibenzo-18-crown-6 improved the yield significantly and switching of solvent to acetone removed the any need of chromatography-based purification. The product (**4.2**) precipitates out of reaction mixture and can be triturated with methanol to remove dibenzo-18-crown-6. Saponification and sulfonation required no change from the conditions already described in Chapter 2. This route afforded Scaffold **F** in 38 % overall yield over 3 steps starting from **2.4**.

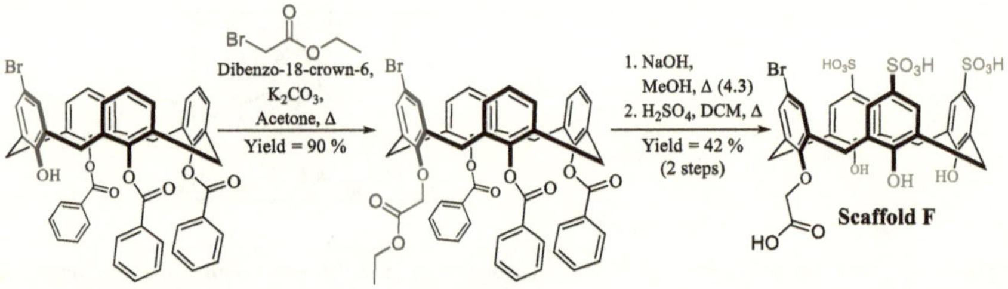

Scheme 4.1. Optimized synthesis of scaffold F integrates well into the already optimized route for making Scaffold A.

4.4. Comparison of Scaffold F vs Scaffold A

We next used NMR to investigate host-guest complexation between Scaffold **F** and methylated lysine. We found that Scaffold **F** binds Kme3 roughly 6-fold weaker than Scaffold **A** (K_d = 1.6 mM versus 0.25 mM, Figure 4.5). Based on our previous experience, we find this unsurprising. But, much to our delight, Scaffold **F** is similarly selective for Kme3 over K. This is a major improvement over our previous attempts at lower rim functionalization, which had resulted in a loss of both affinity and Kme3 selectivity.[176] ^{1}H and ^{13}C NMR suggest a cone conformation for Scaffold **F**, similar to that of like Scaffold **A**.[273,274] One difference is that the methylenes of Scaffold **F** are well resolved doublets while the methylenes of Scaffold **A** are broad coalesced peaks. This suggests vastly different conformational dynamics for the two scaffolds. Our new Scaffold **F** appears to be comparatively more frozen in a single cone conformation and does not undergo macro-ring inversion on the NMR time scale. This preorganization might be the reason for the loss in affinity shown by Scaffold **F**. We think Scaffold **A** is more flexible and accepting of guest as big as Kme3 while Scaffold **F** has a narrower cavity. Nevertheless, the selectivity for Kme3 over unmethylated K is retained.

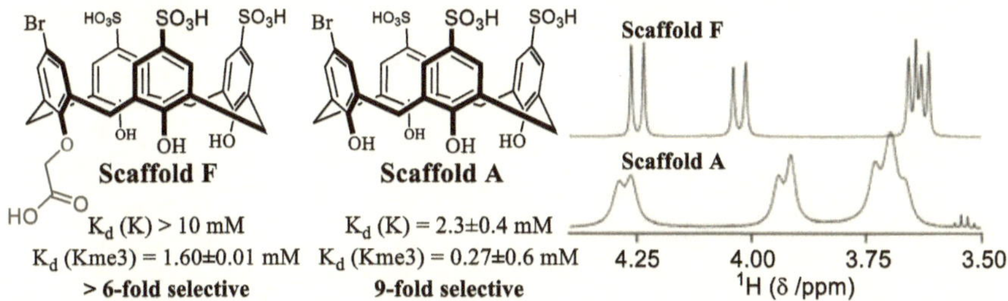

Figure 4.5. Scaffold F vs Scaffold A, a comparison of binding constants and their conformational properties at NMR time scale. Binding constants were estimated by NMR titration (500 MHz) in 40 mM Sodium phosphate buffer at pH 7.4. Methylene region of 1H NMR (recorded in D2O at 500 MHz) shows difference in rigidity between scaffolds.

NMR titrations agree with an end-on encapsulation of Kme3 by Scaffold **F**, like that of Scaffold **A**. Big shifts in the signals for N-methyl, ε- and δ- protons are seen upon encapsulation (Figure 4.6.a). Comparatively, lysine protons show minimal shifts in solution with Scaffold **F** (Figure 4.6.b). Table 4.1 summarizes these complexation-induced shifts upon addition of one equivalent of Scaffold **F**. Kme3 and K bind Scaffold **F** using two very different modes of binding that can be immediately seen in these data. For Kme3, the Δδ is highest for N-methyl protons followed by ε-, δ-, γ- and β- protons. This suggests an end-on encapsulation where the Kme3 sidechain is buried deep inside the cavity of Scaffold **F**, with the N-methyls deepest within the cavity. Comparatively, the δ- proton of lysine shows marginally bigger shift than the ε- proton suggesting a side-on complexation where the lysine sidechain sags into the cavity rather than digging deep. The ammonium, for lysine, stays near the upper-rim sulfonates where it is likely H-bonded. For both lysine and trimethyllysine, the α- protons barely register any shift indicating that they are anchored far away from the cavity. Scaffold **A** shows chemical shift changes for Kme3 and K. Taken together, the data in Table 4.1 and Figure 4.6 support our hypothesis that the lower rim of a calix[4]arene can be carboxymethyl substituted without causing big changes in guest binding modes.[173,175]

Table 4.1. Complexation-induced NMR shift for different CH protons of Kme3 and K after the addition of 1 equivalent of Scaffold F in 40 mM Sodium phosphate buffer (pH 7.4).

Protons	Δδ for Lysine (K)	Δδ for (Kme3)
N-methyl	-	0.54
ε- CH_2	0.11	0.33
δ- CH_2	0.12	0.21
γ- CH_2	0.05	0.05
β- CH_2	0.05	0.02
α- CH	0.01	0.00

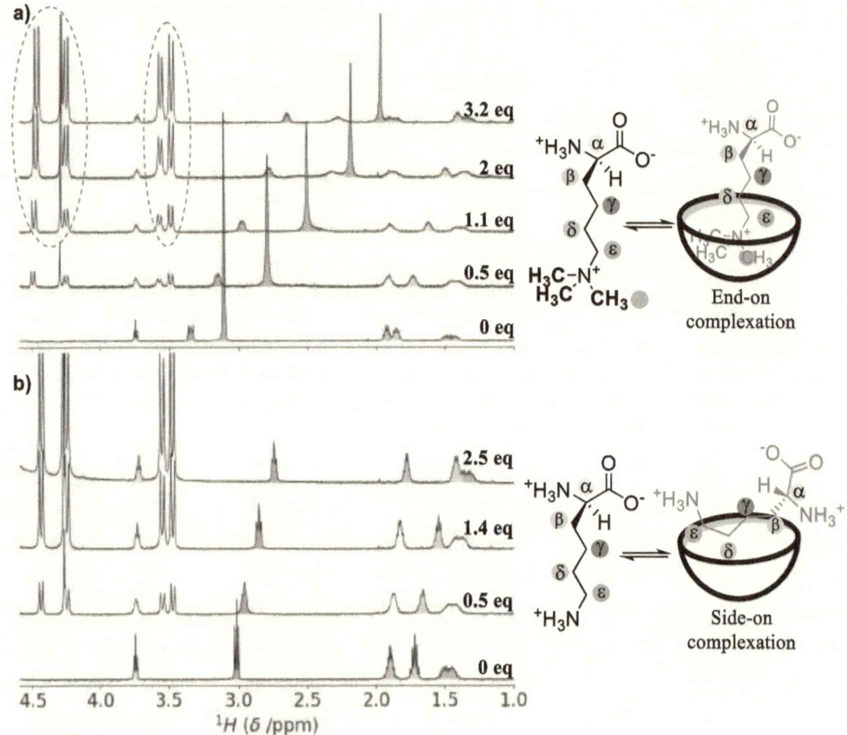

Figure 4.6. NMR titration of Scaffold F with a) Kme3 and b) K. These titrations were performed in 40 mM Sodium phosphate buffer at pH 7.4 (pD 7.0) on a 500 MHz NMR spectrometer. A 5-10 mM sample of Scaffold F was titrated into 1-2 mM solution of amino acid in buffer. The identity of key CH peaks is colour-coded on the chemdraws beside the plot. A schematic structure of complex based on NMR chemical shifts is also shown for each amino acid. Red dashed circles show the position of Scaffold F methylene peaks. More details are shown in section 4.8.3.

4.5. Combining Scaffold F with upper-rim modifications from Chapter 2

With the confirmation that Scaffold **F** binds Kme3 selectively over K and with the same mechanism as Scaffold **A**, we moved forward with our goal to combine upper and lower rim substitutions. Suzuki coupling as described in Chapter 2 was adapted for Scaffold **F** without any change. Three fragments from Figure 2.5 were chosen, viz **O2**, **P5** and **H1**. Their coupling to Scaffold **F** gave us **FO2**, **FP5** and **FH1** (Scheme 4.2). The choice of fragments was made based on the binding strength of their Scaffold **A** counterparts, **AO2**, **AP5** and **AH1**, all of which bind methylated peptide with impressive strength ($K_d < 1$ μM) and >30-fold selectivity over unmethylated peptide.

Scheme 4.2. Suzuki coupling of Scaffold F with selected boronic acid fragments chosen from Figure 2.5 in Chapter 2.

Table 4.2. Comparison of K_d (nM) obtained for Scaffold F with their corresponding Scaffold A analogs.[a]

Peptide	AO2	FO2	AP5	FP5	AH1	FH1
H3K4	4100±200	n.d. [b]	4300 ±300	7100±2600	1300±300	n.d. [b]
H3K4me1	1100±700	n.d. [b]	1300±700	600±230	360±10	< 5 [c]
H3K4me2	290±30	< 10 [c]	440±80	190±50	31±2	9.8±0.5
H3K4me3	110±20	< 10 [c]	140±10	720±40	35±2	18+4
H3K9me3	80±20	< 10 [c]	310±70	>10^5	93±4	< 5 [c]
H3K4ac	7000±200	n.d. [b]	7700±900	>10^5	12000±700	n.d. [b]
H3R2me2a	920±150	< 10 [c]	600±40	340±65	160±10	9.1±0.8
H3R2me2s	1600±400	< 10 [c]	1600±200	1300±460	760±120	20±3

a) All binding constants were determined using IDA as described in Section 2.6. Values tabulated above are all in nM units and are the average of two replicates. Errors are reported as standard error (average of two titrations). b) The fluorescence responses in these titrations were saturated even at the lowest titrant concentration and could not be fit to a 1:1 binding isotherm. c) The fluorescence response had reached plateau at the lowest titrant concentration and only an upper limit could be estimated.

Table 4.2 summarizes all the binding data obtained from Scaffold **F** compounds and compares them with their Scaffold **A** analogs. In general, Scaffold **F** compounds bind methylated peptides with stronger affinity than their Scaffold **A** counterparts. This is in stark contrast to the parent scaffold comparison (*vide supra*) where Scaffold **A** is a stronger host than Scaffold **F**. However, there are troubling trends across the panel. **FO2** and **FH1** show intense response with non-methylated peptides. Within the concentration range of our assay, the fluorescence responses for all titrations with **FO2** jumped at first and then saturated immediately, and we couldn't fit them to any binding isotherm. One interpretation of these data is that all peptides bind very strongly to **FO2** (K_d <10 nM), but we are hesitant to interpret these data in this way. Such broad and strong binding would be unprecedented, and there is nothing about **FO2** that suggests that it should be so different than the other hosts. At the time of this project, the structure activity relationship discussed in Section 2.8 was just beginning to be understood. *Ortho* substituted fragment and heterocycles have given us curious results with Scaffold **A** (Chapter 2). Technical replicates of **FO2** gave the same anomalous binding profile every time. With **FH1**, on the other hand, we had trouble getting consistent titration results (especially for the direct titration with lucigenin). Heterocycle-based substituents on this calixarene framework are difficult to study with IDA and are prone to degradation. Although we haven't confirmed it, we believe that a combination of these factors might be the reason behind the observed binding profile of **FH1**.

FP5 has the best binding curves of all Scaffold **F** compounds. At first glance, it appears to be a candidate Kme2 selective host. However, the lack of binding to H3K9me3 is puzzling. We expect H3K9me3 to bind weaker than H3K4me3 due to presence of one less positive charge, but complete disruption of binding suggests that it is not a candidate to be the pan-specific *N*-methyl peptide binder that we desire. We found, when attempting detailed characterisation of **FP5**, that it shows complicated aggregation behavior in D_2O (see experimental section), something that we did not account for in our IDA study.

Only a much more detailed and direct assay like ITC or NMR can confidently elucidate the finer details of the peculiar behavior of Scaffold **F** compounds but those are out of scope of this book and were not attempted.

4.6. Shortcomings

Scaffold **F** compounds represent an interesting step forward in calixarene-based supramolecular hosts, but their utility towards the goal of this project is questionable after analysis of data from Table 4.2. Our hypothesis that selecting Scaffold **A** fragments based on their host-guest properties and remaking them on Scaffold **F** will result small but manageable changes is demonstrably false. Our inability to predict host-guest binding properties demands another library-based approach directed specifically towards Scaffold **F**. The low yield of the final sulfonation steps has also not been solved. Scaffold **F** takes one more step than Scaffold **A** and, even though the extra step is high yielding (Scheme 4.1), the overall yield for the multi-step synthesis of these potential affinity reagents is poor (2-5%). These factors prevented us from testing

Scaffold **F** compounds as enrichment agents. While we can't conclude anything about their performance as enrichment agents without testing, starting another daunting and resource-consuming library project geared towards Scaffold **F** from scratch will have to wait until we figure out a quicker route to synthesize them.

4.7. Conclusion

Although we haven't confirmed any strong candidate for use as enrichment agent in this project, we can draw some useful conclusions about this new area of calix[4]arene chemistry. Notably, lower rim carboxymethyl substitution is achievable in combination with upper-rim modifications and is overall a highly versatile substitution. We have confirmed that a single such substitution can freeze **PSC** into various forms of guest-binding cone conformations. This preorganization can affect the host-guest properties of calix[4]arene in complex ways that require more detailed study to be understood properly. Our choice of upper rim substitution was unfortunate, but it doesn't rule out the potential of Scaffold **F** as an enrichment agent if a more efficient synthesis can be developed.

4.8. Supporting information

4.8.1. General consideration

Proton NMR (^{1}H NMR) were recorded on a Bruker Avance 500 MHz spectrometer at 23°C and processed with Topspin 3.6.1. Proton chemical shifts are reported in parts per million (ppm) downfield from tetramethylsilane and are referenced to residual proton in the NMR solvent. Data are represented as follows: chemical shift, multiplicity (s = singlet, d = doublet, t = triplet, q = quartet, m = multiplet, br = broad), integration, coupling constant in Hertz. Carbon NMR chemical shifts (^{13}C NMR) were recorded at 125 Hz at 23 °C and processed with Topspin 3.6.1. Carbon chemical shifts are reported in parts per million (ppm) downfield from tetramethylsilane and are referenced to the carbon resonances of the solvent except if the solvent is D_2O, in which case a 5% MeOH solution in D_2O was used as an external reference. Infrared (IR) spectra were obtained using a Perkin Elmer Spectrum-2 (ATR-FTIR) spectrometer. Data are represented as follows: frequency of absorption (cm^{-1}), intensity of absorption (s = strong, m = medium, w = weak, br = broad). Melting points were collected on a Gallenkamp Melting Point apparatus. Microwave reactions were performed on a Biotage Initiator microwave in a heavy glass microwave vial.

All sulfonated compounds were purified using RP-HPLC on Phenomenex Luna C18(2) column (4.6 mm x 250 mm) with 5 μm particle and 100 A pore size. The preparative RP-HPLC purifications were performed using a Shimadzu Prominence LC system using a gradient of acetonitrile in water as mobile phase. Both mobile phase eluents were spiked with 0.1% TFA. The compounds were detected using a DAD set to 280 nm. Analytical LC-MS characterization was done on a Waters Acquity-H UPLC-MS, using a

BEC C18 column with 1.7 μm particle size. Like the preparative HPLC, a gradient of acetonitrile in water was used as mobile phase. Both mobile phase eluents were spiked with 0.1% TFA. The compounds were detected using a DAD set to 280 nm and an inline mass spectrometer.

4.8.2. Synthetic methods for compounds in Scheme 4.1

Compound **4.2.** In a round bottom flask equipped with a magnetic stir bar, compound **2.4** (800 mg, 1 mmol, 1 eq.), K_2CO_3 (270 mg, 2 mmol, 2 eq.) and dibenzo-18-crown-6 (720 mg, 2 mmol, 2 eq.) were suspended in 50 mL acetone. The flask was equipped with a condenser. Ethyl bromoacetate (1340 mg, 0.88 mL, 8 mmol, 8 eq.) was added to this mixture and the reaction was refluxed overnight after which a white precipitate can be seen to have formed. The reaction is cooled, and the precipitate and supernatant separated. The precipitate is washed with ~150 mL of ice cold methanol to give 390 mg of highly pure **4.2**. The supernatant is reduced to a third in volume under low pressure. Ice cold methanol is added to produce more precipitate which is filtered and washed with more ice cold methanol to give 550 mg of crude product (with 5-10 % starting material **2.4**). The second precipitate can be recrystalized using DCM:methanol to give 370 mg of pure **4.2**. The total yield is 90%. Mp: >250 °C (dec). **IR (ATR, cm⁻¹):** 1759m, 1722s, 1600w, 1449s, 1315w, 1255s, 1200m, 1170s, 1093m, 1078m, 1056s, 1022m, 784w, 751m, 704s, 685w, 669w, 558w. **¹H NMR (CDCl₃, 500 MHz):** δ(ppm)= 1.41 (t, 3H, J = 7.3 Hz), 3.51 (d, 2H, J = 15.1 Hz), 3.6 (s, 4H), 3.9 (d, 2H, J = 15.1 Hz), 4.35 (q, 2H, J = 7.1 Hz), 4.47 (s, 2H), 6.63 (m, 7H), 6.82 (s, 2H), 7.24 (d, J = 7.6 Hz, 2H), 7.47 (m, 4H), 7.66 (two overlapping t, 5H), 7.78 (t, J = 7.3 Hz, 2H), 7.88 (d, J = 7.8 Hz, 4H). **¹³C{¹H} NMR (CDCl₃, 125 MHz):** δ(ppm)= 14.5, 36.7, 37.2, 61.2, 69.2, 115.7, 124.8, 125.4, 127.9, 128.2, 128.7, 129.0 (2 overlapping carbons as determined by HSQC), 130.5, 130.8, 131.0 (2 overlapping carbons as determined by HSQC), 131.1, 132.2, 132.6, 133.1, 133.5, 133.7, 133.7, 133.9, 136.0, 148.0, 148.3, 155.7, 164.3, 164.6, 168.2 **HR-ESI-MS:** Calculated for $C_{53}H_{41}BrO_9Na$ [M+Na⁺] = 923.1826, Found 923.18204.

Compound 4.3. In a round bottom flask equipped with a magnetic stir bar, compound **4.2** (600 mg, 0.66 mmol, 1 eq.) and sodium hydroxide (1330 mg, 33.3 mmol, 50 eq.) are suspended in 30 mL methanol and the mixture is refluxed overnight. The reaction mixture is cooled and 150 mL of 1 M HCL is added. The resulting precipitate is filtered and washed with multiple rounds of hot hexane (to remove benzoic acid). The solid is air dried to give 260 mg of an off-white powder in 70 % yield. Mp: >250 °C (dec). **IR (ATR, cm^{-1}):** 3306br, 2932w, 1737m, 1465s, 1448s, 1261m, 1245m, 1191s, 1058m, 751s, 707m, 671m, 664w, 577m. **^{1}H NMR (CDCl$_3$, 500 MHz)**: δ(ppm)= 3.48 (d, 2H, J = 13.5 Hz), 3.49 (d, 2H, J = 13.8 Hz), 4.27 (d, 2H, J = 13.8 Hz), 4.28 (d, 2H, J = 13.5 Hz), 4.86 (s, 2H), 6.72 (t, 1H, J = 7.3 Hz), 6.74 (t, 2H, J = 7.6 Hz), 7.02 (d, 2H, J = 7.7 Hz), 7.04 (dd, J = 6.9, 1.3 Hz, 2H), 7.08 (dd, 2H, J = 7.5, 1.3 Hz), 7.20 (s, 2H), 9.2 (br s, 2H), 9.6 (br s, 1H). **^{13}C{^{1}H} NMR (CDCl$_3$, 125 MHz)**: δ(ppm)= 31.7, 31.9, 72.6, 119.6, 121.8, 122.4, 126.9, 128.4, 128.6, 128.8, 129.1, 129.6, 132.8, 135.9, 148.9, 150.3, 150.7, 170.6. **HR-ESI-MS:** Calculated for C$_{30}$H$_{26}$BrO$_6$ [M+H^{+1}] = 561.0907, Found 561.0742.

Scaffold F. Compound **4.3** (100 mg, 0.124 mmol, 1eq.) was dissolved in minimal amount of DCM in a microwave vial equipped with a magnetic stir bar. Concentrated H$_2$SO$_4$ (0.2 mL, 20 eq.) was added and the vial was sealed. The reaction was then stirred at 60 C for 3 hours. During this time a pinkish precipitate can be observed to be forming in the vial. After 3 hours, the supernatant is decanted and checked for left over starting material using TLC. The precipitate is suspended in minimal amount of EtOAc and taken into centrifuge tubes. Ice cold Et$_2$O is added to the tubes are they are centrifuged. Supernatant is discarded and precipitate is subjected to two more rounds of suspending in Et$_2$O and centrifugation. The final precipitate is air dried to obtain 110 mg of crude product as a pinkish powder.

The crude contains 85% Scaffold F and 15 % of byproducts (partially sulfonated), as judged by HPLC, and can be used forward without purification. The crude can be purified over HPLC in order to obtain analytical samples. **IR (ATR, cm^{-1}):** 3313br, 2943w, 1744m, 1594m, 1475m, 1451w, 1440w, 1199w, 1145w, 1118s, 1030s, 896w, 785m, 671m, 650m, 625s, 5558w. **^{1}H NMR (CDCl$_3$, 500 MHz):** δ(ppm)= 3.67 (d, 2H, J = 13.8 Hz), 3.70 (d, 2H, J = 13.8 Hz), 4.07 (d, 2H, J = 13.8 Hz), 4.29 (d, 2H, J = 13.8 Hz), 4.76 (s, 2H, overlapping with HDO peak, detected using HSQC), 6.50 (br s, 2H), 7.58 (s, 2H), 7.64 (s, 2H), 8.12 (br s, 2H). **^{13}C{^{1}H} NMR (D$_2$O, 125 MHz):** δ(ppm)= 29.7, 30.7, 71.1, 118.6, 126.2, 126.7, 126.8, 128.7, 129.2, 129.4, 131.9, 134.8, 136.1, 136.2, 148.9, 151.3, 152.1, 172.3. **HR-ESI-MS:** Calculated for C$_{30}$H$_{24}$BrO$_{15}$S$_3$ [M-H^{-1}] = 798.9466, Found 798.91750.

4.8.3. NMR titration of free amino acids with Scaffold F

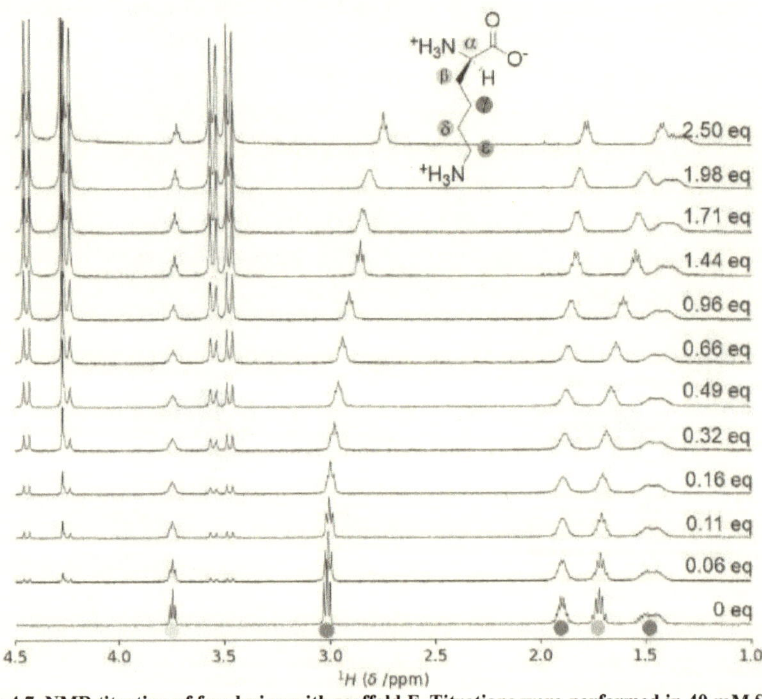

Figure 4.7. NMR titration of free lysine with scaffold F. Titrations were performed in 40 mM Sodium phosphate buffer at pH 7.4 (pD 7.0) on a 500 MHz NMR spectrometer. A 6 mM sample of Scaffold F was titrated into 2.04 mM solution of lysine (both were dissolved in the same buffer). The equivalents of scaffold F with respect to Kme3 is noted on the right side of each spectra. Protons of Kme3 are colour coded as in Figure 4.6.

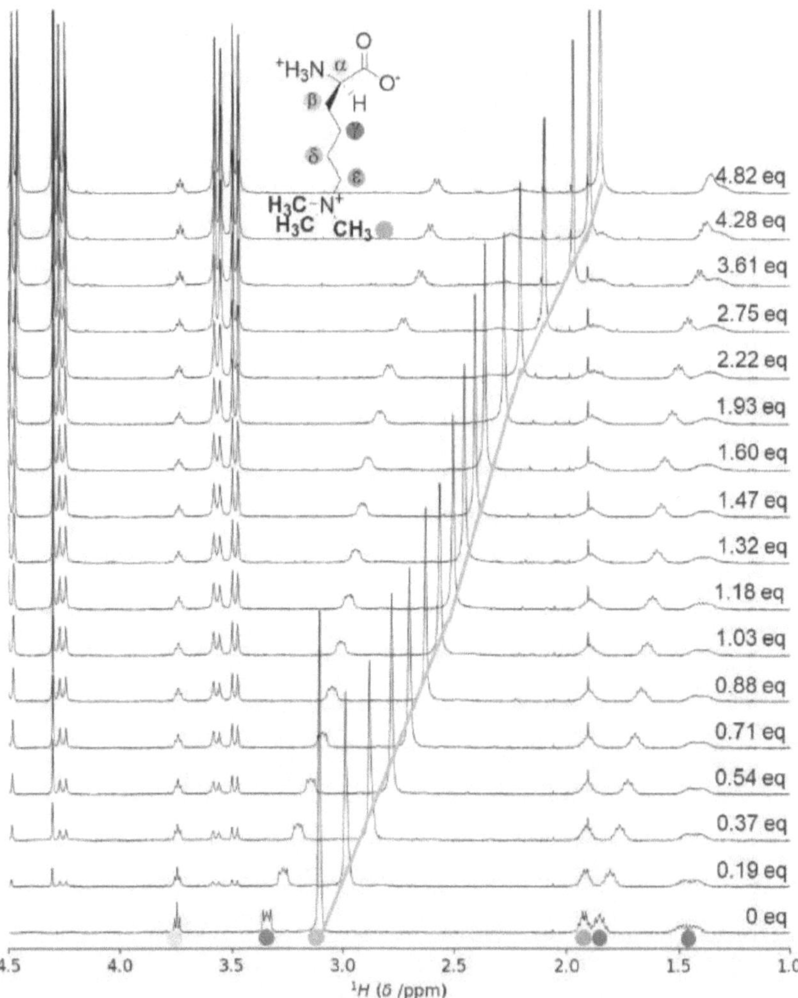

Figure 4.8. NMR titration of Kme3 with scaffold F. These titrations were performed in 40 mM Sodium phosphate buffer at pH 7.4 (pD 7.0) on a 500 MHz NMR spectrometer. A 9 mM sample of Scaffold F was titrated into 1 mM solution of Kme3 (both were dissolved in the same buffer). The equivalents of scaffold F with respect to Kme3 is noted on the right side of each spectra. Protons of Kme3 are colour coded as in Figure 4.6 and the movement of *N*-methyl is traced.

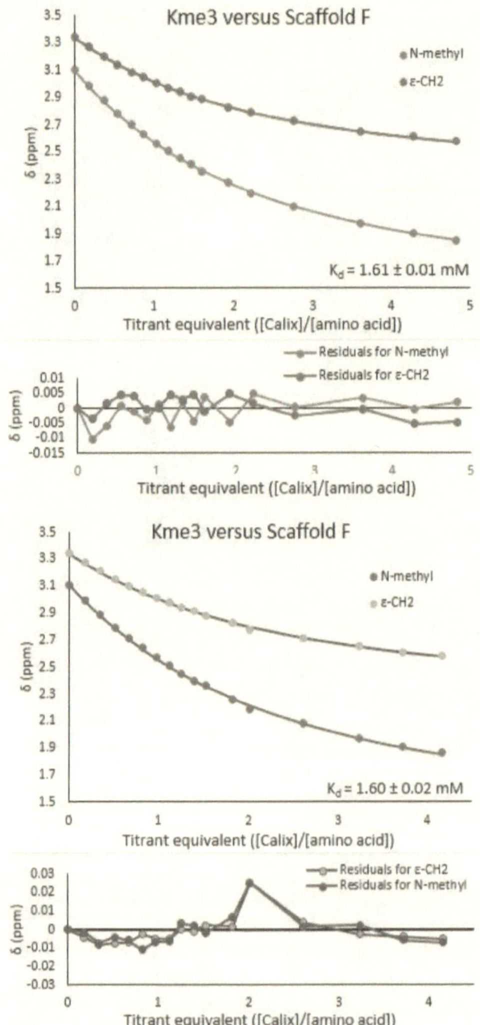

Figure 4.9. Two replicate binding curves obtained from titration of Scaffold F into Kme3. Titrations were performed in 40 mM Sodium phosphate buffer at pH 7.4 (pD 7.0) on a 500 MHz NMR spectrometer. A 9 mM sample of Scaffold F was titrated into 1 mM solution of Kme3 (both samples were made in the same buffer). Binding isotherms were fitted using a 1:1 model at the web portal supramolecular.org.[229] Shifts of both *N*-methyl and ε-CH₂ were fitted. Errors are reported as standard error estimated from the covariance of fit parameters.

4.8.4. General synthesis of compounds FO2, FP5 and FH1.

Scaffold **F** (50 mg, 0.062 mmol 1 eq.), R-B(OH)$_2$ (4 eq.), tetrabutylammonium bromide (TBAB, 10 mg, 0.031mmol, 0.5 eq.), sodium carbonate (25 mg , 0.24 mmol, 4 eq.) and Pd(OAc)$_2$ (2.6 mg, 20 mol%) are dissolved in 3 mL diH$_2$O inside a microwave vial and irradiated at 150 °C for 5 min. 1 mL of 1 M thiourea is added and the reaction is stirred for another hour at 70 °C. This results in precipitation of palladium as a complex with thiourea which can be filtered off. The filtrate can be reduced in volume *in vacuo*, if necessary, and then purified over HPLC.

FO2. Compounds **FO2** was isolated as 23 mg of a white powder with 45% yield. ^{1}H NMR (D$_2$O, 500 MHz): δ(ppm)= 3.08 (s, 3H), 3.51 (d, 2H, J = 13.3 Hz), 3.54 (d, 2H, J = 13.8 Hz), 4.02 (d, 2H, J = 13.3 Hz), 4.14 (d, 2H, J = 14.1 Hz), 4.59 (s, 2H), 5.19 (br s, 1H), 5.50 (br s, 1H), 6.24 (d, 1H, J = 8.1 Hz), 6.38 (d, 1H, J = 7.8 Hz), 7.03 (s, 2H), 7.34 (s, 2H), 7.48 (s, 2H), 7.56 (s, 2H). ^{13}C{^{1}H} NMR (D$_2$O, 125 MHz): δ(ppm)= 30.7, 30.8, 55.5, 71.3, 112.3, 120.4, 126.3, 126.4, 127.7, 127.9, 128.4, 128.4, 128.8, 129.5, 130.6, 132.2, 135.1, 135.8, 136., 149.6, 151.1, 151.5, 152.9, 155.0, 172.2. **HR-ESI-MS:** Calculated for C$_{37}$H$_{30}$O$_{16}$S$_3$ [M-2H^{-2}] = 413.03535, Found 413.0.3412

FP5. Compounds **FP5** was isolated as 28 mg of a brownish white powder with 50 % yield. The compound shows intense aggregation in pure D$_2$O. The spectra tabulated here is missing one resonance which could not be located in D$_2$O even with 2D NMRs. This signal (para hydrogen on biphenyl) is broadened due to intermediated exchange. It can be detected at 7.43 ppm upon addition of 5 % acetonitrile-d3 as a broad triplet (spectra shown in appendix). **^{1}H NMR (D$_2$O, 500 MHz)**: δ(ppm)= 3.66 (d, 2H, J = 13.9 Hz), 3.75 (d, 2H, J = 13.9 Hz), 4.21 (d, 2H, J = 13.9 Hz), 4.39 (d, 2H, J = 13.9 Hz), 4.76 (s, 2H, overlapping with HDO peak, detected using HSQC), 5.24 (br s, 2H), 6.73 (d, 2H, J = 6.6 Hz), 6.99 (d, 2H, J = 6.6 Hz), 7.06 (d, 2H, J = 6.6 Hz), 7.18 (br s, 2H), 6.48 (s, 2H), 7.73 (s, 2H),. **^{13}C{^{1}H} NMR (D$_2$O, 125 MHz)**: 30.8, 31.0 (br), 71.6, 125.1, 125.4 (br), 126.0, 126.3, 126.4, 126.7, 126.8, 127.1, 127.3, 128.2, 128.6, 129.0, 132.6, 135.5, 136.2, 136.4, 137.6, 138.6, 149.7, 151.0, 153.1, 172.2. **HR-ESI-MS:** Calculated for C$_{42}$H$_{32}$O$_{15}$S$_3$$^{-2}$ [M-2H^{-2}] = 436.04571, Found 436.04388

FH1. Compounds **FH1** was isolated as 6 mg of a white powder with 13 % yield. NMR characterisation of **FH1** proved to be difficult. **FH1** shows broad peaks due to aggregation and many resonances were not observed. Combined with extremely low yield of its synthesis and inherent instability of heterocycle coupled *p*-sulfonatocalix[4]arenes, we were never able to make enough **FH1** for a detailed NMR study. The compound was characterized successfully using UPLC-MS (purity and identity) and HRMS (identity). The ^{1}H NMR is too broad to report as numerical values here, but is depicted along with the other spectra in the appendix. **HR-ESI-MS:** Calculated for C$_{38}$H$_{28}$O$_{16}$S$_3$$^{-2}$ [M-2H^{-2}] = 418.02752, Found 418.02614.

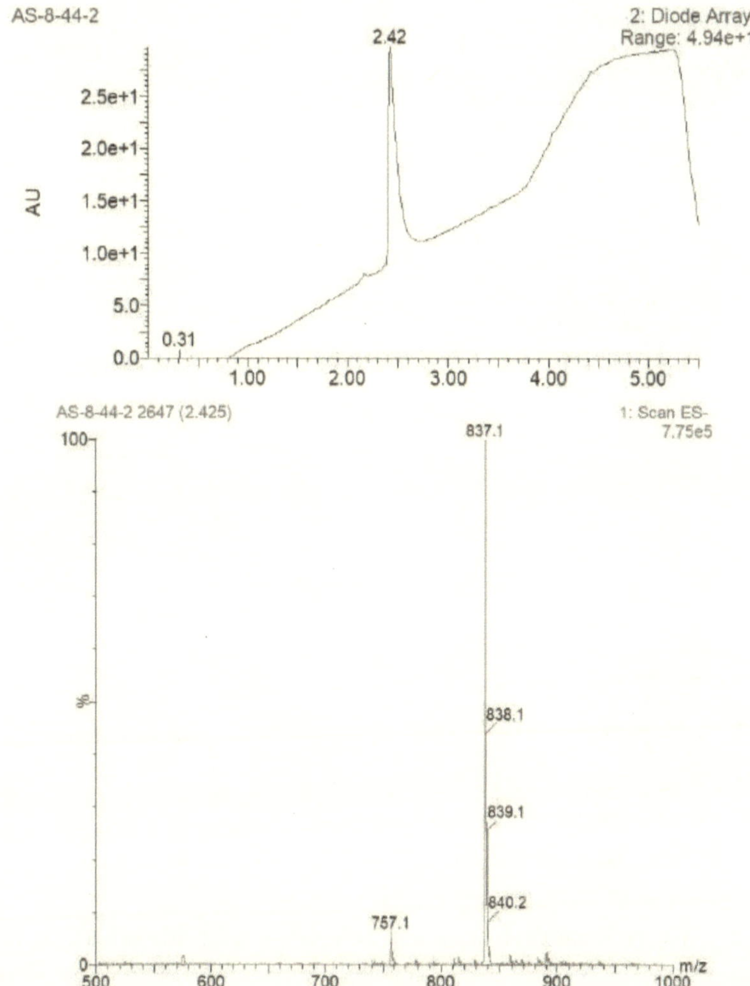

Figure 4.10. UPLC-MS trace for compound FH1 obtained on Waters Acquity-H UPLC. (Top) UV Chromatogram (DAD) of an analytically pure sample of FH1 injected on a BEC C18 column. (Bottom) LR-MS of the eluted peaks shows singly charged molecular ion (m/z = 837.1) as the dominant peak. A small peak for a desulfonated fragment can also be seen at 757.1.

4.8.5. Indicator displacement assay of (performed as described in Section 2.9.6)

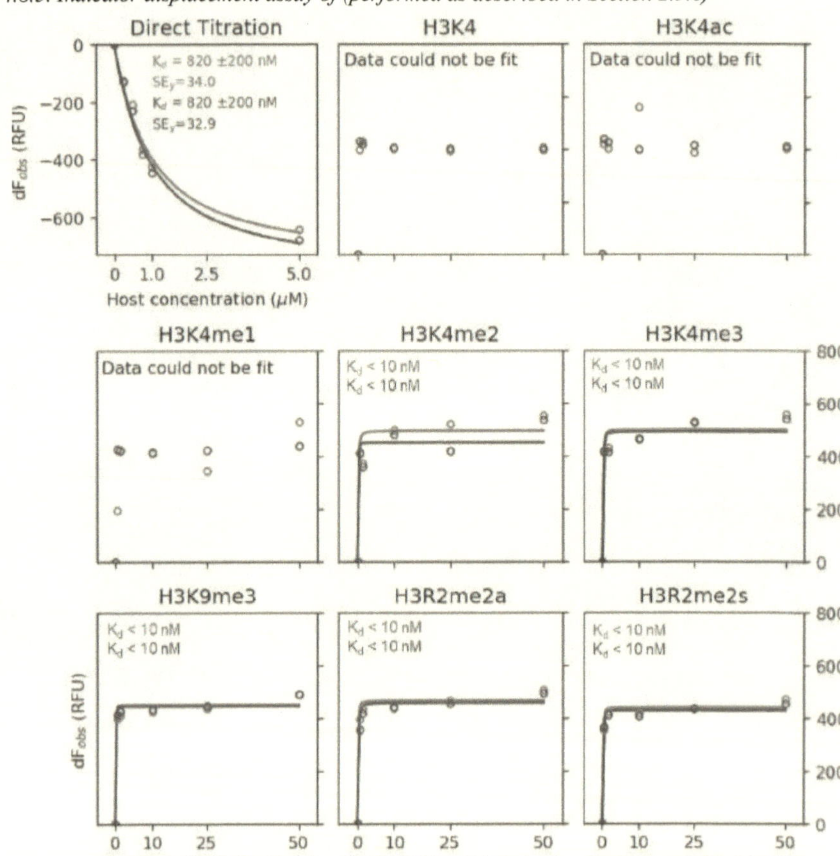

Figure 4.11. IDA of FO2 with 8 different peptides (performed as described in section 2.9.6). Errors are reported as average standard error of fit parameters, calculated from covariance matrix of nonlinear regression. All data set reach the plateau of a binding curve at the first titration point itself. Binding curves for titration with H3K4, H3K4ac and H3K4me1 could not be fitted.

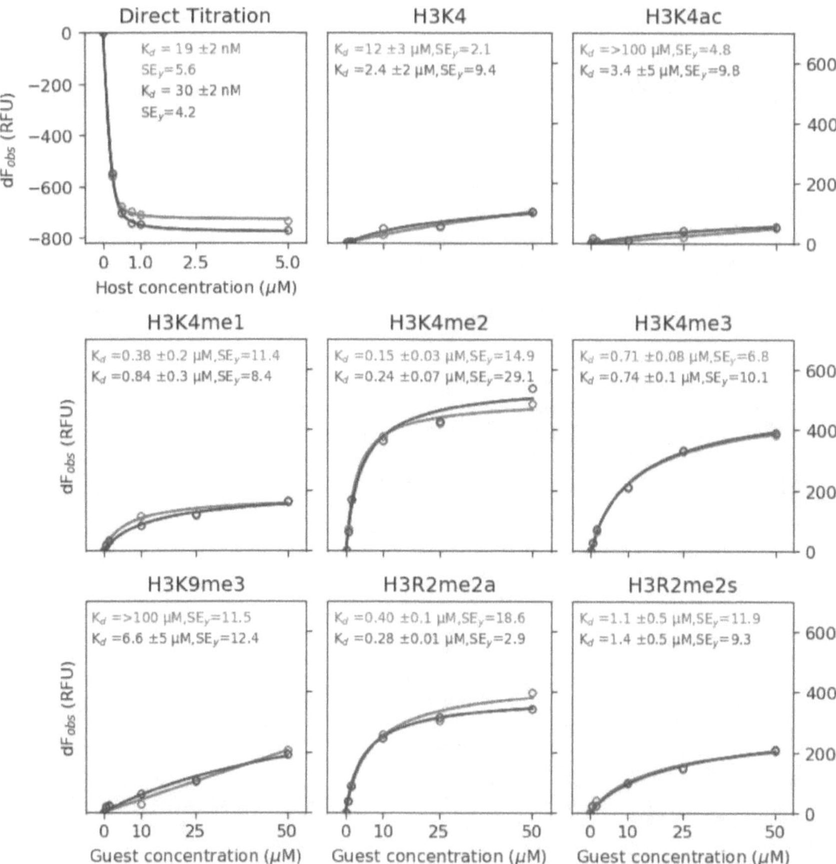

Figure 4.12. IDA of FP5 with 8 different peptides (performed as described in section 2.9.6). Errors are reported as average standard error of fit parameters, calculated from covariance matrix of nonlinear regression.

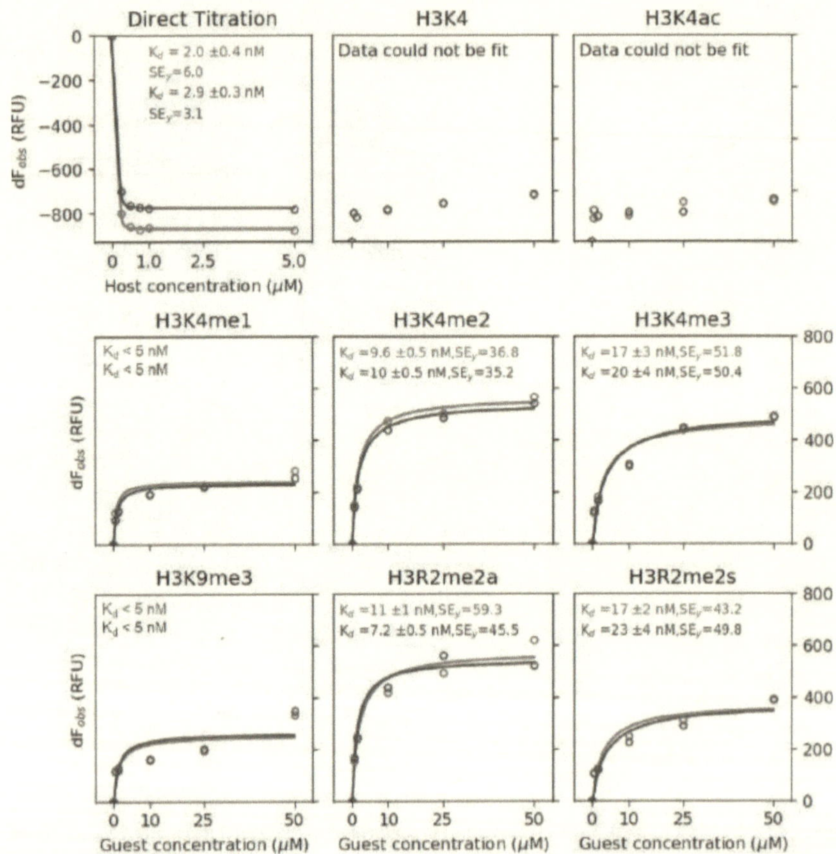

Figure 4.13. IDA of FH1 with 8 different peptides (performed as described in section 2.9.6). Errors are reported as average standard error of fit parameters, calculated from covariance matrix of nonlinear regression. Binding constants for H3K4 and H3K4ac could not be calculated within this concentration range.

Chapter 5. A single sulfonate ester substitution at the lower rim of calix[4]arene leads to an *N,N*-dimethyllysine selective host

Publications

The content of this Chapter are being prepared as a manuscript entitled,

Selective binding of a lower lysine methylation state: an *N,N*-dimethyllysine selective host molecule and its use in methyl proteomics, Alok Shaurya, Graham A.E. Garnett, Melissa J. Starke, Mark C. Grasdal, Charlotte C. Dewar, Aman K. Dheri, Euan L.S. Thomson, and Fraser Hof.

Contributions

Compound **5.1** was first envisioned by GG and synthesized by MS. AS optimized the synthesis and did the IDA, ITC and NMR binding studies, and conceived of, carried out, and analyzed data from all experiments related to calixarene conformations. Utilization of **5.1** in affinity enrichment protocols were first shown by GG and MS and optimized by MG and CD. ET supervised yeast cell culture and sampling. MG performed the lysis, enrichment and analysis of proteomics data with help from CD and ET. The manuscript and supplementary information were written by AS and edited by FH. FH supervised the work.

5.1. Foreword

Despite desirable structural features and supramolecular properties, scaffold **F** compounds required an extra synthetic step to be synthesized which prevented us from testing them as enrichment agents. The biggest lesson learned from all our efforts in the previous Chapters is that we need to minimize the synthetic steps if we are to use the calixarene products in applied science. This Chapter presents our work on a novel lower-rim sulfonate ester derived *p*-sulfonatocalix[4]arene that was first created as an accidental byproduct. It requires just one step to prepare from commercially available material. Its binding properties, conformation dynamics and utility as an enrichment agent for PTM-targeted proteomics are discussed.

5.2. Motivation

The work from Chapter 4 was successful in bringing the lower rim of *p*-sulfonatocalix[4]arenes back to our attention. We were happy to learn that there are substitutions that can be installed at the lower rim without disrupting the binding cavity of the calix[4]arene. However, the testing of these calix[4]arenes as enrichment reagents was severely hampered by the long synthetic scheme that was required. In search for a perfect host which can perform to our liking in multiple different conditions we ended up heavily decorating our host. The synthetic chemistry developed along the way is useful but arduous. Most of the steps were optimized to high yield but two key steps proved to be a big bottleneck.

First, the sulfonation step could never be optimized to high yields. A maximum 60% yield was obtained, but some replicates of this reaction could yield as little as 30% product. The obtained product is also highly polluted with salts (likely sodium and potassium sulfate) and some partially sulfonated calix[4]arenes which brings the overall yield even lower (Figure 5.1.a). Second, coupling reactions were efficient only if a clean sulfonation product was used. For all compounds in Chapter 2 and 4, crude sulfonation products were moved forward without purification. This would result in a mixture of products, fully and partially sulfonated, which were purified using an HPLC. While this strategy worked for the scale used in library synthesis (20-40 mg crude), at bigger scales (>200 mg) the purification became arduous. The only solution was to purify the sulfonation product before attempting any coupling reaction which added an extra, time consuming, HPLC steps to our already long scheme. Even after careful purification, coupling reactions with scaffold **E** was never clean. Side reactions at the lower rim were inevitable and were another cause for poor yields (Figure 5.1.b).

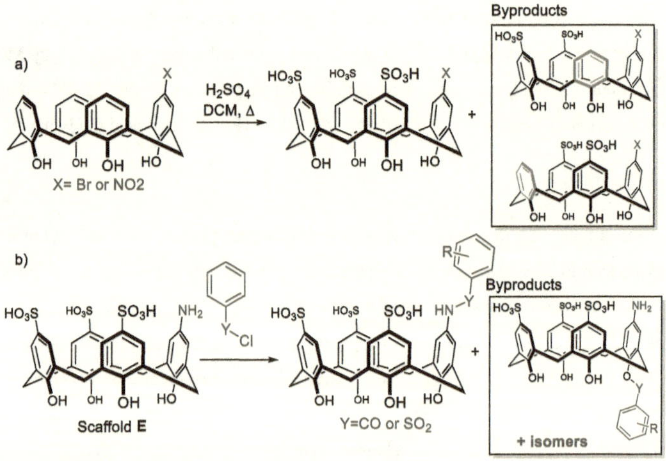

Figure 5.1. Bottleneck reactions that severely hampered our yields. a) Sulfonation, b) Amide/sulfonamide coupling using scaffold E.

Figure 5.1.b also shows the key observation that led to the work described in this Chapter. Esterification of lower rim phenols was an ever-present side reaction throughout Chapter 2. This was detrimental to the goals of Chapter 2 but we conceived the idea that such reactions can be utilized to install carboxylic acid containing fragments at the lower rim of calix[4]arene. Such fragments were already present in our inventory (Figure 2.5). With the new knowledge of single lower rim functionalization that maintained supramolecular properties of sulfonated calix[4]arene (Chapter 4) we decided to invest time towards this hypothesis using sulphonyl chloride fragments. Carboxylic esters were decided against due to their well established tendency to undergo hydrolysis at the slightly basic pH values that are convenient for proteomics studies.

5.3. Design and synthesis

There are three key properties that we are keen to see in our hosts before we move forward with enrichment studies, 1) water solubility, 2) strong affinity towards methylated peptides (low micro- to nanomolar dissociation constants) and 3) presence of an additional functional group (like a carboxylic acid) which allows us to immobilize them to a solid phase. We realized that, instead of beginning our synthetic chemistry from water insoluble and completely unsubstituted calix[4]arene **2.1**, starting from commercially available p-sulfonatocalix[4]arene (**PSC**) itself gives us first two of our desired properties. **PSC** is water soluble and binds our test set of methylated peptides with sub-micromolar binding constants (Table **2.2**). All we need is a chemical handle to be able to immobilize it on a solid support. We chose fragment **1** (4-(chlorosulfonyl)benzoic acid) from Chapter 2 (Figure 2.5) for this task. This is the same fragment that was used to synthesize **E1** in Chapter 3, albeit by coupling it at the upper rim. For this project we choose to react the sulfonyl chloride to the lower rim phenols instead.

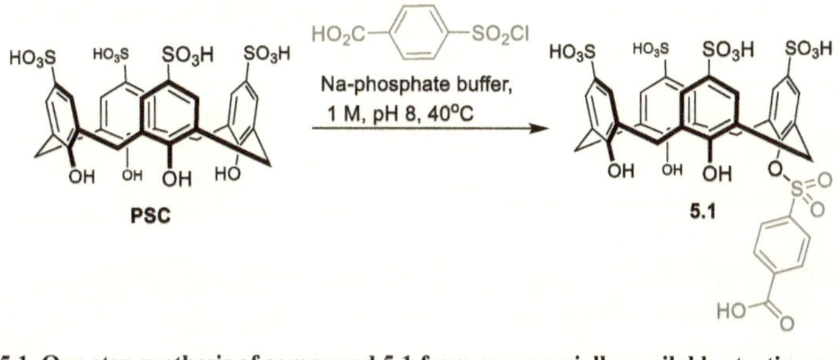

Scheme 5.1. One step synthesis of compound 5.1 from commercially available starting material.

Reaction of commercially available **PSC** with 4-(chlorosulfonyl)benzoic acid in aqueous buffer provided the mono-substituted compound **5.1** through reaction of one out of four lower-rim phenols (Scheme 5.1). Formation of small of amount of disubstituted byproducts was observed in early trials but was minimized by slow addition of the sulfonyl chloride reagent. An isolated yield of 60% after HPLC purification is reliably achieved for this selective mono-functionalization. In the absence of the possibility of upper rim functionalization there are no isomeric byproducts that can be produced (Figure 5.1.b). This was a huge, and much-needed, improvement over all the previous schemes described in this book. A 60% yield would be considered moderate, but the reaction was optimized easily at bigger scales without any change in chemical profile. A 200 mg reaction had the same yield as a 20 mg reaction and would routinely yield more than 120 mg product which would be enough for preparation of multiple batches of enrichment agent. The reaction is over in 4-6 hours and multiple batches of reaction can be done in parallel and stored indefinitely before purification. While sulfonate esters have a reputation for being hydrolytically unstable, we found that the product is exceptionally bench stable at variety of pH in both solid and solution phase. An aqueous solution was stored at room temperature and monitored by ^{1}H NMR periodically. It showed no evidence of decomposition over >3 months.

5.4. Surprise dimethyllysine selectivity of 5.1

We established the binding behavior of **5.1** with eight different PTM peptides as in previous Chapters. To our surprise, **5.1** was found to bind H3K4me2 peptide with K_d of 180 nM, and 6-30 fold selectivity over the other two Kme3 peptides in our set (Table 5.1). Comparatively, **PSC** is 6-14-fold selective for Kme3 containing peptides over Kme2 peptide, favouring the higher methylation state in the way that is typical for sulfonated calixarenes. K_d values for unmethylated and Kme1 containing peptides are the same for **5.1** and **PSC**, suggesting a plausible size-based discrimination between larger guests by **5.1**.

Table 5.1. K_d values for binding of 1 and PSC with 8 different PTM bearing peptides[a], determined using indicator displacement assay[b] as described in Chapter 2.

Peptides[a] (overall charge)	K_d for PSC (μM)[c]	Selectivity for Kme2 peptide[d]	K_d for 5.1 (μM)[c]	Selectivity for Kme2 peptide[d]
H3K4 (+3)	4.1 ± 0.4	4.9	3.9 ± 0.5	21.6
H3K4me1 (+3)	2.2 ± 0.2	2.7	2.1 ± 0.9	11.6
H3K4me2 (+3)	0.83 ± 0.1	1	0.18 ± 0.02	1
H3K4me3 (+3)	0.56 ± 0.4	0.7	1.0 ± 0.1	5.5
H3K9me3 (+2)	1.4 ± 0.1	1.7	5.3 ± 1.2	29.4
H3K4ac (+2)	17 ± 4	20.0	19 ± 4	110.0
H3R2me2a (+3)	2.3 ± 0.2	2.8	2.0 ± 0.7	11.1
H3R2me2s (+3)	5.7 ± 0.3	6.8	4.9 ± 0.9	27.2

[a] All peptides sequences are same as described in Section 2.6 [b] IDA was performed in 10 mM Sodium Phosphate buffer (pH 7.4). Plots for each titration is shown in Section 5.9.2 [c] Average of two titrations. Errors reported as standard deviation calculated from the covariance of fit parameters. [d] Selectivity is defined as K_d(Peptide)/K_d(H3K4me2)

To gain more insight into the binding selectivity, we performed ITC and NMR titrations. ITC experiments with **5.1** partly supported the Kme2 selectivity seen in IDA studies. H3K4me-x (x=2,3) peptides were titrated into a solution of **5.1** at 30°C (Table 5.2). Compound **5.1** is reproducibly 1.3-fold selective for the Kme2 peptide under these conditions. We attribute the decrease in observed selectivity in the ITC experiment to small amounts of non-specific binding under the much higher concentrations required in ITC vs. those used near the ideal dilute limit in fluorescence titrations (5 µM **5.1** and 50 µM peptides in IDA vs 50 µM **5.1** and 1-2 mM peptides in ITC).

Table 5.2. Thermodynamic parameters obtained from ITC titration[a] of H3K4me$_x$ (x=2, 3) peptides with compound 5.1.

Peptide	ΔH (kJ/mol)[b]	TΔS (kJ/mol)[b]	ΔG (kJ/mol)[c]	K_d (nM)[d]
H3K4me2	-52.8 ± 0.4	-18.9 ± 0.6	-33.9 ± 0.2	1460 ± 100
H3K4me3	-40.0 ± 0.2	-6.7 ± 0.3	-33.3 ± 0.14	1810 ± 100

[a] ITC titrations were performed at 303 K in 10 mM sodium phosphate buffer (pH 7.4). Each peptide (1-2 mM) was titrated into 50 µM 1. MicroCal VP-ITC software was used for data collection and curve fitting (Section 5.9.3) [b] Average of two ITC runs. Errors reported are average error from the curve fitting program [c] Back calculated from ΔH and ΔG. [d] Calculated from ΔG.

NMR titrations show that 5.1 binds peptides through selective engagement of methylated side chains. No significant upfield shifts occur for the peaks corresponding to any amino acids other than K4me-x (x=2,3) and R2 (Figure 5.2). Such upfield shifts are characteristic of encapsulation with arene-rich hydrophobic cages and prove that complex between host 5.1 and peptides are mainly due to encapsulation of hydrophobic cations in the side chain of lysine and arginine. Interestingly the alanine Cα-H protons show downfield shift upon encapsulation. The alanine sidechain, while hydrophobic, is too small to be encapsulated within the cavity of 5.1 and is apparently engaged by an external, deshielding region of the host. The dimethyllysine side chain engages the pocket of 5.1 in a unique way. The resonance for the N-methyl protons from H3K4me2 peptide are in intermediate exchange between two bound states (NMR time scale) in the presence of 1 and broaden significantly at calixarene concentrations higher than 0.5 equivalents with respect to peptide. The N-methyl protons from H3K4me3, on the other hand, stay sharp and appear to be in fast exchange throughout the titration, as is normally observed for any Kme3-calixarene complex.

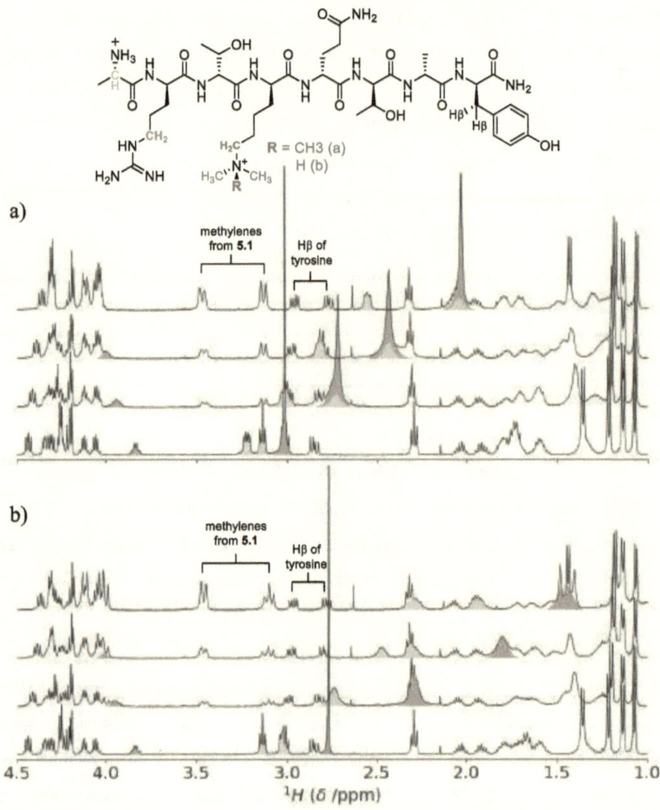

Figure 5.2. Titration of 1 with peptides (a) H3K4me3 and (b) H3K4me2 in 10 mM Na-phosphate buffer (pD 7.5). The peptides are held at constant concentration of 0.5 mM while the concentration of calixarene is varied from 0.0, 0.5, 1.0 and 2.0 equivalent in each plot (bottom to top). The general structure of the peptide is shown above, and important signals are colour coded or labelled.

5.5. Conformational dynamics of 5.1 reveal its uniqueness

Comparison of **5.1** to **PSC** by ¹H-NMR showed vastly different conformational dynamics for the two macrocycles. The ¹H NMR data is as expected. The methylene ¹H peaks of **5.1** are four sharp doublets (J ~12-14 Hz), showing slow inversion through the annulus on the ¹H NMR time scale (> 0.8 and 1.1 ms[275] for either methylene pair at 300 MHz at 298 K in D_2O). This slow inversion is expected for any calixarene with a lower-rim substituent too large to pass through the annulus. **PSC** is in the C_{4v} cone conformation, and has the expected rapid intraannular inversion for a calix[4]arene that is unsubstituted at the lower rim, shown by a single broad resonance for its germinal CH$_2$ protons in both D$_2$O and phosphate buffer (Figure 5.3.b).

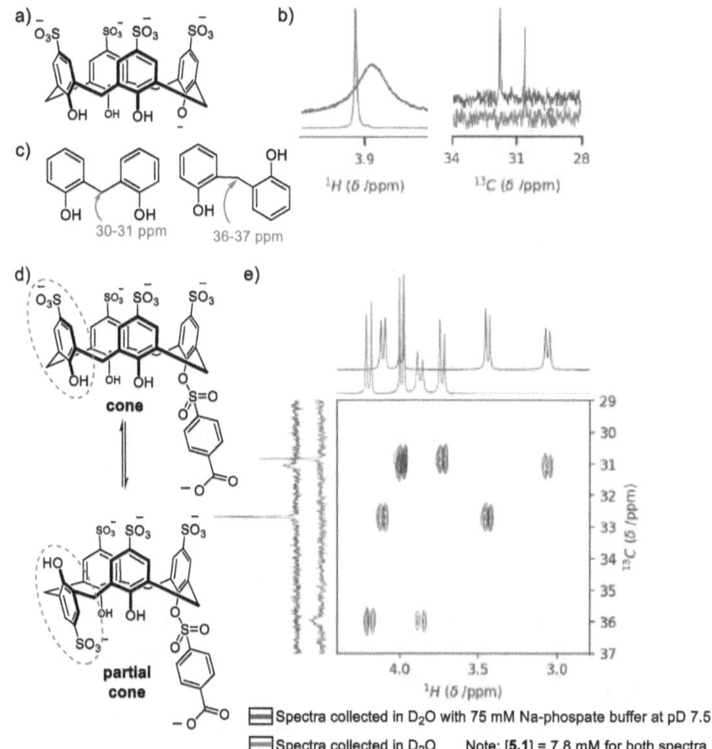

Figure 5.3. ^1H and ^{13}C NMR show the difference in conformational dynamics of PSC and 1. a) PSC in its expected C_{4v} cone conformation. b) Broad methylene signal ^1H NMR and ^{13}C chemical shift at ~31 ppm corroborate an overall cone conformation for PSC in both D$_2$O and buffered D$_2$O. (c) ^{13}C chemical shift of calix[4]arene methylene carbons are well correlated with the dihedral angles between neighbouring phenols. Values are taken from reference [273]. d) Compound 5.1 has different conformational flexibility from that of PSC, including significant interconversion between cone and partial cone conformations, as supported by e) HSQC of 5.1 with ^1H and ^{13}C spectra projected along respective axes. Certain ^{13}C peaks have significant broadness which suggest an equilibration between two conformers – see main text for a detailed explanation.

The traits of the two symmetry-inequivalent methylene ^{13}C resonances in **5.1** reveal its unusual conformation and dynamics. One methylene carbon is seen as a sharp signal in ^{13}C NMR, while the second is very broad, and can only be observed with confidence using HSQC (Figure 5.3.d). This shows that the two symmetry-inequivalent halves of **5.1** have different conformational dynamics. The symmetry and dynamics of ^{1}H and ^{13}C NMR resonances are best explained by **5.1** existing as a mixture of cone and partial cone conformations, each of which has a single internal mirror plane and overall C_s symmetry (Figure 5.3.c). The chemical shifts of the ^{13}C resonances in pure D_2O (31.4, 35.3 ppm) are diagnostic of a partial cone,[273] which has a distorted, smaller binding surface (relative to the fully open cone conformation) that might explain the surprising Kme2 selectivity of **5.1**. The mixture of one sharp and one broad methylene $^{13}CH_2$ resonance remains in buffered D_2O (Figure 5.3.d), although the chemical shift changes suggest different degrees of population of the partial cone conformation.[218,273]

5.6. Unique features that drive the altered selectivity

NMR and ITC combine to shed light on the key features of H3K4me2-calix[4]arene complex. At first glance, it is difficult to conclude anything from the NMR due to the broad resonances of the encapsulated dimethyllysine signal (Figure 5.2.b). But the thermodynamic parameters obtained from ITC are remarkably different for H3K4me3 and H3K4me2, and the data for H3K4me2 seem to match well with the unusual dynamics observed by NMR (Table 5.2). Binding of H3K4me2 with **5.1** suffers from a large entropic penalty ($T\Delta S = -18.9 \pm 0.6$ kJ/mol). Comparatively, the entropic penalty for H3K4me3 is moderate ($T\Delta S = -6.7 \pm 0.3$ kJ/mol). Despite this, the overall enthalpic compensation for H3K4me2 is big enough to make **5.1** selective for H3K4me2. It is not possible to explain these parameters using only "a tight fit between host and guest" as an explanation. Instead, we believe that the partial-cone conformation of **5.1** plays a more important role than just restricting the cavity size of calix[4]arene. In this conformation, the anti phenol's oxygen (red in Figure 5.4.a) gets close to the cavity of calix[4]arene and can interact with the encapsulated guest molecules directly. We believe that dimethyllysine of H3K4me2 is able to form a hydrogen bond with this phenol oxygen in partial cone conformation (Figure 5.4.a). This would lower the rotational degrees of freedom dimethyllysine can have inside the cavity and result in the more negative $T\Delta S$ that we observe. Comparatively, H3K4me3 sidechain cannot form any such hydrogen bond and suffers only a moderate loss of entropy upon binding. Another part of this model is supported by both NMR and ITC. The high, favourable enthalpy of H3K4me2 binding can arise from the additional strong hydrogen bond between host and guest. This interaction would result in restricted dynamics for Kme2 that Kme3 does not experience, supported by the unusual broadness of the N-methyl and ε-methylene signals of H3K4me2's lysine side chain.

While the picture depicted in Figure 5.4.a is consistent with all experimental data, it by no means is the only binding mode. The cone conformation of **5.1** can still encapsulate lysine side chains. Its partial cone conformation forms another negatively charged portal lined with aromatic rings at its lower rim (Red line, Figure 5.4.b). The complexation induced shift in NMR are characteristic of a more enveloping host-guest complex and suggest upper rim interactions to be dominant. However, cationic ammonium guests, like lysine side chains, must be able to interact with lower rim partial cone conformation too (Red line, Figure 5.4.b). It is difficult to isolate the contribution of each mode towards overall affinity. We merely propose that Figure 5.4.a represents the binding mode that tips the overall selectivity towards H3K4me2.

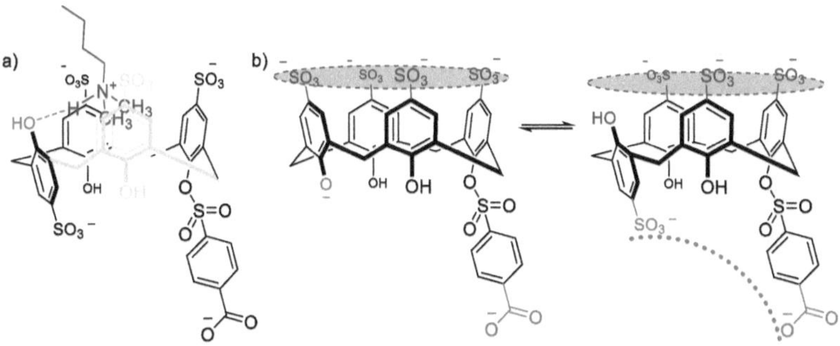

Figure 5.4.a) A chemdraw representation of the most prominent binding mode for dimethyllysine as supported by NMR and ITC. b) The cone-partial cone equilibrium of 5.1 generates multiple negatively charged portals lined with aromatic groups at both upper and lower rims (shown in blue and red respectively). The observed binding strength contains contributions from all of these.

Figure 5.4.b also helps to explain the discrepancy between IDA and ITC values. Compound **5.1** is significantly less selective for H3K4me2 based on ITC data (Tables 5.1 and 5.2). We speculate that the non-specific binding to the lower rim is likely not reported on by IDA at all. This is because the fluorescent signal in IDA comes indirectly from displacement of lucigenin which interacts primarily through the upper rim sulfonates.[276] However, non-specific electrostatic interactions at the lower rim are going to strongly influence ΔH, which is seen in the ITC data. We believe that values obtained from ITC, which is a direct assay, are more accurate depiction of peptide-calix[4]arene complex and as they capture a much more complete picture.

5.7. Immobilization and utilization in proteomics experiment

We immobilized and used solid-supported **5.1** in a proteomics enrichment experiment similar to E1 in Chapter 3. We conjugated **5.1** to amine-functionalized agarose beads (Affi-gel 102, Bio-Rad) by EDC coupling to create **5.1-agarose** (Figure 5.5). We sought to assess its potential on cell lysates from real-world samples. A cell lysate from a commercial fermentation using brewer's yeast (*S. cerevisiae*) was obtained from a local brewery, the protein extracted, the sample proteolyzed (Section 5.9.5), and the resulting mixture of peptides was run onto **5.1-agarose** and then eluted with a high salt solution using protocols similar to those reported in Chapter 3. This generated in a single experiment a matched input control, unretained fraction, and retained fraction. We analyzed each sample using a standard proteomics method as in Chapter 3. Peptide sequences were determined by LC-MS/MS fragmentation patterns, matched to known sequences from the *S. cerevisiae* genome. Lysine methylation sites and states were identified and located on individual residues by PTM mass differences observed during fragmentation; we based further analyses on those marks assigned with >90% confidence by the Scaffold proteomics analysis software (Section 5.9.6). We found dramatic enhancement of both di- and trimethyllysine-containing peptides from a complex cell lysate (Figure 5.5.d).

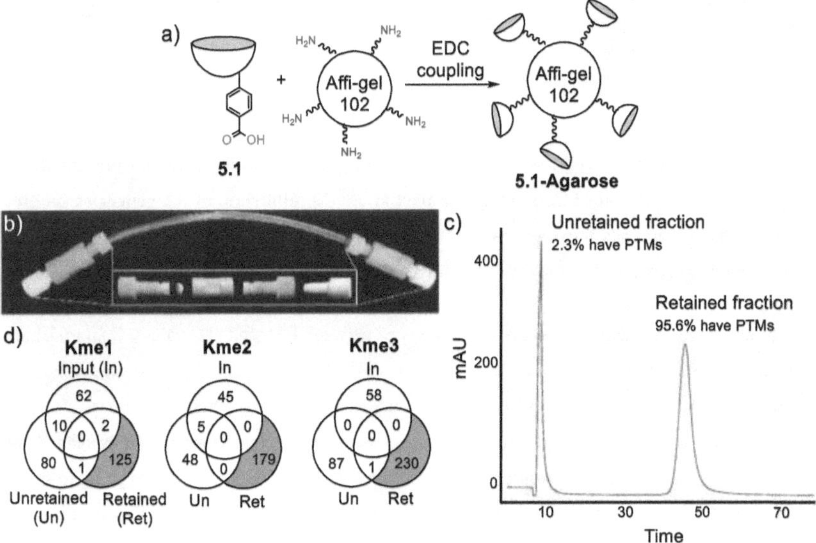

Figure 5.5. Compound 5.1 powers discovery of methyllysine marks. (a) Compound 5.1 was coupled to amine containing Affi-Gel 102 beads using EDC. (b) The calixarene-modified gel was then packed into capillaries. (c) The resulting capillary were used to enrich proteolyzed sample from cell extract of *S. cerevisiae* (d) Distribution of methylated peptides seen in the samples before (input) and after enrichment (unretained and retained) is summarized in the Venn diagram.

Affinity separation with immobilized **5.1** allows identification of many new Kme sites that would not otherwise be observed if only the unenriched sample (input control) was being analysed. The unretained fraction contained a relatively small proportion of methyl PTMs (<3% of the total number of peptides identified), while there was a very high proportion (>95%) of peptides with PTMs among the total set of peptides in the retained fraction. The number of Kme2 sites observed increased from 50 to 179 after enrichment, and the number of Kme3 sites increased from 58 to 231. The Venn diagrams of identified methyl marks show that essentially all Kme2 and Kme3 marks seen in the retained fraction are not observed at all in the input controls, similar to results obtained in Chapter 3. Detailed lists of all proteins and their lysine methylation sites observed are reported in Tables 5.3, 5.4 and 5.5 (Section 5.9.7). Interestingly, the number of methylated proteins identified increases in both unretained and retained fraction (116 and 111 respectively) compared to input sample (98 different methylated proteins). It appears that partitioning of input sample based on methylation improves the visibility of low-level proteins in both fractions. A combined analysis of each fraction, thus, increases coverage of detected proteins much more than non-enriched input sample.

5.8. Conclusion

In conclusion, **5.1** is novel host that is Kme2 selective in solution, and that enables broad-spectrum enrichment of methylated peptides from a cell lysate. The one-step preparation of **5.1** is a significant improvement over the complex, multi-step syntheses required for other Kme-binding macrocycles shown in this book and published in literature. The Kme2 selectivity arises unexpectedly from a remote, lower-rim modification. This selectivity likely arises from the strong influence of lower-rim modification on host conformation. We have also demonstrated the utility of **5.1-agarose** as an enrichment agent in a real-world proteomics sample. Unlike **E1** (Chapter 3), immobilized **5.1** is able to successfully enrich cell lysates, a much more complex sample of the sort that is routinely used in proteomics studies. Immobilization through the lower rim appears to have made the calix[4]arene cavity much more accessible and is a better option than immobilization through upper rim.

5.9. Supporting information

5.9.1. Synthesis of compound 5.1

PSC (200 mg, 0.27 mmol, 1 eq.) was dissolved in sodium phosphate buffer (1 M, pH 8). To this solution, 36 mg of (4-chlorosulfonyl)benzoic acid (0.162 mmol, 0.6 eq.) was added and the mixture was stirred for 1 hour at 40°C (This temperature was maintained to prevent the phosphate salts from crashing out). At this point, UPLC showed 25% reaction progress. After this, small portions of (4-chlorosulfonyl)benzoic acid (12-18 mg, 0.05-0.081 mmol, 0.2-0.3 eq) were added at 2-hour intervals and reaction monitored using UPLC-MS until <5% starting material was visible. During this time, the pH of the reaction mixture was regularly checked and maintained between 7.5 and 8.0. The total amount of (4-chlorosulfonyl)benzoic acid added was 114 mg (0.517 mmol, 1.9 eq) added in 6 portions. After completion, a few drops of saturated $FeCl_3$ solution were added to precipitate most of the phosphate salt. The mixture was centrifuged, and the supernatant collected. The precipitate was washed twice with water. The combined supernatant and washes were then subjected to preparative RP-HPLC () using a 30-minute gradient of acetonitrile in water (starting from 10% acetonitrile to 90% acetonitrile). The compound was obtained as a white powder with yields of 60-65%. **^{1}H NMR (500 MHz D_2O, concentration – 7.8 mM):** δ (ppm) = 3.81 (d, 2H, J = 14.3 Hz, Ar-CH$_2$-Ar), 3.95 (d, 2H, J = 16.1 Hz, Ar-CH$_2$-Ar), 4.07 (d, 2H, J = 14.3 Hz, Ar-CH$_2$-Ar), 4.28 (d, 2H, J = 16.1 Hz, Ar-CH$_2$-Ar), 6.81 (br, 2H, Ar-H), 7.39 (d, 2H, J = 1.6 Hz, Ar-H), 7.44 (d, 2H, J = 1.7 Hz, Ar-H), 7.49 (br d, J = 7.8 Hz, 2H, Ar-H), 7.68 (s, 2H, Ar-H), 7.74 (s, 2H, Ar-H); **^{13}C{^{1}H} NMR (75 MHz D_2O, referenced externally using 10% methanol in D_2O, concentration – 40.7 mM):** δ (ppm) = 31.5, 35.4 (br), 126.9, 127.1, 127.4, 127.5, 128.1, 129.5, 129.6, 129.9, 130.9, 135.7, 136.0, 136.2, 137.2, 138.5, 142.1, 146.8, 151.7, 153.6, 169.0; **IR (ATR, cm^{-1})** = 3361 (br), 1695 (m), 1595 (w), 1448 (w), 1380 (w), 1109 (s), 1031 (s), 621 (s), 529 (s) cm^{-1}; **HR-ESI-MS:** Calculated for $C_{35}H_{26}O_{20}S_5^{2-}$ [M-2H]$^{2-}$= 462.9816, Found 462.97945.

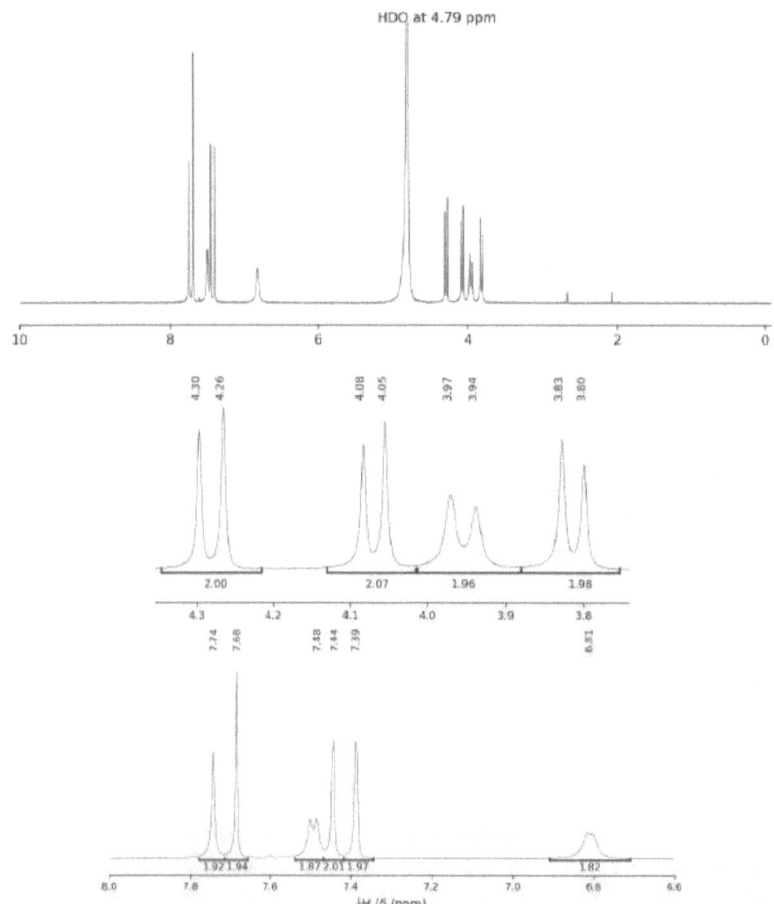

Figure 5.6: 500 MHz ^{1}H NMR trace of compound 5.1 (top) in D$_2$O. Blown up trace in aliphatic (middle) and aromatic (bottom) region are also shown. The sample concentration of 5.1 was 7.8 mM.

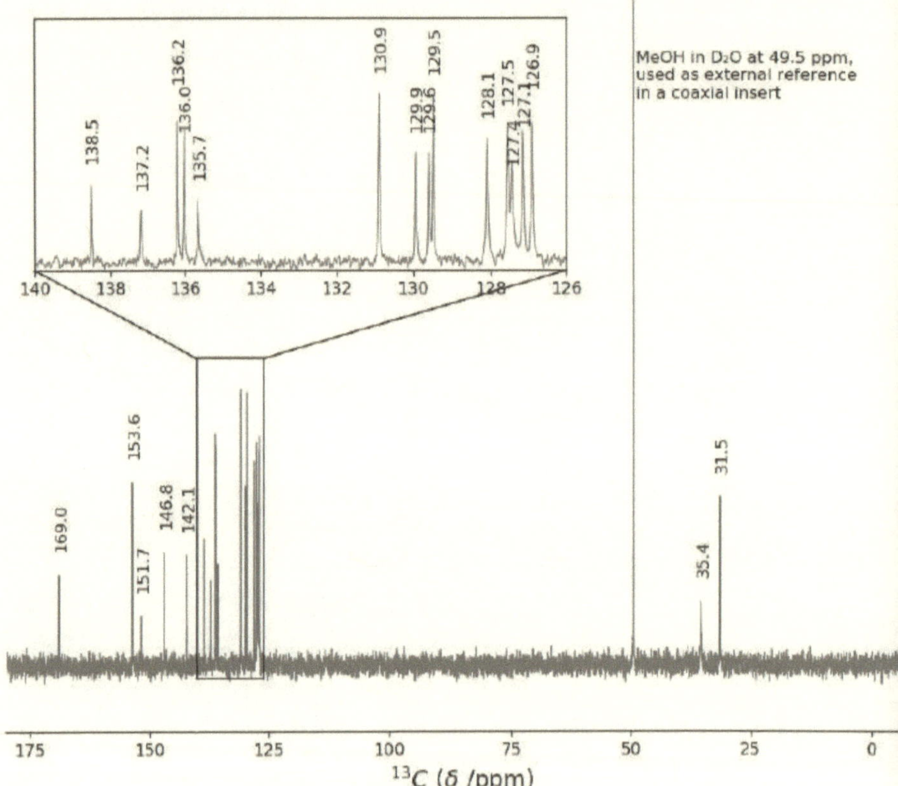

Figure 5.7: ^{13}C{^{1}H} **NMR trace of compound 5.1 collected in D2O at 75 MHz frequency. The samp[le] concentration of 5.1 was 40.7 mM. Inset shows the blown up version of aromatic region (126-140 ppm) [of] this NMR.**

*5.9.2. IDA curves for **5.1** with 8 different PTM containing peptides.*

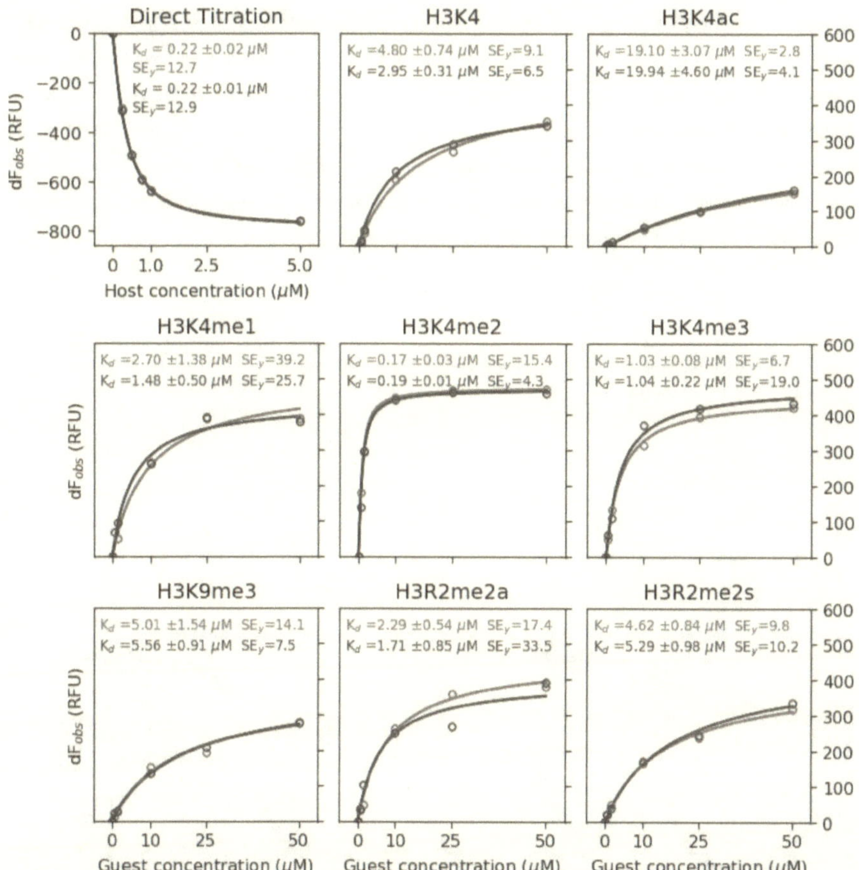

Figure 5.8. Plots for IDA of compound 5.1 (host). The first plot (upper left) is the direct titration of lucigenin with 5.1. The next 8 plots are competitive titrations of 5.1-Lucigenin complex with 8 different peptides bearing different modified amino acids. The concentration of lucigenin was fixed at 250 nM for all titrations while the concentration of 5.1 was fixed at 0.5 μM in the competitive titrations. The experiment was done in a 96-well plate as described in Section 2.9.6. Duplicates are shown as red and blue data sets. Errors reported as standard error derived from covariance of fit parameters.

5.9.3. Isothermal calorimetry (ITC)

ITC was performed using Microcal VP-ITC (GE Healthcare). Titration were performed in 10 mM sodium phosphate buffer at pH 7.5 at 303 K. Concentration of peptides were calculated using A_{280} while that of **5.1** was calculated using qNMR. All solutions were thoroughly degassed before experiment. In all cases, solution of **5.1** was loaded in the ITC cell while peptide solution was loaded into the syringe. The full titration involved 50 injections of 5 μL (except first injection which was 2 μL) at constant interval of 300 seconds. Binding curves were produced using supplied Origin software and fit using a 1-sites binding model. The first point of titration was discarded. Each titration was performed in duplicate and the final data reported is average of two runs. ITC data and curve fit are shown below.

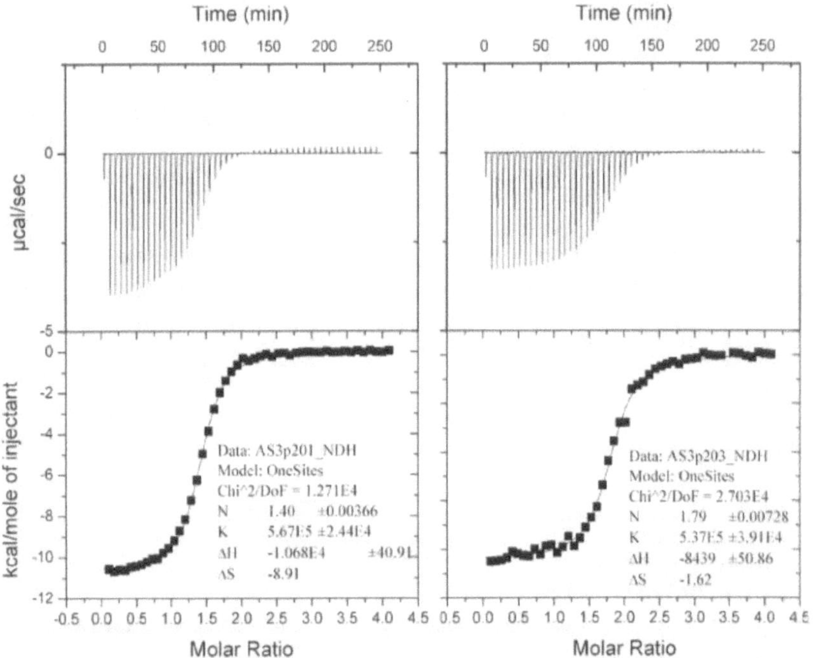

Figure 5.9. Replicate ITC curves for titration of 5.1 and H3K4me3 peptide. The concentration of peptide was 2.1 mM (calculated using A_{280}) while concentration of 5.1 was kept at 100 μM (determined using qNMR).

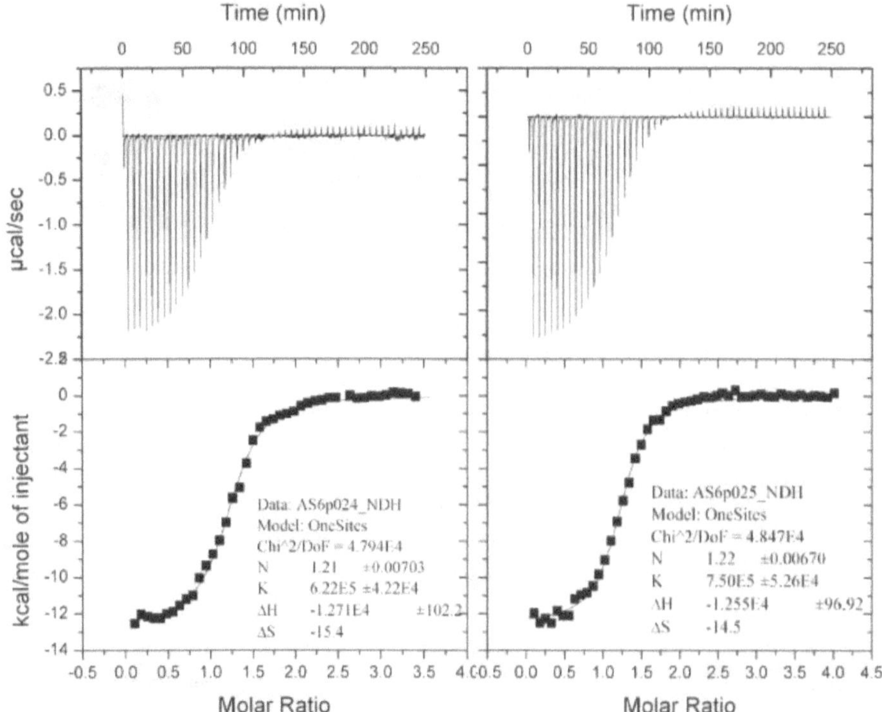

Figure 5.10. Replicate ITC curves for titration of 5.1 and H3K4me2 peptide. The concentration of peptide was 1 mM (calculated using A_{280}) while concentration of 5.1 was kept at 50 µM (determined using qNMR).

5.9.4. Immobilization of compound 1 on Affi-Gel resin and preparation of enrichment columns

Calixarene **5.1** was coupled to Affi-Gel-102 (primary amine functionalized, cross-linked agarose resin; Bio-Rad) using EDC as the coupling agent. Reaction pH was kept close to 5.0 and the extent of coupling was judged by monitoring changes in concentration of **5.1** in the supernatant using HPLC. Once satisfactory amount of immobilization of **5.1** was achieved on the resin, it was packed into clear PFA tube (1/16" x 1/8", I.D. x O.D.) as described elsewhere.[277] Both ends of the capillaries were closed with fittings (IDEX, part. no. XP-335-CP), ferrules (IDEX, part. no. P-300NX-CP), frits (VICI-Jour part. no JR-1150-10P-5) and union (IDEX, part. no. P-703-01) before attaching it to an AKTA Prime FPLC. The column was then washed overnight with 50 mM sodium phosphate + 2 M NH_4Cl solution at 0.3 mL/min and then at 0.4-0.6 mL/min until the resin was evenly packed (visual inspection). Care must be taken to avoid formation of air bubbles, gaps and/or headspace in the column.

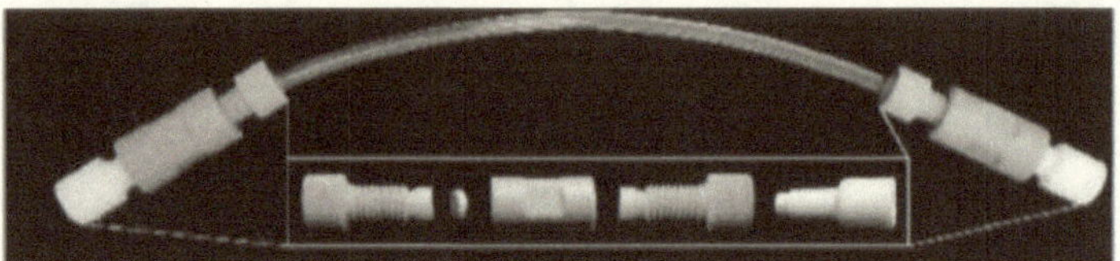

Figure 5.11. Fully assembled enrichment column with 1-Affi-Gel-102 packed inside

5.9.5. Yeast cell sampling, lysis, digestion and enrichment

Yeast cell sampling: Yeast cells were sampled from the fermentation tanks at Phillip's Brewing & Malting, Victoria, BC Canada.

Yeast cell lysis: Cells were washed with 50 mM ammonium bicarbonate buffer and counted with hemocytometer. 1.6×10^9 cells were collected and lysed in 1 mL of 50 mM ammonium bicarbonate using an equal volume of 0.5 mm acid-washed glass beads (Sigma-Aldrich) by vortexing for 8 cycles of 1-minute vortex followed by 1 minute on ice. The supernatant was collected and sonicated by undergoing 3 cycles of sonication at 40 amps x 15 seconds with 30 seconds on ice between each. The extract was then centrifuged at 20,000G for 30 min at 4°C to pellet debris. The supernatant was collected, and the protein concentration was evaluated using a Bradford (BioRad reagent, BSA standard) assay with 50 mM ammonium bicarbonate as a diluent/blank.

Lysate preparation: Crude lysate was denatured and reduced with 3.5 M Guanidine hydrochloride and 50 mM dithiothreitol (Sigma-Aldrich) for 30 minutes at 65°C. The reduced lysate was cooled to room temperature, then alkylated with 100 mM iodoacetamide (Sigma-Aldrich) for 30 minutes in the dark at room temperature. The lysate alkylation reaction was quenched in a 4:1:3 volume ratio of methanol, chloroform, and water. Following this, interfacial protein precipitation was induced by centrifugation at 3700 rpm for 15 minutes. The top layer (aqueous methanol) was removed and replaced with same volume of methanol. The sample was the vortexed for 1 minute and centrifuged again at 3700 rpm for 15 min. The supernatant was discarded. The protein pellet was dried, then resuspended in 50 mM NaOH (150 µL). The sample was placed in an ultrasonic water bath for ca. 20 minutes. The protein pellet was further diluted with 100 µL of 50 mM ammonium bicarbonate buffer (pH 7.8) and 750 µL of dH$_2$O. A further 10-minute sonication concluded protein solubilization. The residual precipitant was pelleted by centrifuging at 3,700 rpm at 4°C for 45 minutes. The solubilized protein solution (supernatant) was assayed with Bio-Rad Protein Assay Dye Reagent (Bio-Rad).

Digestion and Buffer Exchange: 1 mg of solubilized proteins were digested at a 1:100 ratio of protein to GluC protease (Promega, catalog no V1651) for 18 hours at 37°C while shaking. Next, the digest was buffer exchanged with PD minitrap G10 desalting column (GE Healthcare, catalog no. 28918010). Following the buffer exchange, the peptides were eluted into 50 mM phosphate buffer (pH 7.4).

Methylated peptide enrichment: The peptide sample was spiked with a methylated internal standard peptide. 100 µl of the spiked sample was aliquoted and saved as the input sample. The spiked sample was then injected into AKTA Prime Plus connected to the calixarene-functionalized affinity column (Section 5.9.4). The column was washed first with binding buffer (50 mM phosphate buffer pH 7.4), for 30 minutes and then eluted with binding buffer spiked with 2 M NH$_4$Cl salt (introduced in a 30-minute gradient). The resulting chromatogram is shown in Figure 5.5.c. Unretained and retained peptide fractions were independently pooled. The input, unretained, and retained fractions were assayed with Pierce[TM] Quantitative Fluorometric Peptide Assay (Thermo Fischer) and sent to the UVic-Genome BC Proteomics Centre (www.proteincentre.com) for proteomic analysis.

5.9.6. Proteomics workflow and analysis

All proteomics samples were submitted for proteomics analysis at the UVic/Genome BC Proteomics Centre. All details below are as provided by the analytical service. The lists of identified peptides and PTMs were then manually collected and tabulated (Tables 3.1-3.3).

Desalting and sample cleanup: Each sample was desalted using an Oasis HLB µ-elution plate SPE (30 µm) (Waters part#186001828BA). Volume corresponding to ~30 µg of peptide was loaded. Following binding and washing, peptides were eluted with 180 µL (80% v/v Acetonitrile, 0.1% v/v Formic acid), speed vacuumed to near dryness and rehydrated with 20 µL 2% Acetonitrile, 0.1% formic acid in water.

LC-MS/MS parameters: The peptide mixture (5 µL) was separated by on-line reverse phase chromatography using a Thermo Scientific EASY-nLC 1000 system with a reverse-phase pre-column Magic C18-AQ (100 µm I.D., 2.5 cm length, 5 µm, 100 Å) and an in-house prepared reverse phase nano-analytical column Magic C-18AQ (75 µm I.D., 15 cm length, 5 µm, 100 Å, Michrom BioResources Inc, Auburn, CA), at a flow rate of 300 nL/min. The chromatography system was coupled on-line with an Orbitrap Fusion Tribrid mass spectrometer (Thermo Fisher Scientific, San Jose, CA) equipped with a Nanospray Flex NG source (Thermo Fisher Scientific). Solvents were A: 2% Acetonitrile, 0.1% Formic acid; B: 90% Acetonitrile, 0.1% Formic acid. After a 300 bar (~ 8µL) pre-column equilibration and 300 bar

(~ 3μL) nanocolumn equilibration, samples were separated by a 140-minute gradient (0 min: 5%B; 110 min: 35%B; 10 min: 45%B; 10 min: 100%B; hold 10min: 100%B).

The Orbitrap Fusion: Instrument parameters (Fusion Tune 3.0 software) were as follows for orbitrap (OT-MS) iontrap (IT- MS/MS) with HCD fragmentation: Nano-electrospray ion source with spray voltage 2.45 kV, capillary temperature 275 °C. Survey MS1 scan m/z range 380-2000 profile mode, resolution 120,000 FWHM@200m/z, one microscan with maximum inject time 50 ms. The Siloxane mass 445.120024 was used as lock mass for internal calibration. Data-dependent acquisition Orbitrap survey spectra were scheduled at least every 3 seconds, with the software determining "Top-speed" number of MS/MS acquisitions during this period. The automatic gain control (AGC) target values for FTMS and MSn were 400,000 and 5,000 respectively. The most intense ions charge state 2-7 exceeding 50,000 counts were selected for HCD and CID ion trap MSMS fragmentation with detection in centroid mode. Monoisotopic Precursor Selection (MIPS) was enabled and Dynamic exclusion settings were: repeat count: 2; repeat duration: 10 seconds; exclusion duration: 10 seconds with a 10ppm mass window. The ddMS2 IT HCD scan used a quadrupole isolation window of 1.6 Da; IonTrap rapid scan rate centroid detection first mass 100m/z, 1 microscan, 50ms maximum injection time and stepped collision energy 40% ± 3. The ddMS2 IT CID scan used a quadrupole isolation window of 1.6 Da; IonTrap rapid scan rate centroid detection, auto-normal scan mass range, 1 microscan, 250ms maximum injection time and normalized collision energy = 35%.

Software: Raw files were created by XCalibur 4.1.31.9 (Thermo Scientific) software and analysed with Proteome Discoverer 2.2.0.388 software suite (Thermo Scientific). Parameters for the Spectrum selection to generate peak lists of the CID spectra were; activation type: CID; s/n cut-off: 1.5; total intensity threshold: 0; minimum peak count: 1; precursor mass: 350-5000 Da.

Database: The sample peak lists were submitted to a Mascot 2.5.1 server UP_yeast_Canon (76,618 sequences; 33,759,464 residues) database search for yeast samples as follows: precursor tolerance 5 ppm; MS/MS tolerance 0.8 Da; GluC enzyme 2 missed cleavages; FT-ICR instrument type; fixed modification: Carbamidomethylation (C); variable modifications: acetyl(K), methyl (K, R), dimethyl (K, R), trimethyl (K). Percolator settings: Max delta Cn 0.05; Target FDR strict 0.01, Target FDR relaxed 0.05 with validation based on q-Value.

Peptide, protein and PTM identification: Scaffold (version Scaffold_4.8.7, Proteome Software Inc., Portland, OR) was used to validate MS/MS based peptide and protein identifications. Peptide identifications were accepted if they could be established at greater than 95.0% probability by the Peptide Prophet algorithm (Keller, A et al. *Anal. Chem.* **2002**; 74(20): 5383-92) with Scaffold delta-mass correction. Protein identifications were accepted if they could be established at greater than 90.0% probability and contained at least 1 identified peptide. Protein probabilities were assigned by the Protein Prophet algorithm (Nesvizhskii, Al et al. *Anal. Chem.* **2003**; 75(17): 4646-58).

UniProtKB Mapping: Identifications of methyl containing proteins were mapped using UniProtKB (The UniProt Consortium *Nucleic Acids Research.* **2019**; 47(D1): D506-D515). Accession numbers of statistically significant identifications were exported from Scaffold software, then exported into the UniProtKB Retrieve/ID mapping tool. Non-redundant matches were checked for known post-translational modifications with the iPTMnet database of UniProtKB.

5.9.7. Identified methylation sites and proteins in each fraction.

The methylated proteins identified in each fraction are listed in Tables 5.3–5.5, along with their Uniprot accession code and methylated lysine position. The letter code X represents cases where the methylation site could not be determined accurately because multiple amino acid could have been methylated. A few entries are now obsolete on Uniprot and were instead accessed using Uniparc.

Table 5.3. Methylated proteins with identified lysine methylation sites observed in input fraction.

S.no.	Uniprot accession number	Protein name	Location	Kme3 sites found	Kme2 sites found	Kme1 sites found
1	Q751X9_ASHGO	AFR696Cp	nucleus, cytoskeleton	K710		
2	B3LTG7_YEAS1	Nucleolar complex-associated protein 3	nucleus (nucleolus)	K89	K95	
3	Q6FMP6_CANGA	Uncharacterized protein	nucleus (inner membrane)	K171		
4	C7GM33_YEAS2	GAL10p	unknown	K283		
5	E7NGG6_YEASO	ATPase GET3	Endoplasmic reticulum	K76		K69
6	B3LM19_YEAS1	Uncharacterized protein	unknown	K261		K258
7	G2WNG1_YEASK	K7_Uaf30p	unknown	K73		
8	Q6FX09_CANGA	Uncharacterized protein	unknown	k230		K239,K240
9	E7QJP0_YEASZ	Jjj1p	unknown	K455, K457	K463,K464	K460,K461,K462, K463,K464
10	G2WEZ3_YEASK	K7_Yta7p	unknown	K1246, K1249		
11	G2WP16_YEASK	K7_ATG29p	unknown	K173,K176		K180
12	B3LJV2_YEAS1	Protein disulfide-isomerase MPD1	unknown	K297,K304,K305	K300, K301, K313	K304,K305
13	C7GN53_YEAS2	Mediator of RNA polymerase II transcription subunit 4	nucleus	K227,K232		
14	DNA2_YEAST	DNA replication ATP-dependent	nucleus	K19, K20		K8,K19,K20

		helicase/nuclease DNA2				
15	A6ZVJ5_YEAS7	Threonyl-tRNA synthetase	Cytoplasm	K28,K29	K13,K17	K12
16	Q757J6_ASHGO	AER017Cp	unknown	K71		
17	C8Z6D3_YEAS8	Glt1p	unknown	K623		
18	Q754F2_ASHGO	AFR118Wp	unknown	K28		
19	E7M1R3_YEASV	Elongation factor 1-alpha	cytoplasm	K315		
20	Q750I5_ASHGO	AGL034Cp	cytoplasm	K273	K281	
21	SPT7_YEAST	Transcriptional activator SPT7	nucleus	K672	K678,K679	K678,K679,K693
22	E7KK18_YEASL	Skt5p	unknown	K663,K688,K689,K690		K684
23	A6ZU13_YEAS7	5'-3' exoribonuclease 1	cytoplasm	K112,K113		
24	SDO1L_YEAST	Restriction of telomere capping protein 3	nucleus, cytoplasm	K73		
25	Q756S4_ASHGO	AER180Wp	preribosome	K37		
26	G2WDZ1_YEASK	K7_Cue3p	unknown	K223,K230,K579	K578	
27	E7Q9G3_YEASB	YOR287C-like protein	nucleus (nucleolus)	K93		K95
28	E7NNP8_YEASO	Uip4p	unknown	K258		K248,K251
29	C8ZC92_YEAS8	Tef4p	unknown	K228		K225,K230
30	MYO1_YEAST	Myosin-1	Cytoskeleton	K873		
31	MPS1_ENCCU	Probable serine/threonine-protein kinase MPS1 homolog	unknown	K169,K172,K173	K149	
32	E7Q765_YEASB	Chs5p	unknown	K645,K647	K648	K651,K652,K654
33	Q754F6_ASHGO	AFR114Wp	mitochondrion, cell wall	K268		
34	Q6FWI9_CANGA	Uncharacterized protein	cytosol	K191		
35	Q6FR15_CANGA	Uncharacterized protein	mitochondrion	K80,K84		K90
36	G2WIC0_YEASK	K7_Ptr2p	membrane	K581		
37	E7KRT2_YEASL	Glycylpeptide N-tetradecanoyl transferase	unknown	K6	K4	
38	Q7RAQ7_PLAYO	Saccharomyces cerevisiae ORF 2310-related	unknown	K304,K306,K310,K328	K298,K303	K297
39	E7LWQ7_YEASV	Ltv1p	unknown	K111	K108	
40	E7NNV4_YEASO	YPL088W-like protein	unknown	K254		
41	B5VJM4_YEAS6	YGR281Wp-like protein	membrane	K551	K555	
42	IRC19_YEAS6	Increased recombination centers protein 19	unknown	K114		
43	G2WDZ5_YEASK	K7_Arc1p	unknown	K146		
44	Q6FNI8_CANGA	Uncharacterized protein	nucleus (nucleolus)	K290	K289	
45	ESF1_ASHGO	Pre-rRNA-processing protein ESF1	nucleus (nucleolus)	K463,K470	K464	
46	EF1A_YEAST	Elongation factor 1-alpha	cytoskeleton	K315		
47	B3LSL9_YEAS1	Aldose reductase	unknown	K290,K291		

48	Q75BC4_ASHGO	ADL357Cp	mitochondrial matrix, nucleus	K794	
49	G2WJ15_YEASK	K7_Swi6p	unknown	K626	
50	A6ZPZ2_YEAS7	GDP/GTP exchange factor	unknown	K1007	
51	E7Q0Q0_YEASB	Rpl23ap	ribosome	K83,K87	
52	ZIP1_YEAST	Synaptonemal complex protein ZIP1	nucleus	K288	K286,K289
53	Q754P1_ASHGO	AFR031Cp	cytoplasm	K456	
54	C7GSM6_YEAS2	Coronin	unknown	K559,K563, K565	K563,K565
55	Q6FS29_CANGA	Uncharacterized protein	unknown	K370	
56	EF1A_ASHGO	Elongation factor 1-alpha	Cytoplasm	K315	
57	Q6FRC5_CANGA	Uncharacterized protein	cellular bud neck	K891	
58	Q752R1_ASHGO	Dihydrolipoyl dehydrogenase	Mitochondrion	K166	
59	Q753R1_ASHGO	AFR352Cp	unknown	K582,K584	
60	VATF_YEAST	V-type proton ATPase subunit F	Vacuole	K46	K34
61	C8ZE49_YEAS8	40S ribosomal protein S1	Cytoplasm (ribosome)	K54	
62	E7LV11_YEASV	Eno1p	cytosol	K240,K242	K242
63	Q75AF0_ASHGO	ADL033Wp	plasma membrane	K1574	
64	E7NJV3_YEASO	Ltv1p	unknown	K108,K111	
65	SMI1_ASHGO	KNR4/SMI1 homolog	unknown	K609	
66	Q755D6_ASHGO	AFL113Cp	nucleus	K733	
67	B3LJA0_YEAS1	5'-3' exoribonuclease	nucleus	K487	
68	FAS2_YEAST	Fatty acid synthase subunit alpha	cytosol (fatty acid synthase complex)	K559	
69	E7LX31_YEASV	Glg1p	unknown	K569	K562,K572
70	ARF1_YEAST	ADP-ribosylation factor 1	Golgi apparatus	K126	
71	Q6FLJ8_CANGA	Uncharacterized protein	nucleus		K76
72	Q75E11_ASHGO	AGOS_ABL138Wp	nucleus, cytosol		K9
73	A6ZY61_YEAS7	MutS-like protein	unknown		K786
74	E7Q9A9_YEASB	Sas5p	Nucleus		K11
75	A6ZXE6_YEAS7	V-ATPase V1 sector subunit A	unknown		K410
76	A2R3S6_ASPNC	Aspergillus niger contig An14c0180, genomic contig	unknown		K72
77	H2B2_YEAST	Histone H2B.2	Nucleus		K11,K16,K17, K21
78	E7Q8T1_YEASB	YNR021W-like protein	unknown		K387
79	A6ZNB5_YEAS7	Helicase in mitochondria	unknown		K140,K143
80	Q6FW47_CANGA	Uncharacterized protein	cytoplasm		K258
81	Q6FSV3_CANGA	Uncharacterized protein	unknown		K361
82	E7NNU3_YEASO	Sse1p	unknown		K359

83	A6ZZ18_YEAS7	Zinc finger transcriptional activator	Nucleus	K152,K154
84	Q75AD4_ASHGO	ADL017Cp	clathrin adaptor complex	K245
85	B5VLA4_YEAS6	YJL074Cp-like protein	chromosome	K160
86	B3LID9_YEAS1	Pyruvate decarboxylase isozyme	unknown	K301,K303
87	GRP78_ASHGO	Endoplasmic reticulum chaperone BiP	Endoplasmic reticulum lumen	K550
88	Q752C8_ASHGO	Folic acid synthesis protein fol1	Mitochondrion	K114
89	Q75B04_ASHGO	ADL119Wp	unknown	K96
90	E7QMG4_YEASZ	Fsp2p	unknown	K326
91	E7M1E0_YEASV	Sse1p	unknown	K331
92	E7Q591_YEASB	Dal1p	unknown	K29
93	Q758I6_ASHGO	AEL224Wp	unknown	K299,K307
94	Q75DF8_ASHGO	ABR068Cp	cell wall, cytosol, membrane	K72
95	RL7_ASHGO	60S ribosomal protein L7	cytosolic large ribosomal subunit	K3
96	G2WEB5_YEASK	tRNA(His) guanylyltransferase	unknown	K37
97	Q74ZR3_ASHGO	AGR135Cp	unknown	K233,K239
98	A6ZTG0_YEAS7	tRNA (guanine(37)-N1)-methyltransferase	Mitochondrion matrix, nucleus, cytoplasm	K84

Table 5.4. Methylated protein with identified lysine methylation sites observed in retained fraction.

S.no.	Uniprot accession number	Protein name	Location	Kme3 sites found	Kme2 sites found	Kme1 sites found
1	B3LJY7_YEAS1	Class V myosin	cytoskeleton	K1038,K1041, K1042		
2	G2WA51_YEASK	K7_Snt1p	nucleus	K352,K359,K361		
3	E7KR12_YEASL	Tef4p	unknown	K38,K40,K43		
4	E7QAE9_YEASB	Rlf2p	unknown	K157		
5	E7NH18_YEASO	Yen1p	unknown	K515,K516,K522	K528,K547	K528,K537, K543,547
6	Q6FQI3_CANGA	Uncharacterized protein	nucleus	K217,K218,K220, K221,K223,K224, K226,K232,K233, K235,K240,K242	K236,238	
7	E7Q5F6_YEASB	YJL123C-like protein	unknown	K70		K78
8	Q75FB3_ASHGO	Structural maintenance of chromosomes protein	nucleus	K174		K179
9	SP381_YEAST	Pre-mRNA-splicing factor SPP381	nucleus	K239	K241,K245	K241,K245, K246
10	E7NFJ2_YEASO	Glt1p	unknown	K1562,K1564		K1566
11	Q6FTF2_CANGA	Uncharacterized protein	membrane	K733,K734		
12	E7KST6_YEASL	FK506-binding protein	nucleus (nucleolus)	K242,K246	K249,K250	K249,K250

137

#	Accession	Protein	Localization			
13	A6ZT62_YEAS7	Regulator of Ty1 transposition	unknown	K980		K988
14	E7Q0M2_YEASB	Yro2p	membrane	K311,K316,K317, K318,K320	K305,K307, K309,K310, K311,K313, K320,K321, K323,K324, K326,K327	
15	Q6FQ98_CANGA	Uncharacterized protein	nucleus, preribosome	K10	K5,K12,K13	
16	Q74ZW9_ASHGO	Structural maintenance of chromosomes protein	nucleus	K519,K520,K524	K519,K520	
17	EF3_ASHGO	Elongation factor 3	unknown	K1011,K1022, K1023	K1020,K1021	K996,1003, 1009, 1010
18	A2QGD0_ASPNC	Similarity to hypothetical myb-related protein	unknown	K127,K129		
19	TOM1_ASHGO	Probable E3 ubiquitin-protein ligase TOM1	nucleus	K935		K944
20	E7KPI9_YEASL	Dcd1p	unknown	K7,K15		
21	Q75AK5_ASHGO	ADL078Cp	cytosol (large ribosomal subunit)	K123	K121	K113,K116, K117,K120
22	Q6FNA8_CANGA	Uncharacterized protein	unknown	K205		K224
23	E7Q7L3_YEASB	Gis4p	unknown	K703,K704		K706
24	G2WEH5_YEASK	U3 small nucleolar RNA-associated protein 22	nucleus (nucleolus)	K802,K808		
25	A2QAB0_ASPNC	Similarity to hypothetical membrane protein YOR206w	unknown (membrane?)	K187	K184	K185
26	Q758T5_ASHGO	Serine/threonine-protein phosphatase 2A 56 kDa regulatory subunit	cytoskeleton, nucleus	K64,K66,K70, K84,K54	K78,K80	K61
27	E7QLC0_YEASZ	Mei5p	unknown	K182,K183		K182,K183
28	E7KK02_YEASL	Methionine aminopeptidase 2	cytoplasm	K46,K47,K48, K49	K45	K41,K42,K43
29	E7Q6Y1_YEASB	Eukaryotic translation initiation factor 3 subunit J	Cytoplasm	K15,K24	K12,K20,K21, K23,K26	
30	E7Q9L9_YEASB	YOR338W-like protein	unknown (Nucleus?)	K155,K160,K170, K177,K179	K155,K157, K177,K179	K177,K179
31	E7Q5I4_YEASB	Aly2p	unknown	X351		
32	E7KTW3_YEASL	Alr1p	membrane	K374,K387	K372	K385
33	Q74ZH8_ASHGO	AGR229Wp	unknown	K656,K657	K651,K653	
34	Q6FN04_CANGA	Uncharacterized protein	nucleus (nucleolus), small-subunit processome	K109,K114,K133	K124,K125	K103,K104
35	E7Q4F4_YEASB	YGR273C-like protein	unknown	K158,K159	K157	K151
36	Q75EE4_ASHGO	AAR137Wp	cytoskeleton, nucleus	K391,K446,K447, K448	K396,K446,K4 47	K389,K446, K447
37	B3LNS1_YEAS1	tRNA (adenine(58)-N(1))-methyltransferase non-catalytic subunit TRM6	nucleus	K474		K473
38	E7LZB4_YEASV	RNA cytidine acetyltransferase	nucleus (nucleolus)	K956,K958,K959, X955	K595, X955	K952

#	Accession	Protein	Localization			
39	Q75EC8_ASHGO	AAR146Wp	extracellular region/cell surface	K382,K383	K381	
40	Q75AU0_ASHGO	ADL170Cp	nucleus (nucleolus), 90S preribosome	K391,K392	K394,K395	K399
41	E7Q5J4_YEASB	Nsp1p	nucleus (nuclear pore)	K66,K78,K85		
42	Q6FNJ2_CANGA	Uncharacterized protein	membrane	K376,K377,K381, K403,K404,K406, K407	K376,K391, K393,K394, K397,K406, K407	K377,K381, K387
43	E7QL17_YEASZ	Mum3p	unknown	K302	K320	K320
44	C7GPR8_YEAS2	Transcription elongation factor Spt6	nucleus	K1344		
45	G2WBJ5_YEASK	K7_Spp41p	nucleus	K973		K980
46	Q759M8_ASHGO	ADR248Cp	GID complex	K98,K99		
47	E7NJY6_YEASO	Bud2p	unknown	K980	K972,K975	K984
48	IMH1_ASHGO	Golgin IMH1	Golgi apparatus membrane, cytoplasm	K259,K260,K269, K270		K262,K264
49	E7KQT6_YEASL	Mnn4p	membrane	K1161,K1162, K1163, X1167	K1161,K1162, K1163	K1161,K1162, K1163, X1167
50	B3LQ79_YEAS1	Geranylgeranyltransferase type II alpha subunit	Rab-protein geranylgeranyltransferase complex	K13		K12,K20
51	C8ZC92_YEAS8	Tef4p	unknown	K230,K234	K233	K235
52	PXR1_ASHGO	Protein PXR1	nucleus (nucleolus)	K225,K226,K228, K229,K231,K232	K226,K228, K229,K231, K232	K225,K228, K229,K231, K232
53	C8ZFA5_YEAS8	Cus1p	nucleus	K38	K39,K40	
54	E7NNG5_YEASO	Cam1p	unknown	K145,K146,K157	K146	K158
55	E7LXP1_YEASV	Sik1p	membrane	K87	K91,K92,K102, K109	K93
56	Q6FMZ8_CANGA	Uncharacterized protein	membrane	K130		K133
57	G2WEE3_YEASK	K7_Rsc1p	nucleus	K7457		
58	E7Q765_YEASB	Chs5p	unknown	K664,K665,K666	K664,K665, K666	K643,K644
59	E7KTJ0_YEASL	Inp52p	unknown	K1170,K1178		K1170,K1178
60	A6ZZ14_YEAS7	Conserved protein	unknown	K801,K803,K804		K797
61	Q6FTM3_CANGA	Uncharacterized protein	mitochondria	K135,K152,K155, K158	K165	K136,K137
62	E7LV21_YEASV	Pxr1p	unknown	K174,K175,K177, K178,K180,K181, K183,K184,K186, K187,K188	K191,K193, K194,K197, K198,K200	
63	Q6FJB2_CANGA	Uncharacterized protein	nucleus/ 90S preribosome	K411		
64	B3LU26_YEAS1	Uncharacterized protein	unknown	K21,K31,K37	K38,K41,K42	K27
65	E7NFQ6_YEASO	Dpb4p	unknown	K153	K146,K156	K146
66	Q75DY8_ASHGO	ABL115Wp	nucleus	K29,K30,K33	K34,K37,K40, K46,K49	K41,K44
67	Q7RAQ7_PLAYO	Saccharomyces cerevisiae ORF 2310-related	unknown	K560	K562	K563

No.	ID	Protein name	Localization			
68	EAF6_ASHGO	Chromatin modification-related protein EAF6	nucleus	K11,K13		
69	H2B1_CANAL	Histone H2B.1	nucleus, chromosome	K37	K30,K31,K34	K11,K16,K17
70	E7NEY9_YEASO	Spp381p	unknown	K239,K241	K246	
71	Q6FP17_CANGA	Uncharacterized protein	cytosol, nucleus	K552,K553,K556	K561,K563, K569,K576 ,K580	
72	E7NKE6_YEASO	Ioc2p	unknown	K659,K677,K676	K677	K664,K667, K676
73	ESF1_ASHGO	Pre-rRNA-processing protein ESF1	nucleus (nucleolus)	K515,K525		K526,K528
74	E7QH84_YEASZ	Rrp14p	unknown	K308,K314	K307,K310	
75	N1P3M2_YEASC	Vas1p	cytoplasm	K64	K37,K38,K67, K68,K69,K70	
76	E7Q7Y0_YEASB	Yku70p	nucleus	K590,K591,K595, K596,K598, X592	K595,K596,K5 98	
77	Q6FK28_CANGA	Uncharacterized protein	Cytoskeleton	K171,K175,K177	K167,K168	K171,K175
78	E7Q2G4_YEASB	Mrp20p	ribosome	K50,K53,K56	K53,K56	
79	Q6FIW4_CANGA	Uncharacterized protein	nucleus	K36,K39,K40,K4 2,K46,K47,K51,K 52,K58,K60,K64, K65	K36,K39,K40, K42,K46,K47, K51,K52,K58, K60,K64,K65	K36,K39,K40, K51,K52,K58, K60,K64
80	Q756J7_ASHGO	AER257Wp	nucleus (nuclear chromatin)	K465,K473,K476	K449,K461, K462	
81	G2WMC1_YEASK	K7_Pap2p	unknown	K544,K545,K547		K544,K545
82	E7Q1R3_YEASB	Nop14p	small-subunit processome	K22,K27	K35,K44	
83	Q6FNB3_CANGA	Uncharacterized protein	unknown	K96,K97,K100, K102,K107,K113, K116,K119,K120, K121	K92,K93	
84	Q759A4_ASHGO	ADR373Wp	cytosol (small ribosomal subunit)	K194,K197,K199	K202	
85	G2WKX7_YEASK	K7_Tma23p	unknown		K159,K160, K161,K162, K164	
86	Q75DQ3_ASHGO	ABL036Cp	unknown		K633	
87	A2QRC8_ASPNC	Cluster of Similarity to secretory protein Sec5	exocyst		K528	
88	Q6FS05_CANGA	Phospholipid-transporting ATPase	membrane		K1459	
89	Q757V3_ASHGO	AEL091Cp	unknown		K2,K10	
90	Q758E3_ASHGO	AEL181Cp	nucleus (nucleolus)		K969,K972	
91	E7NEH3_YEASO	Fun12p	unknown		K2	
92	E7LYG8_YEASV	Acetolactate synthase	unknown		K426	
93	A6ZX90_YEAS7	Hexose transporter	membrane		K556	K566
94	Q6FRF3_CANGA	Uncharacterized protein	nucleus		K324,K325	
95	Q6FRX8_CANGA	Uncharacterized protein	nucleus		K751	K756
96	B3LSD5_YEAS1	Meiosis-specific protein SPO13	unknown		K18	
97	Q6FT31_CANGA	Uncharacterized protein	Endoplasmic reticulum		K195	K190,K195, K202
98	E7Q7H4_YEASB	Peptidylprolyl isomerase	unknown		K82,K83,K86, K88,K89	K78

S.no.	Uniprot accession	Protein name	Location	Kme3 sites	Kme2 sites	Kme1 sites
99	HSP7F_ASHGO	Heat shock protein homolog SSE1	Cytoplasm		K601,K606	K597,K599
100	E7NML6_YEASO	Msb4p	unknown		K370	
101	SET2_ASHGO	Histone-lysine N-methyltransferase, H3 lysine-36 specific	nucleus		K655	
102	Q75EK8_ASHGO	AAR071Wp	endoplasmic reticulum membrane			K1005
103	E7KCZ2_YEASA	Pxr1p	unknown		K178,K180	K83,K84,K86, K87,K88,K90, K91,K93,K94, K97
104	H2B2_YEAST	Histone H2B.2	nucleus	K37	K30,K31,K34	K11,K17,K18
105	E7QFU0_YEASZ	Rix1p	unknown			K590,K591
106	Q757B8_ASHGO	AER095Wp	nucleus			K332
107	E7NP23_YEASO	Rrp15p	unknown			K155
108	E7NNT1_YEASO	Hho1p	nucleus			K4,K8,K12, K14
109	E7QIQ0_YEASZ	Prp24p	unknown			K414,K421,K4 222,K423
110	Q6FPR8_CANGA	Uncharacterized protein	unknown			
111	A6ZR22_YEAS7	CCR4-NOT transcriptional complex subunit	unknown			K101

Table 5.5. Methylated proteins and identified lysine methylation sites observed in unretained fraction.

S.no.	Uniprot accession number	Protein name	Location	Kme3 sites found	Kme2 sites found	Kme1 sites found
1	E7Q6D2_YEASB	YKR075C-like protein	unknown	K3,K24	K1,K25	K15,K17,K20
2	Q6FP93_CANGA	Uncharacterized protein	nucleus	K1237,K1238, K1245		
3	Q75EP5_ASHGO	AGOS_AAR034W	unknown	K78,K81,K85,K88, K89,K93,K98,K102	K85,K88,K89, K93	K74,K75,K85, K102,K103
4	G2WM16_YEASK	DNA topoisomerase 2	unknown	K1276,K1277		
5	EF1A_YEAST	Elongation factor 1-alpha	cytoskeleton	K315		
6	E7Q7C9_YEASB	Spp382p	unknown	K339		
7	Q75AE5_ASHGO	ADL028Wp	nucleus	K223,K224,K228		
8	Q74ZY4_ASHGO	Midasin	nucleus	K4355		
9	E7NI30_YEASO	Erg1p	membrane	K196		
10	A6ZXR7_YEAS7	Essential protein involved in intracellular protein transport	Golgi membrane	K1724,K1731		K1718
11	G2WEZ3_YEASK	K7_Yta7p	unknown	K1246,K1249		
12	E7NNV4_YEASO	YPL088W-like protein	unknown	K254		
13	Q75DS7_ASHGO	ABR022Cp	cytoplasm	K519		
14	E7KFP9_YEASA	Peptidylprolyl isomerase	nucleus (nucleolus)	K167		K170,K172
15	G2WEH5_YEASK	U3 small nucleolar RNA-associated protein 22	nucleus (nucleolus)	K625		K629
16	Q6FWI9_CANGA	Uncharacterized protein	cytosolic ribosome, NatA complex	K191		

#	Accession	Protein	Localization			
17	E7NL54_YEASO	Fpr4p	nucleus (nucleolus)	K238		K241,K243
18	Q750U5_ASHGO	AGL156Wp	endosome, HOPS complex	K377	K383	
19	SMI1_ASHGO	KNR4/SMI1 homolog	unknown	K465,K468,K477	K477	K459,K460, K468,K471, K477
20	Q756S4_ASHGO	AER180Wp	preribosome	K37		
21	Q750I5_ASHGO	AGL034Cp	extracellular space, membane, cytoplasm	K273,K277	K281	
22	G2WDZ1_YEASK	K7_Cue3p	unknown	K223,K230		
23	C8ZIY6_YEAS8	Nop4p	unknown	K525,K529,K544	K526,K527	K530,K535
24	E7KSG3_YEASL	Clustered mitochondria protein homolog	Cytoplasm	K8		
25	A6ZZF3_YEAS7	Ebna1-binding protein	unknown	K60,K61		
26	A6ZMY2_YEAS7	Inorganic pyrophosphatase	cytoplasm	K294		
27	C8Z6P2_YEAS8	Hop	unknown	K542	K534	K539
28	MSH3_ASHGO	DNA mismatch repair protein MSH3	nucleus	K315		
29	C7GPH2_YEAS2	Bpt1p	membrane	K1001		
30	Q6FQG5_CANGA	Uncharacterized protein	Endosome	K80	K81,K82	K81,K82,K99
31	A6ZRL4_YEAS7	J-protein (Type III)	unknown	K455,K457,K465,K467	K458,K459,K461,K464	
32	BRE1_YEAST	E3 ubiquitin-protein ligase BRE1	nucleus	K449		K460
33	Q755C6_ASHGO	Signal recognition particle subunit SRP72	Cytoplasm	K3		
34	MYO1_YEAST	Myosin-1	Cytoskeleton (myosin II complex)	K873		
35	E7M1R3_YEASV	Elongation factor 1-alpha	cytoplasm	K315		
36	E7QFJ1_YEASZ	Myo1p	Cytoskeleton (myosin complex)	K863		
37	Q6FPM1_CANGA	Uncharacterized protein	unknown	K602,K614,K616		K593,K594
38	Q6FSW1_CANGA	Uncharacterized protein	nucleus inner membrane	K380		
39	E7LWQ7_YEASV	Ltv1p	unknown	K108	K112	
40	Q754F9_ASHGO	AFR111Cp	unknown	K449		
41	Q6FKB7_CANGA	Uncharacterized protein	unknown	K311,K324,K326,K330	K324,K326,K330	K295,K308, K324
42	G2WCA2_YEASK	K7_Uso1p	Golgi membrane	K1724,K1731		K1718
43	E7NFA6_YEASO	Mgr1p	unknown	K45	K55	
44	A6ZPZ2_YEAS7	GDP/GTP exchange factor	unknown	K19		
45	E7KHP8_YEASA	Nop12p	unknown	K27,K30	K2	
46	N1P624_YEASC	Uso1p	cytoplasm, membrane	K515,K522		K509
47	E7NKJ5_YEASO	Bre2p	nucleus	K425		
48	E7KRT2_YEASL	Glycylpeptide N-tetradecanoyltransferase	unknown	K6	K4	
49	Q6FMP2_CANGA	Uncharacterized protein	unknown	K158		K155,K161
50	B3LRT0_YEAS1	DNA repair helicase RAD3	nucleus	K141,K143	K131	

51	E7KLS7_YEASL	Histone H2B	nucleus, chromosome	K6,7,11,16,17		K37
52	C7GVC5_YEAS2	Sir4p	unknown	K1289		
53	Q75BC0_ASHGO	ADL353Cp	nucleus (nucleolus)	K4,K778	K2	
54	E7KPB2_YEASL	Cic1p	unknown	K303		
55	Q74ZT5_ASHGO	AGR113Wp	unknown	K416,K417		K413
56	Q6FY60_CANGA	Uncharacterized protein	unknown	K29,K43,K47		
57	SS120_YEAST	Protein SSP120	unknown	K222		
58	B5VTZ1_YEAS6	Uncharacterized protein	membrane	K26		K23
59	A7A1D0_YEAS7	Conserved protein	unknown	K172		
60	C8Z4U9_YEAS8	EC1118_1D0_2718p	membrane		K2	
61	C8ZE49_YEAS8	40S ribosomal protein S1	Cytoplasm (ribosome)		K54	
62	Q6FNH4_CANGA	Uncharacterized protein	chromosome		K59	K68,K72
63	SET2_ASHGO	Histone-lysine N-methyltransferase, H3 lysine-36 specific	nucleus, chromosome		K533	K534,K535
64	Q754P7_ASHGO	AFR025Cp	unknown		K27	
65	Q752G9_ASHGO	AFR604Cp	unknown		K633,K635	
66	EF1G1_YEAST	Elongation factor 1-gamma 1	nucleus, cytoplasm		K224,K234	K236
67	Q753X2_ASHGO	AP complex subunit beta	membrane coat		K613	
68	E7Q0Q0_YEASB	Rpl23ap	structural constituent of ribosome		K83,K87	
69	Q75F13_ASHGO	AAL085Cp	unknown		K381	
70	Q758S1_ASHGO	AEL318Wp	Mitochondrion outer membrane		K106	
71	A6ZMF6_YEAS7	Glucose-6-phosphate 1-epimerase	unknown		K149	
72	C8ZD85_YEAS8	Cleavage and polyadenylation specificity factor subunit 2	nucleus		K438	
73	Q753M7_ASHGO	AFR285Cp	ribosome		K28,K32	
74	E7NHC6_YEASO	obsolete	unknown		K29,K33	
75	E7KRC4_YEASL	Atg10p	unknown		K9	K8
76	Q752R1_ASHGO	Dihydrolipoyl dehydrogenase	unknown		K166	
77	Q75A89_ASHGO	U3 small nucleolar RNA-associated protein 22	nucleus (nucleolus)		K531	
78	E7Q2S8_YEASB	Gin4p	unknown		K506	K510
79	E7KE87_YEASA	Iml2p	unknown		K219	
80	EF1A_ASHGO	Elongation factor 1-alpha	cytoplasm		K221	
81	Q75AB1_ASHGO	ADR007Cp	unknown			K220
82	G6PI_ASHGO	Glucose-6-phosphate isomerase	cytoplasm			K20
83	HSP7F_ASHGO	Heat shock protein homolog SSE1	cytoplasm			K467
84	A2R3S6_ASPNC	Aspergillus niger contig An14c0180, genomic contig	unknown			K71,K72
85	A6ZTG8_YEAS7	3-isopropylmalate dehydrogenase	cytoplasm			K4,K5

86	G3P_ASHGO	Glyceraldehyde-3-phosphate dehydrogenase	cytoplasm		K107,K114, K115
87	Q758M4_ASHGO	AEL262Cp	unknown		K554,K558
88	TOM1_ASHGO	Probable E3 ubiquitin-protein ligase TOM1	nucleus (nucleolus)		K944
89	E7QCY2_YEASZ	Cdc37p	unknown		K402,K414
90	YBR4_YEAST	Putative uncharacterized membrane protein YBR064W	Membrane		K86
91	Q755D6_ASHGO	AFL113Cp	nucleus		K10
92	GRP78_ASHGO	Endoplasmic reticulum chaperone BiP	endoplasmic reticulum lumen		K550
93	E7Q8T1_YEASB	YNR021W-like protein	Membrane		K387
94	Q759Z5_ASHGO	ADR128Cp	unknown		K53
95	G3P3_YEAST	Glyceraldehyde-3-phosphate sdehydrogenase 3	Cytoplasm		K303,K306
96	Q757G8_ASHGO	AER045Cp	endoplasmic reticulum, Golgi membrane		K1029,K1031
97	Q6FW47_CANGA	Uncharacterized protein	Cytoplasm		K258
98	Q6FNR6_CANGA	Uncharacterized protein	nucleus, cytoplasm		K220
99	Q6FVV4_CANGA	Uncharacterized protein	unknown		K195
100	G2WJT3_YEASK	Tubulin alpha chain	cytoskeleton		K60
101	Q74ZB2_ASHGO	40S ribosomal protein S30	ribosome		K2
102	UTP10_ASHGO	U3 small nucleolar RNA-associated protein 10	nucleus (nucleolus)		K746
103	Q754V3_ASHGO	AFL032Cp	aminoacyl-tRNA synthetase multienzyme complex		K132
104	B3LID9_YEAS1	Pyruvate decarboxylase isozyme	unknown		K301,K303
105	Q75DL6_ASHGO	ABR002Cp	unknown		K169
106	E7LW29_YEASV	Ura2p	unknown		K2107
107	Q75E11_ASHGO	ABL138Wp	cytosol, nucleus		K9
108	Q6FMP7_CANGA	Uncharacterized protein	intracellular (unknown)		K255
109	Q75DF8_ASHGO	ABR068Cp	cell wall, cytosol,plasma membrane		K72
110	E7NHX6_YEASO	Pil1p	unknown		K247
111	Q6FSV3_CANGA	Uncharacterized protein	unknown		K361
112	E7NFS7_YEASO	YDR161W-like protein	unknown		K160
113	HSP75_ASHGO	Ribosome-associated molecular chaperone SSB1	cytoplasm		K103
114	E7NP23_YEASO	Rrp15p	unknown	K82,K83	K72,K74,K87, K90
115	Q75AP0_ASHGO	Mitochondrial import inner membrane translocase subunit TIM44	Mitochondrion inner membrane		K50
116	Q752C8_ASHGO	Folic acid synthesis protein fol1	mitochondrial envelope		K114

Chapter 6. Concluding remarks

6.1. Impact of this work

The global interest in biologically targeted host-guest chemistry has increased dramatically during this book work.[278–283] This growth is inspired by the huge diversity of functions performed by proteins themselves, that includes recognition, molecular assembly and catalysis, all of which are result of carefully organized combination of non-covalent interactions. The quest to understand and mimic these functions is driven by the desire to contribute in multiple interdisciplinary sub-fields of concern to modern medicine, like biochemical profiling,[284] biomarker detection,[285] drug delivery[286] etc.

Host-guest chemistry in water is extremely challenging, and the biggest impact of this work in that it greatly expands the repertoire of water-soluble calix[4]arene hosts. We have developed methods to regioselectively functionalize sulfonated calix[4]arenes (Chapter 2), and shown that such strategies lead to diversification of host-guest behavior in ways that are difficult to predict, but very desirable for exploration (Chapters 2 and 4). These efforts are also the first towards synthesis of upper rim arylated calix[4]arenes. Our general conclusion is that arylation leads to an improved affinity towards hydrophobic guest molecules, but we have also discovered and commented on the nuances that come with it, like unwanted aggregation and fluorescence. These are valuable lessons which can help future projects to avoid problematic substitutions.

Another major achievement of this work is the identification of a new form of lower-rim substitution in the form of a sulfonate ester (Chapter 5). To the best of our knowledge, no previous attempts to make sulfonate esters at the lower rim have been reported. We were pleasantly surprised by its ease of synthesis and very high stability. The dramatic effect of this novel substitution on the conformation of calix[4]arene was also established. We hope that it inspires new design principles among researcher looking to modify calix[4]arene cavity.

At the start of this book, there were no host molecules that were selective for Kme2. The only published claim came during this work from Waters group at University of North Carolina.[189] Our efforts have led to the second, and arguably more stable and more selective, host that binds Kme2 over Kme3 (compound **5.1**). In addition, the affinity for Kme3 peptides has been driven to new levels (<100 nM; Chapters 2, 4, and 5). Previously published hosts, that are weaker that the ones presented in this book, were shown to be useful reagents to study interactions of histones *in vitro*. We can easily envision utilization of our library molecules for similar causes.

Finally, we pioneered the idea that a small supramolecular host can be used as an enrichment agent for PTM proteomics (Chapters 3 and 5). The best-performing agent for these purposes to date is compound **5.1**. While our proteomics data are preliminary, this host has all the desired properties to be a successful enrichment agent. It can be synthesized in relatively large quantities, is water soluble, immune to degradation in water, and can be immobilized easily. The data in Chapter 5 represent the first demonstration of a supramolecular host being used as an enrichment agent in a real-world proteomics sample. We believe we have established a new route to translate host-guest chemistry to chemical biology.

2.1.   6.2. Future work and consideration

The first and foremost attempt to expand upon the work presented here must be to use our best performing compound (**5.1**) in a a series of more extensive proteomics experiments. For *N*-methyl proteomics, one step would be to use our enrichment approach in a variety of cell lines, with direct comparisons to antibody-based Kme enrichment approaches. Another step is to integrate our enrichment strategies with SILAC-based proteomics in order to provide more quantitative data on methylation levels and the degree of enrichment provided by our approach. Both sets of experiments are underway in the labs of a new collaborator.

Another big bottleneck towards the goals of this book that can be improved upon is the overall yield of final products. In the end, we were successful at developing a calix[4]arene host (compound **5.1**) which could be utilized as an enrichment agent and synthesized in a single step from commercially available reagents. While it can be argued that heavy decoration of the calix[4]arene cavity can always improve its properties, it is difficult to apply these molecules to a real world application without access to large amount of material (i.e. >100 mg). As such, the development of quicker routes to regioselectively functionalized calix[4]arenes would have a huge impact. Such strategies can not only be targeted towards lower rim, as is highlighted in this work, but also at other positions on calix[4]arene skeleton that are not focused here.

HO₃S   HO₃S   SO₃H   SO₃H
OH   OH   OH   HO   C–H   H
HO₃S   HO₃S   SO₃H   SO₃H
OH   OH   OH   HO   C–H   R

Figure 6.1. Methylene carbon of calix[4]arene (highlighted in red) provides an additional position which can be explored to immobilize calix[4]arenes and also alter their binding profile

One such position is the methylene bridge of calix[4]arene (Figure 6.1). Similar to our rationale about lower rim, immobilization of calix[4]arene using the methylene carbons would prevent any obstruction of binding face. Methylene bridge substitution is known to impact calix[4]arene conformations and its host guest properties. We can hypothesize that such substitutions can not only provide an additional chemical handle to use but also be used to alter the supramolecular properties by changing the conformational dynamics. Synthetic methodology for such calix[4]arenes exists, but has not been applied to *p*-sulfonatocalix[4]arene.[287–290] The existing methodologies require multiple synthetic steps to install substituents at methylene carbon which must be improved upon first.

Biochemistry is full of molecules that play a big role *in vivo* but are present in low concentrations evade detection. *N*-methylated of lysine and arginine are one such set of molecules but there are plenty other PTMs and low-abundance biomolecules for which an enrichment technique is lacking. We hope that this book sparks new interest in supramolecular chemistry where researchers challenge their host molecule to act as an enrichment agent for different biomolecules. Such a combined interdisciplinary effort would have huge impact on understanding cellular machinery.[291]

Appendix A – ^{1}H and ^{13}C{^{1}H} NMR spectra

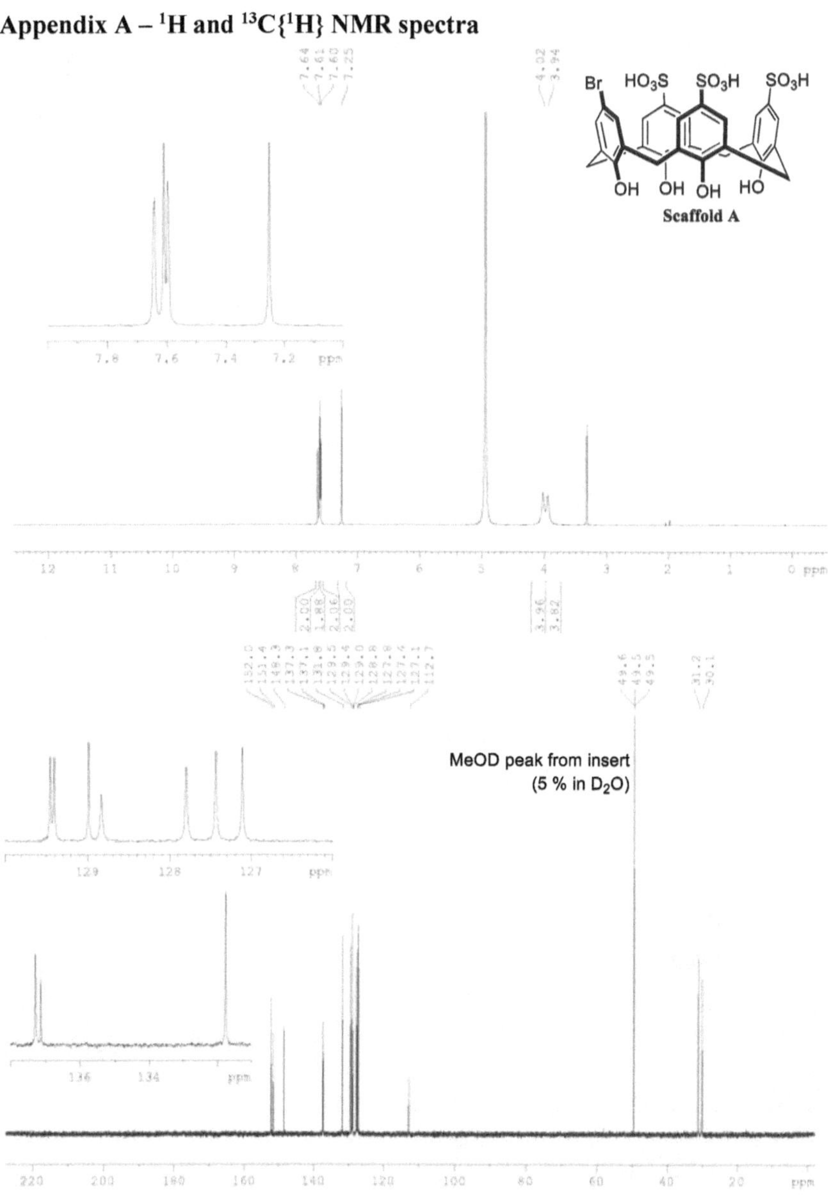

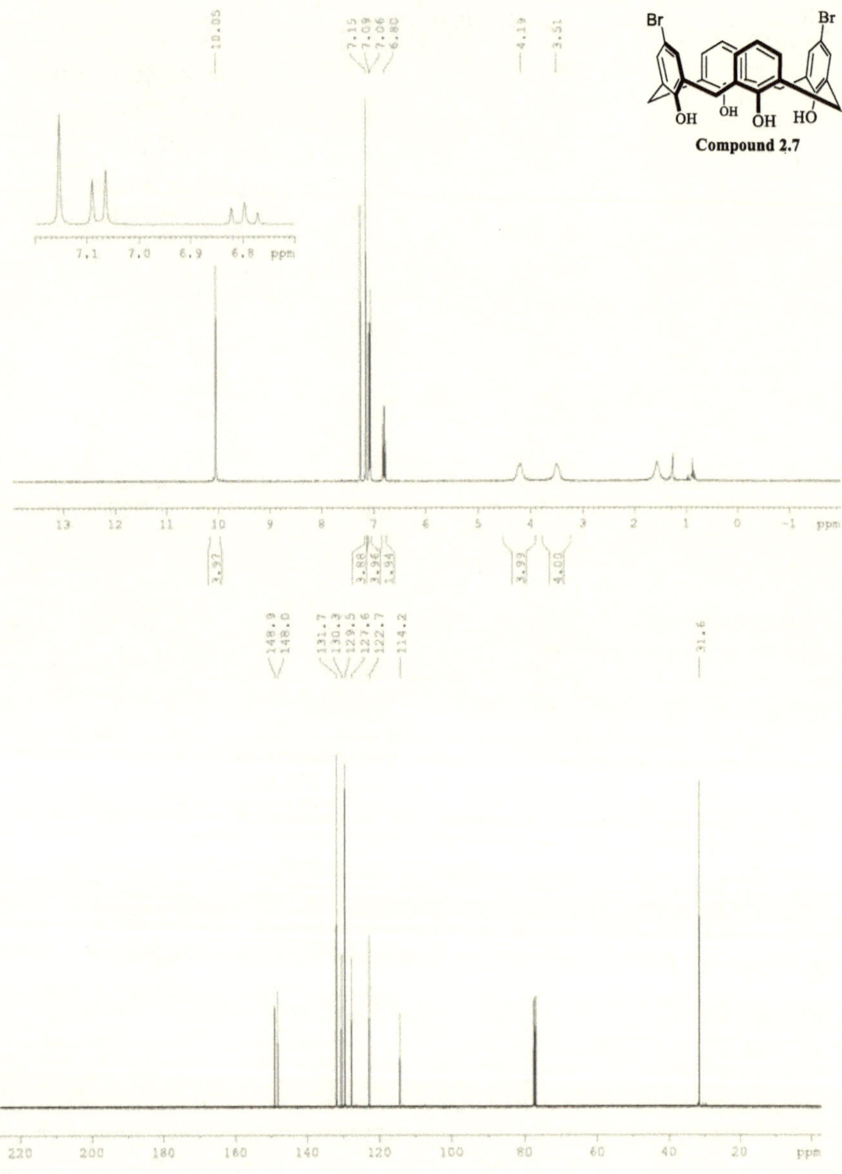

Br
Br
OH
OH
OH
HO
Compound 2.7

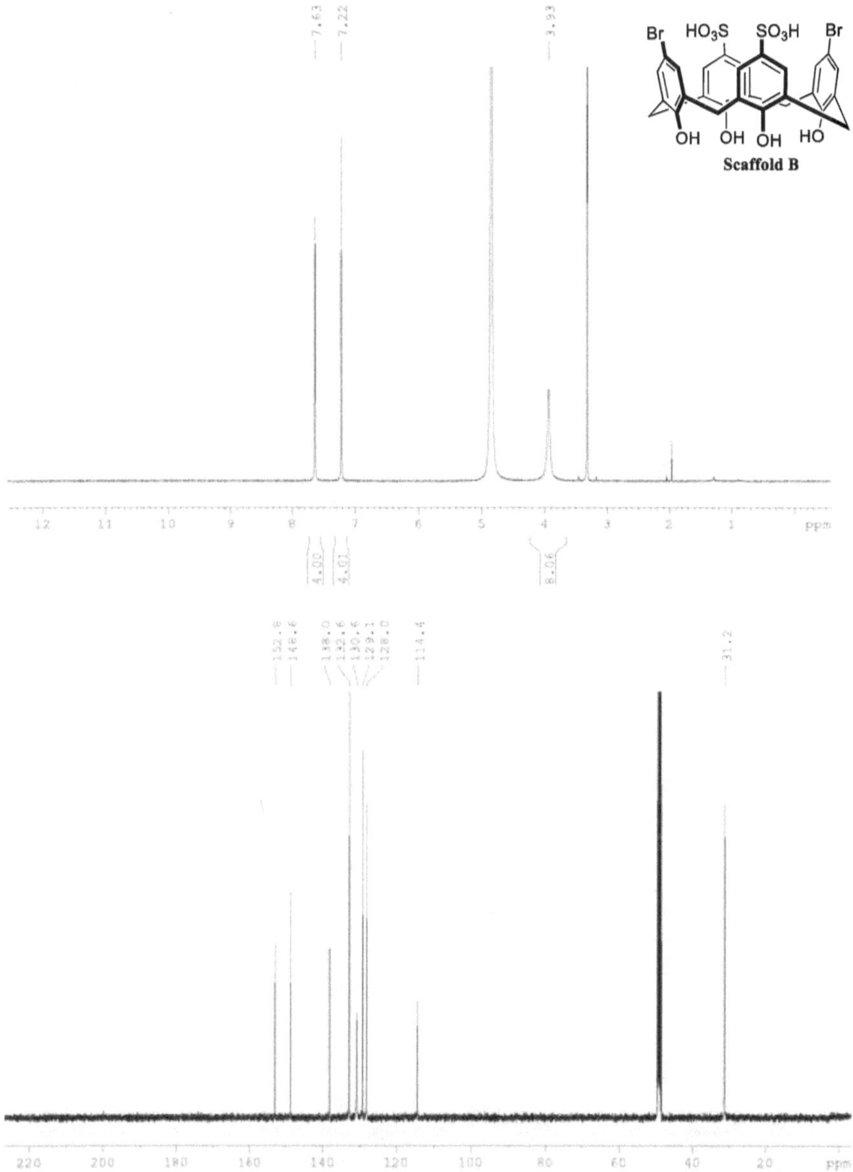
HO3S   SO3H
Br   Br
OH   OH   OH   HO
Scaffold B

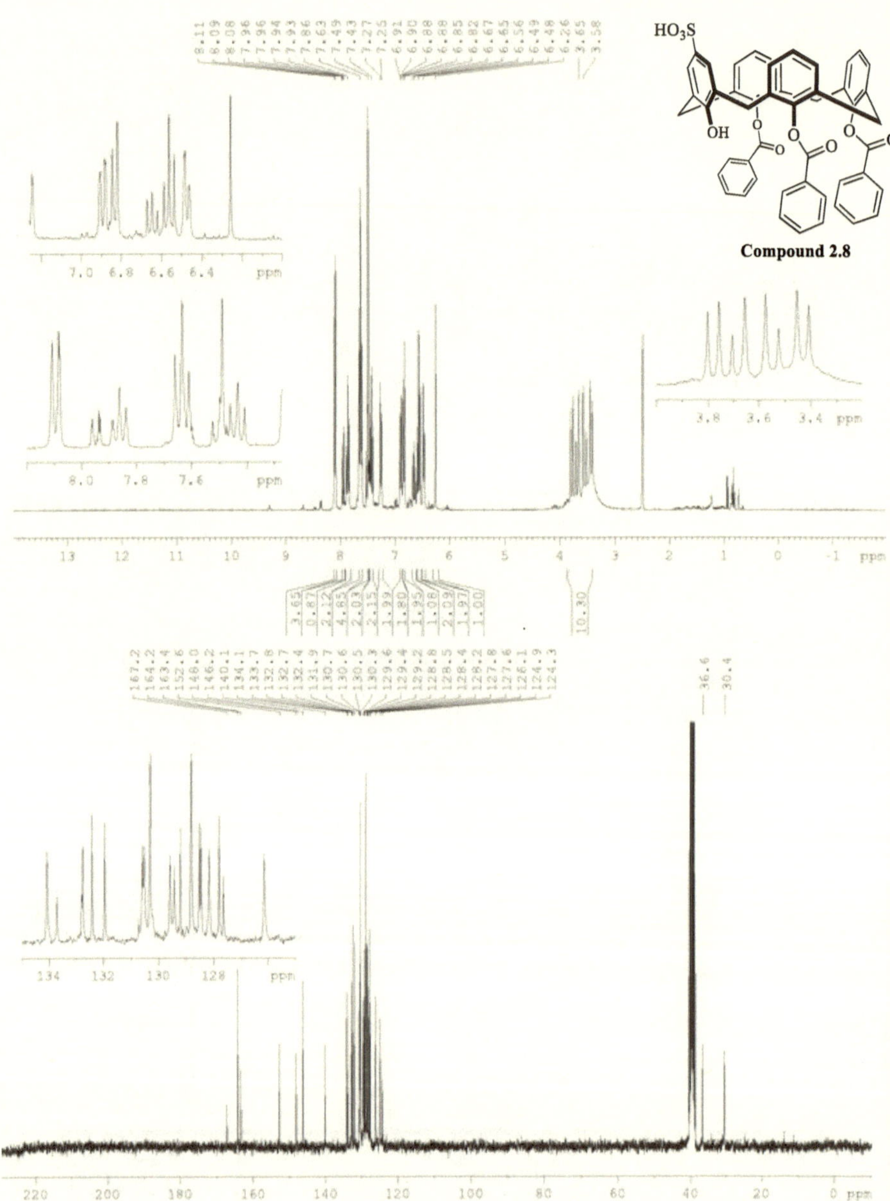

HO3S
OH
Compound 2.8

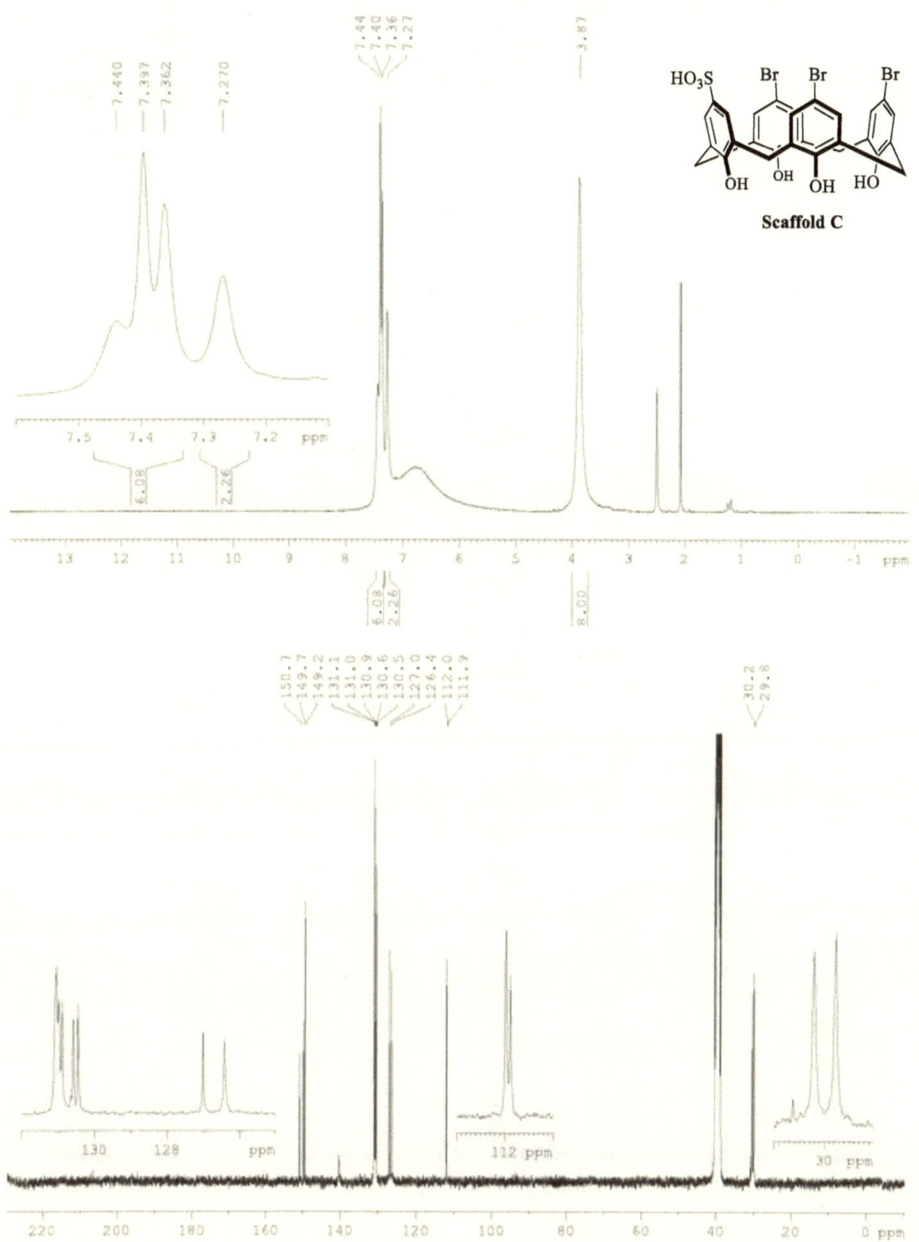

7.440
7.397
7.362
7.270
7.44
7.40
7.36
7.27
3.87
HO₃S
Br
Br
Br
OH
OH
OH
HO
Scaffold C
150.7
149.7
149.2
131.1
131.0
130.9
130.6
130.5
127.0
126.4
112.0
111.9
30.2
29.8

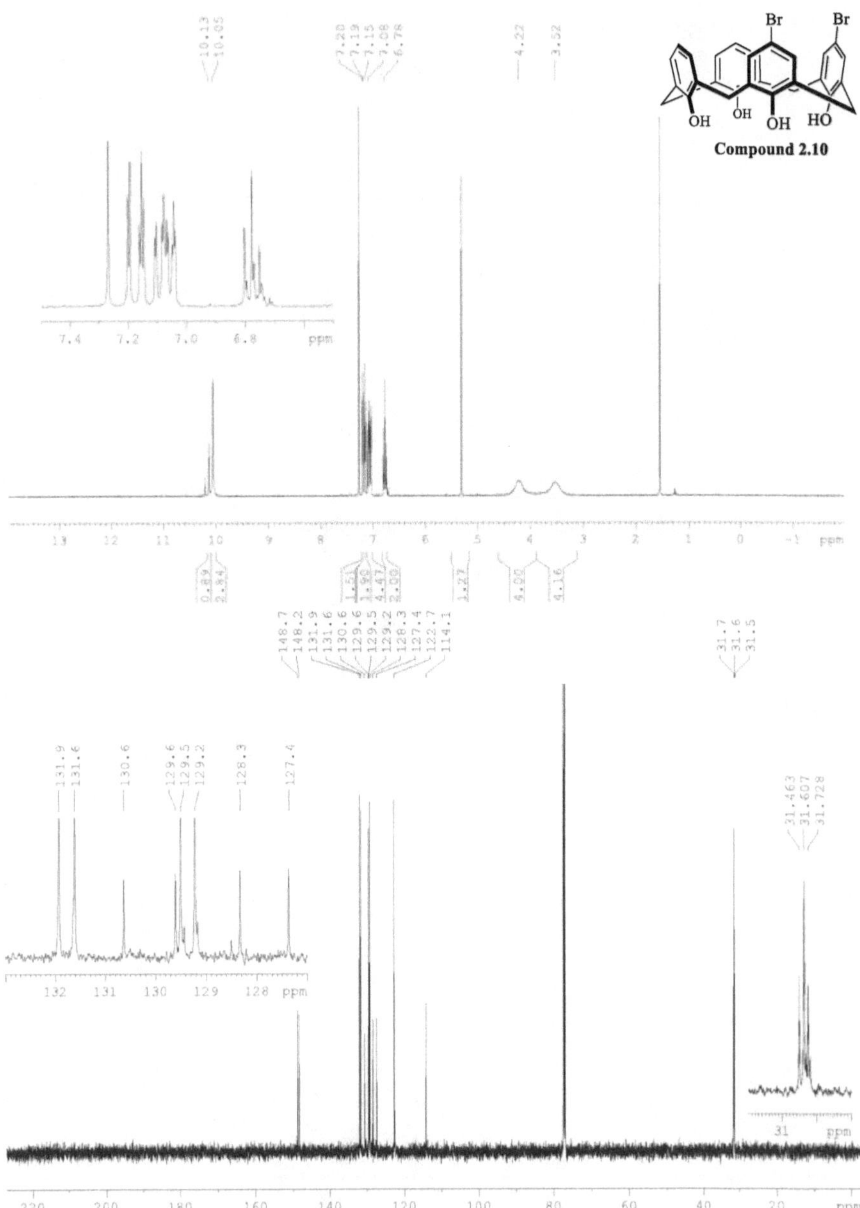
Br
Br
OH
OH
OH
HO
Compound 2.10

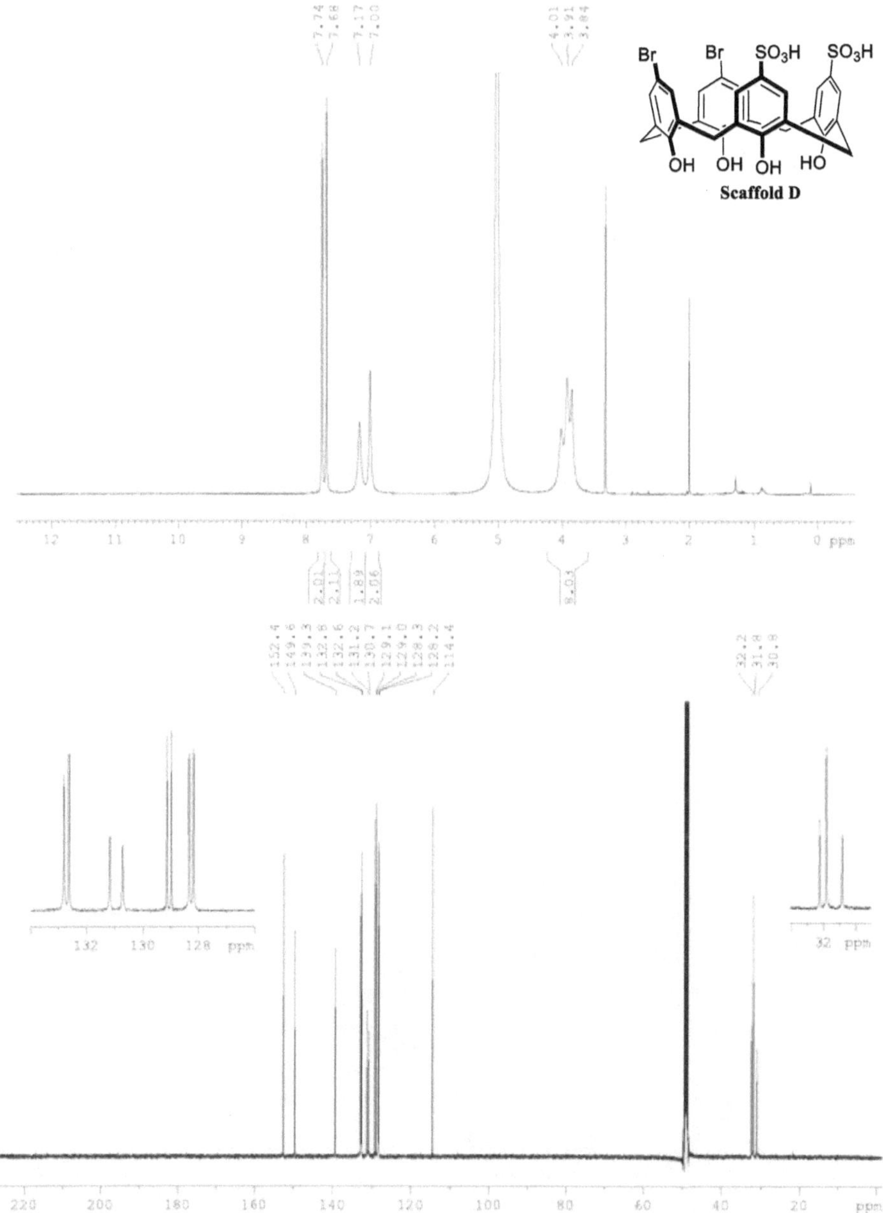

Scaffold D

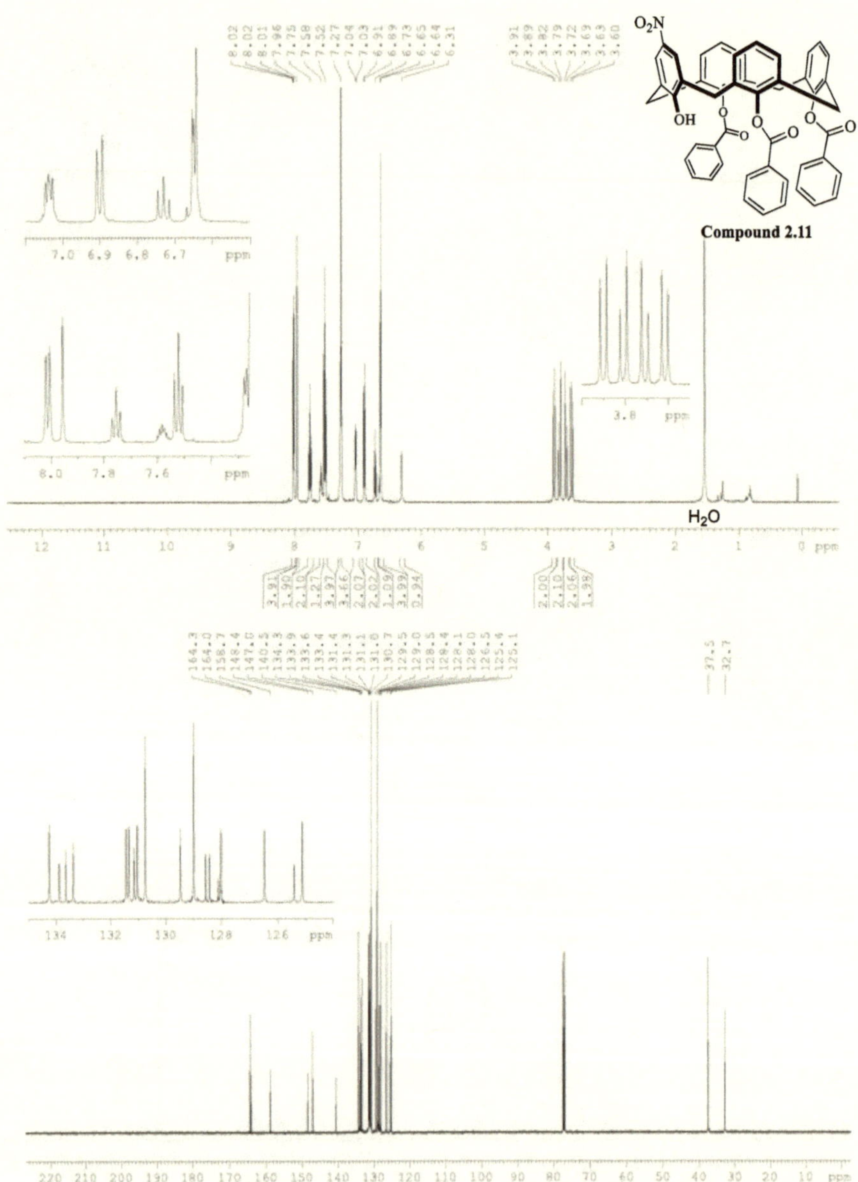
Compound 2.11
H₂O

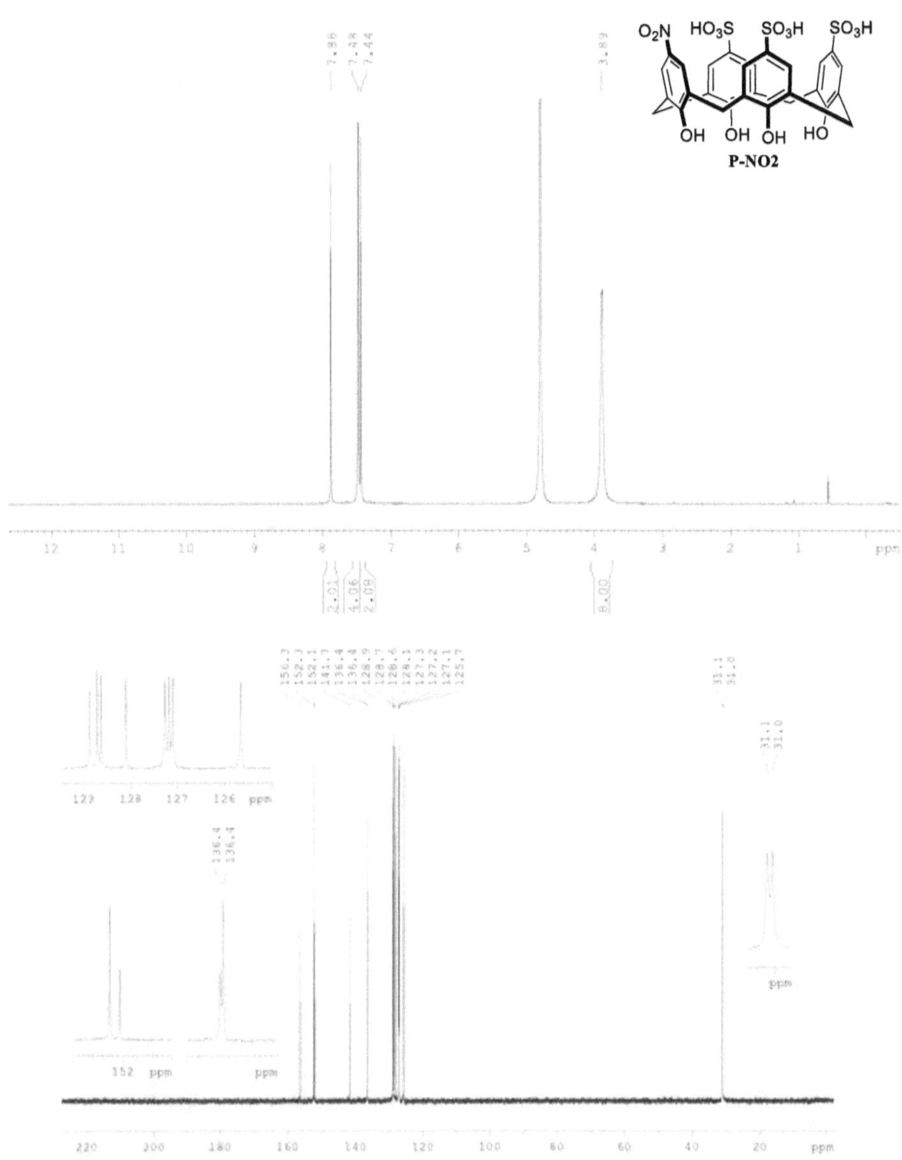
O2N
HO3S
SO3H
SO3H
OH
OH
OH
HO
P-NO2

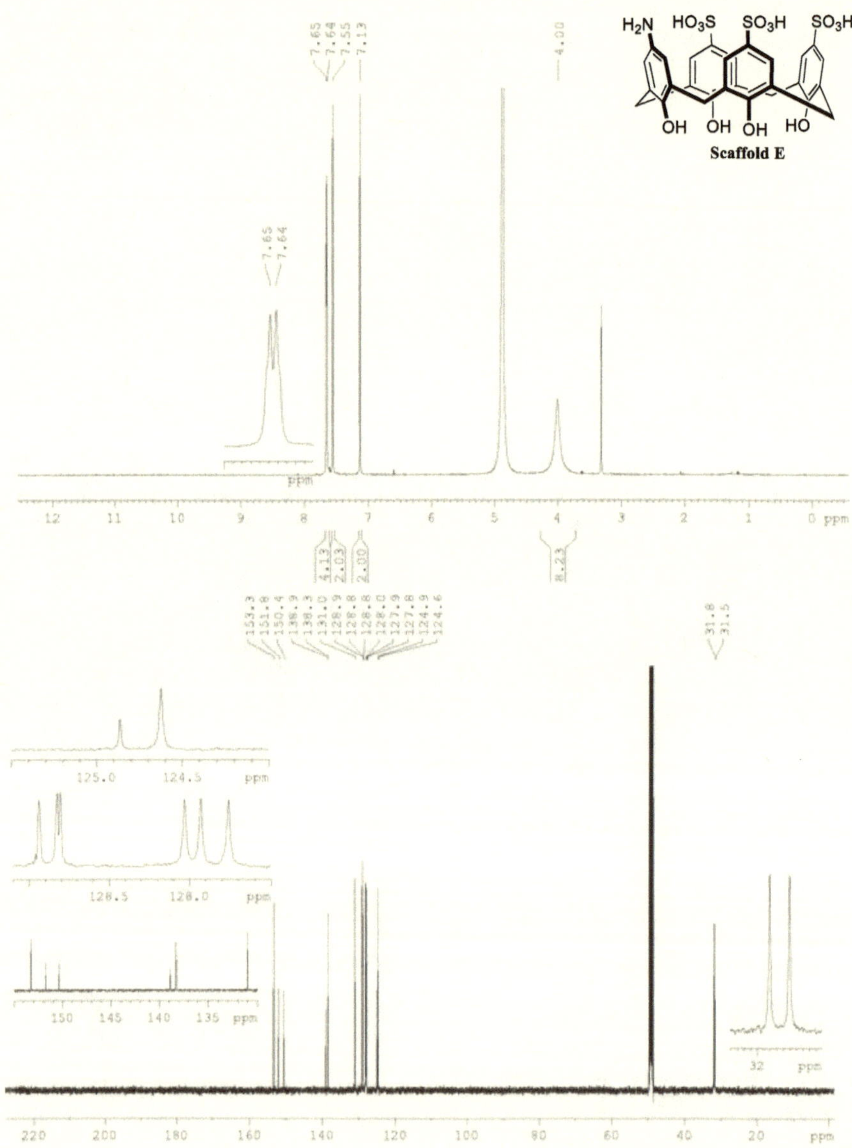

H₂N HO₃S SO₃H SO₃H
OH OH OH HO
Scaffold E

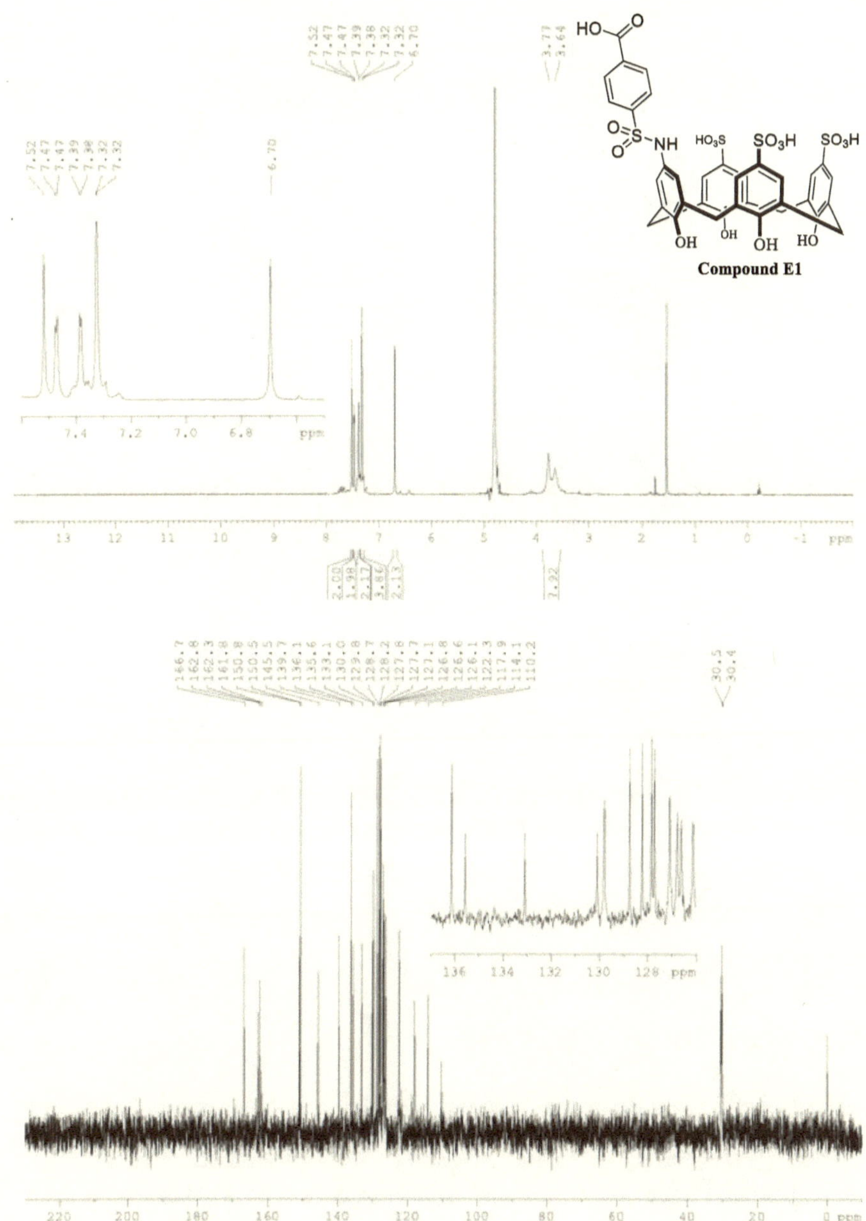

HO₃S   SO₃H   SO₃H
HO
O
OH
OH
OH
HO
Compound E1

S

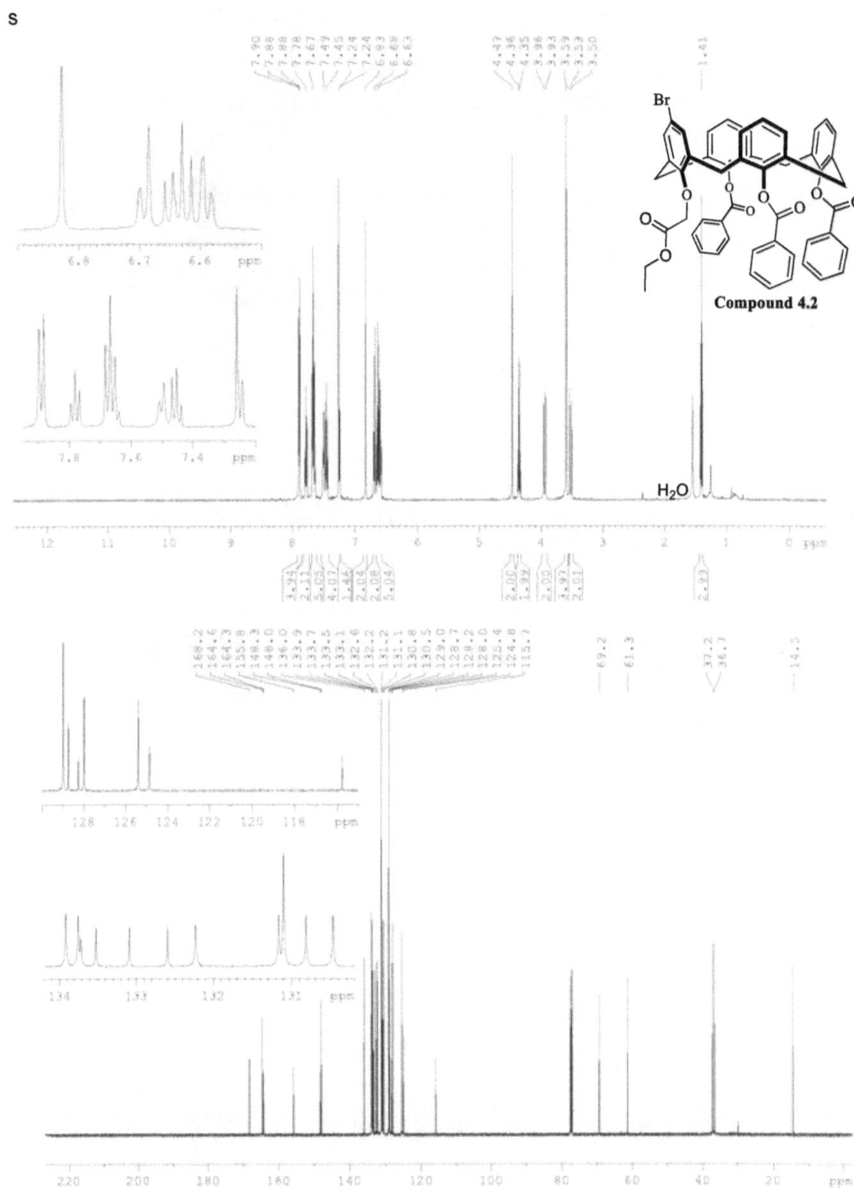

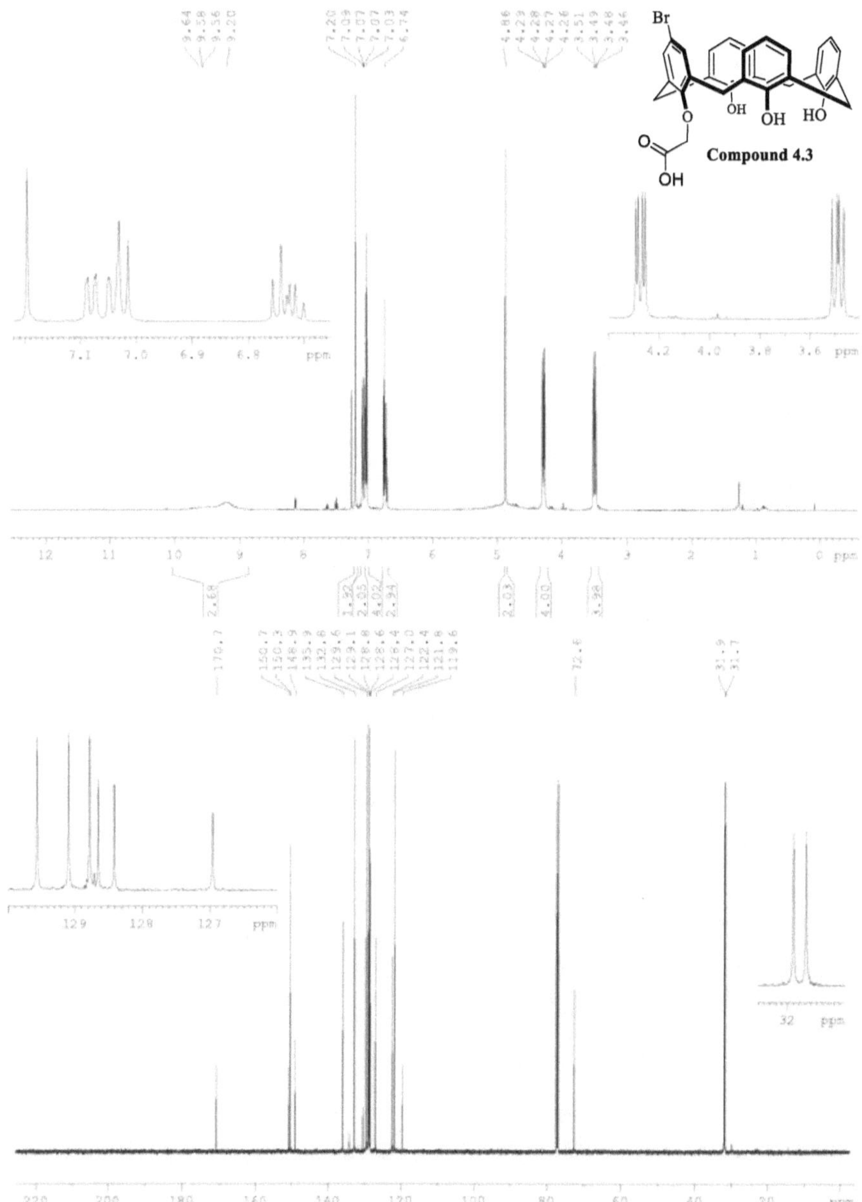
Compound 4.3

Br
HO₃S
SO₃H
SO₃H
OH
OH
HO
Scaffold F
HO
O
O

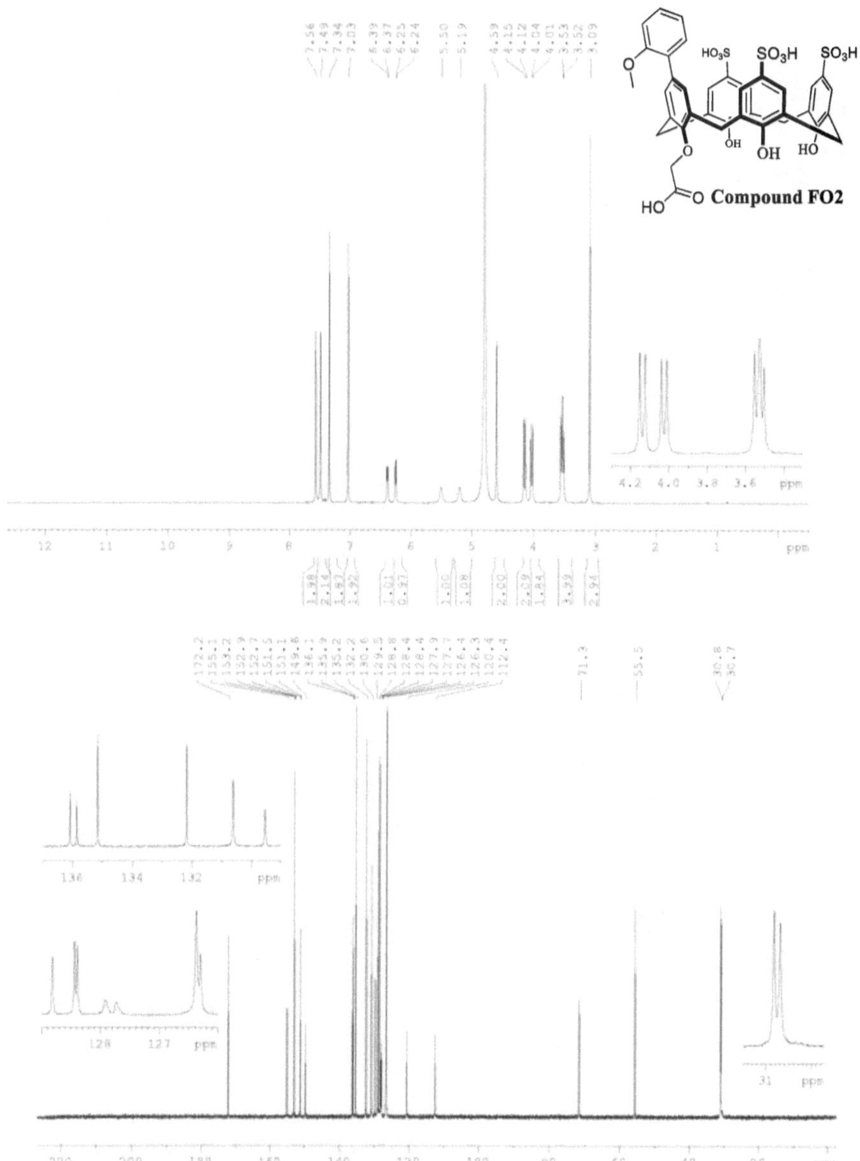

Compound FO2

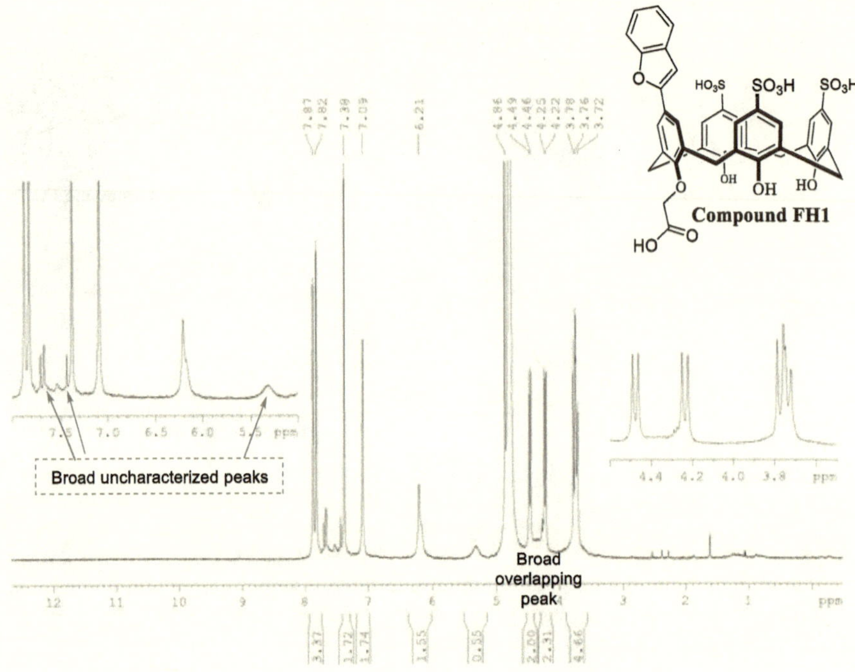

HO₃S   SO₃H   SO₃H
OH   OH   HO
Compound FH1
HO
O
7.87
7.82
7.38
7.05
6.21
4.86
4.49
4.46
4.25
4.22
3.78
3.76
3.72
Broad uncharacterized peaks
4.4 4.2 4.0 3.8 ppm
7.5 7.0 6.5 6.0 5.5 ppm
Broad overlapping peak
12 11 10 9 8 7 6 5 4 3 2 1 ppm
3.87
1.72
1.74
1.55
0.55
2.00
2.31
4.66

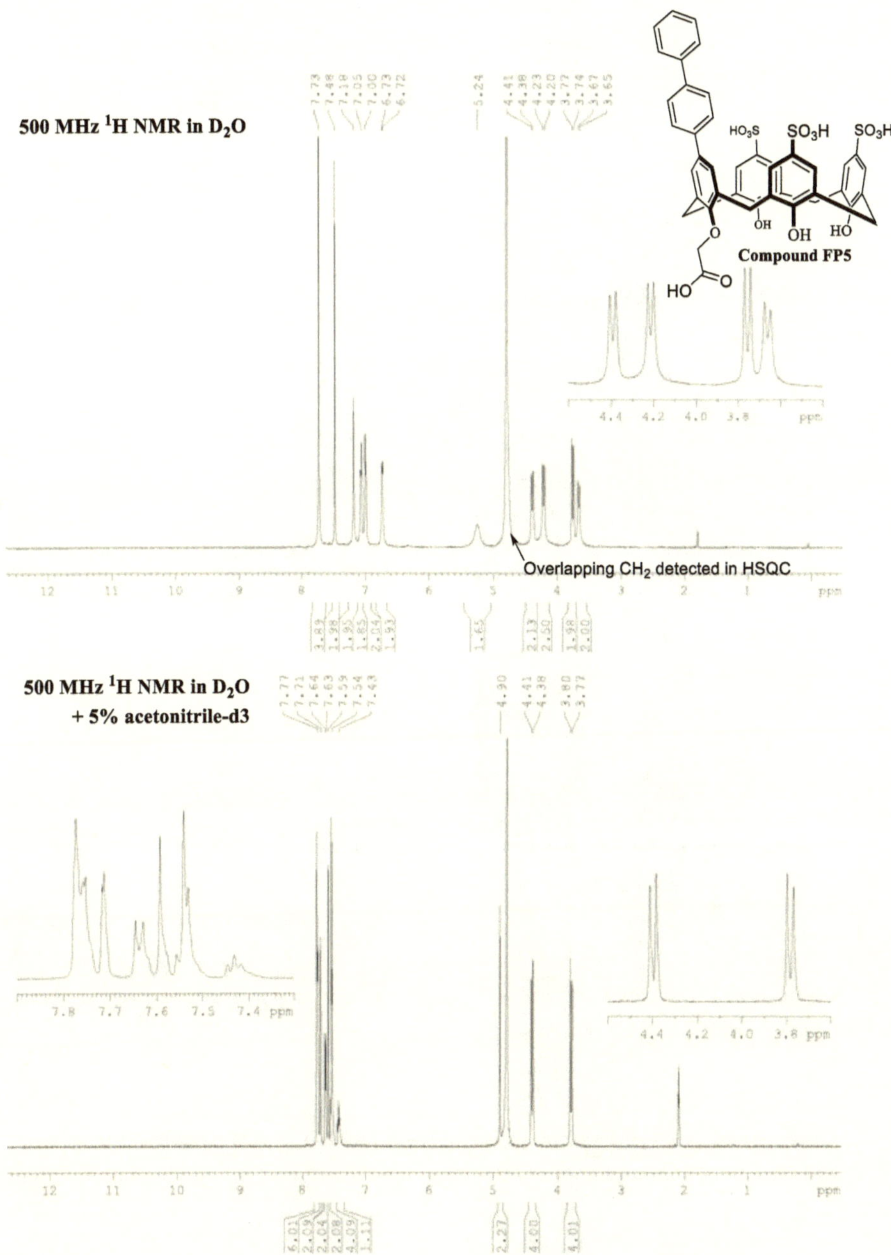
500 MHz ¹H NMR in D₂O
HO₃S
SO₃H
SO₃H
OH
OH
HO
HO
O
Compound FP5
Overlapping CH₂ detected in HSQC
500 MHz ¹H NMR in D₂O
+ 5% acetonitrile-d3

125 MHz ^{13}C NMR in D$_2$O

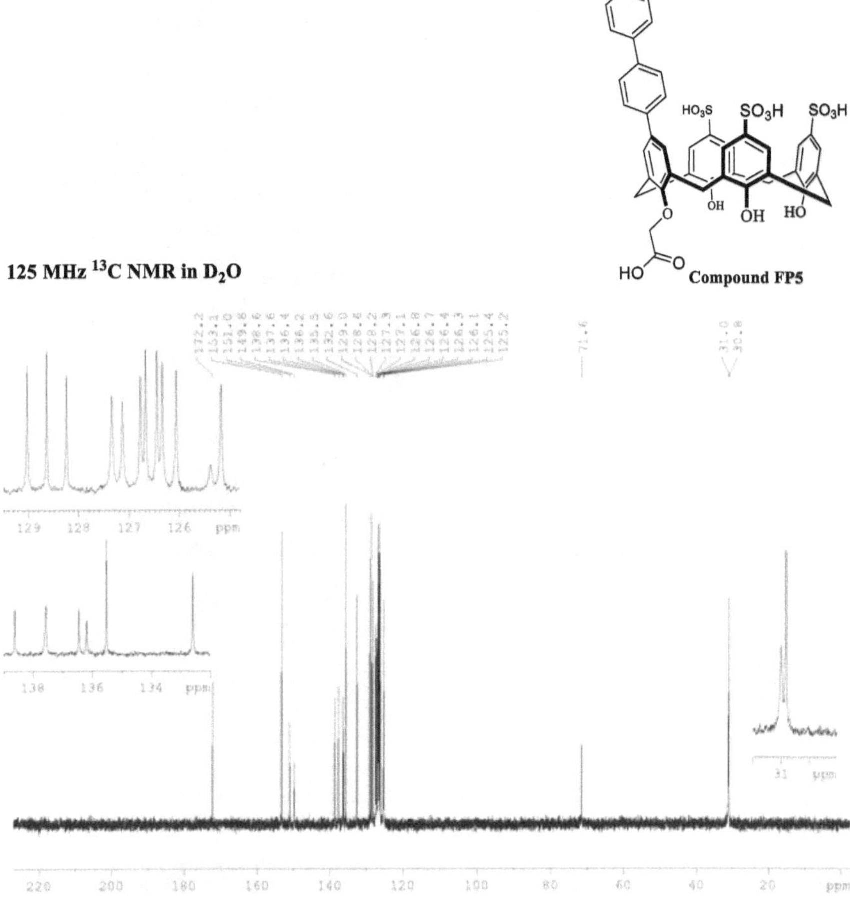

Appendix B – UPLC-MS of test peptides used in this work

Peptide - H3K4

Sequence (N to C terminus) **– ARTKQTAY** (N terminus = free amine, C terminus = amide)

Prominent mass spectrocscopy peaks – 937.58 (M+H⁺), 469.44 (M+2H⁺), 313.50 (M+3H⁺)

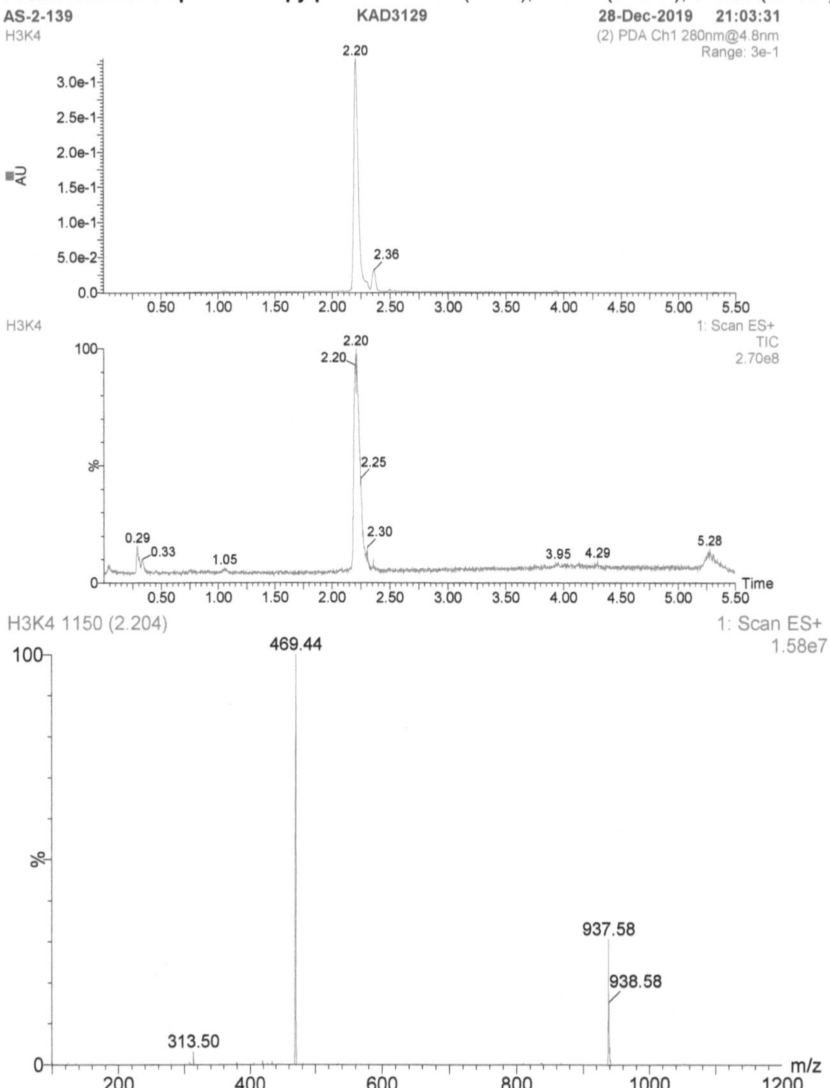

Peptide - H3K9me3

Sequence (N to C terminus) **– TAR(Kme3)STGY** (N terminus =acetyl, C terminus = amide)

Prominent mass spectrocscopy peaks – 1080.49 (M+TFA), 483.94 (M+H$^+$)

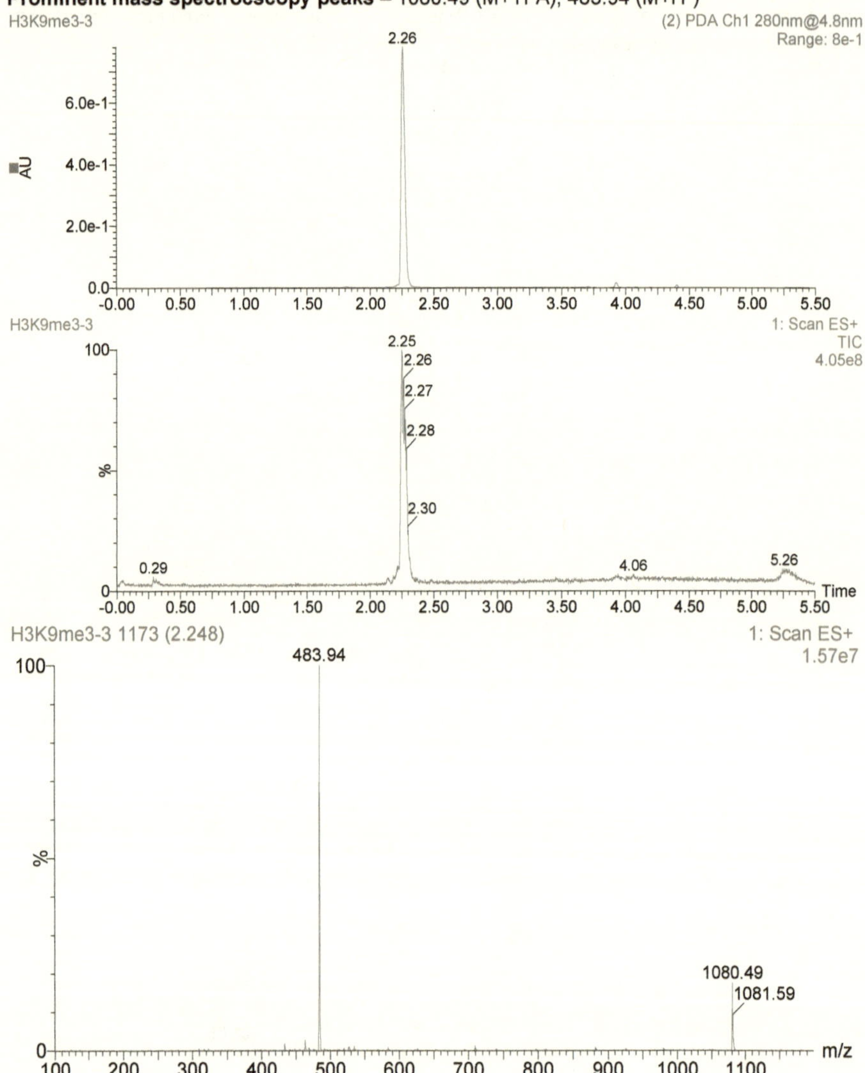

Peptide - H3K4ac

Sequence (N to C terminus) **– ART(Kac)QTAY** (N terminus = free amine, C terminus = amide)

Prominent mass spectrocscopy peaks – 979.58 (M+H⁺), 490.45 (M+2H⁺)

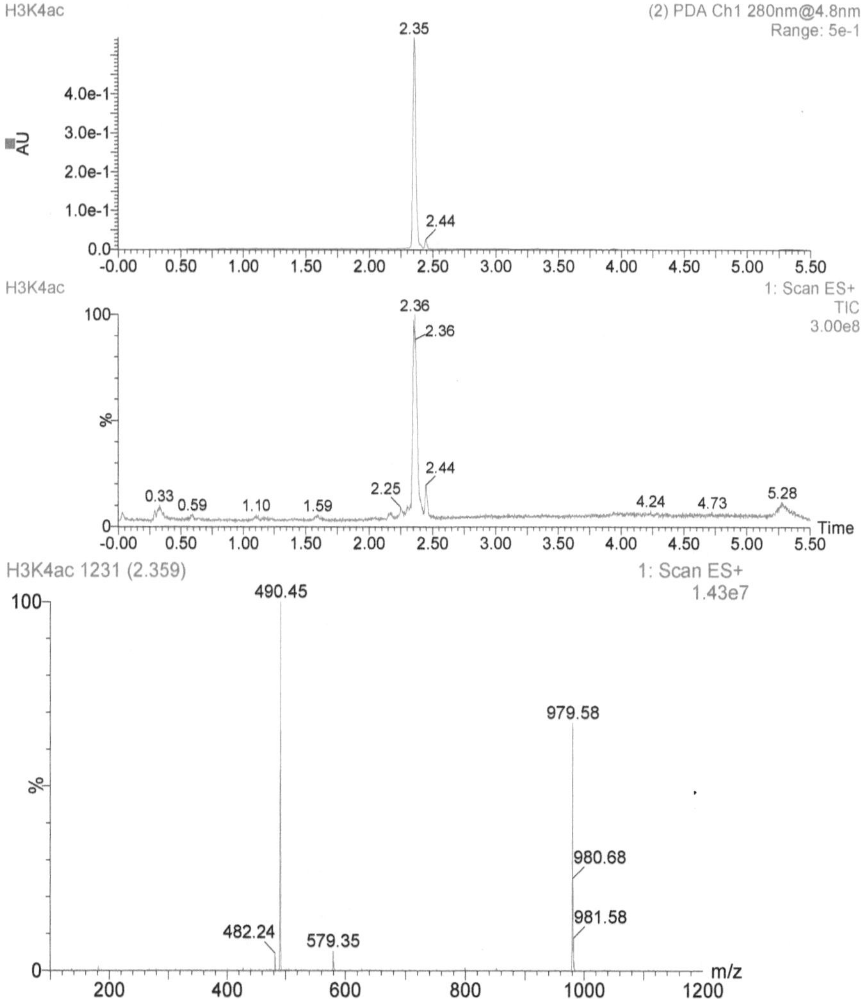

Peptide - H3K4me3

Sequence (N to C terminus)**– ART(Kme3)QTAY** (N terminus = free amine, C terminus =amide)

Prominent mass spectrocscopy peaks – 1093.59 (M+TFA), 490.45 (M+2H[+]), 327.50 (M+3H[+])

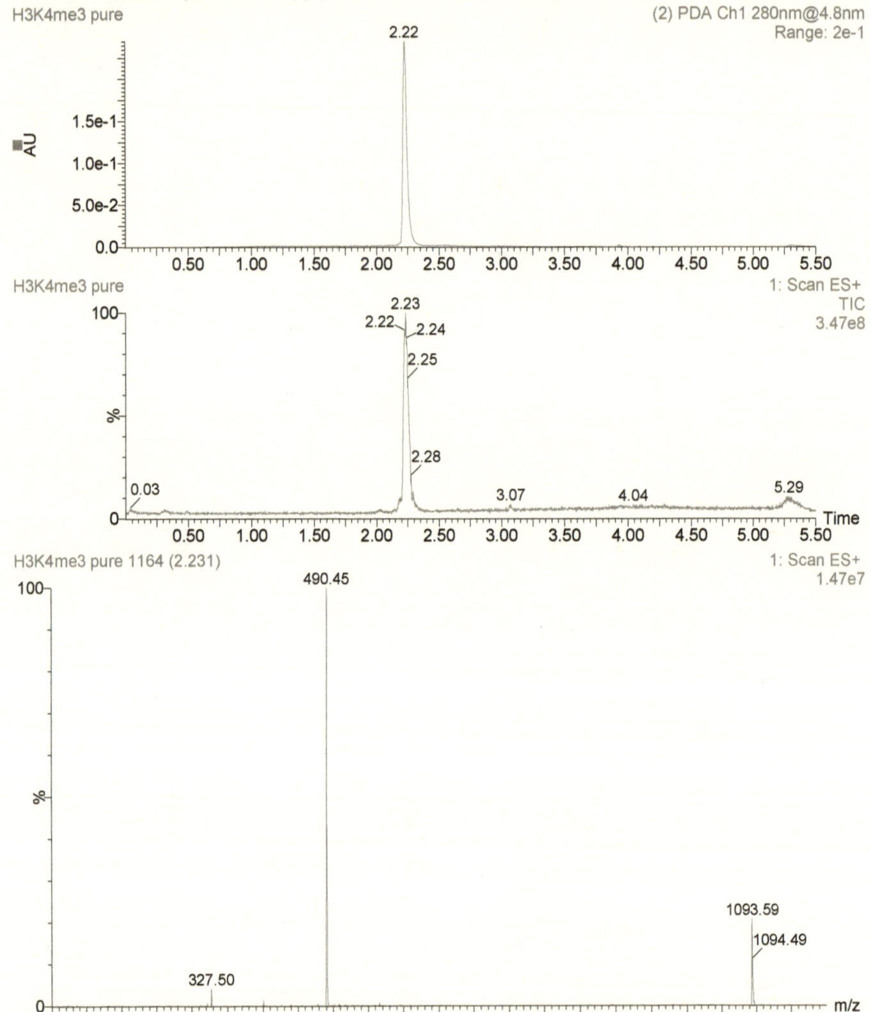

Peptide - H3K4me2

Sequence (N to C terminus) **– ART(Kme2)QTAY** (N terminus= free amine, C terminus= amide)

Prominent mass spectrocscopy peaks – 965.68 (M+H$^+$), 483.44 (M+2H$^+$)

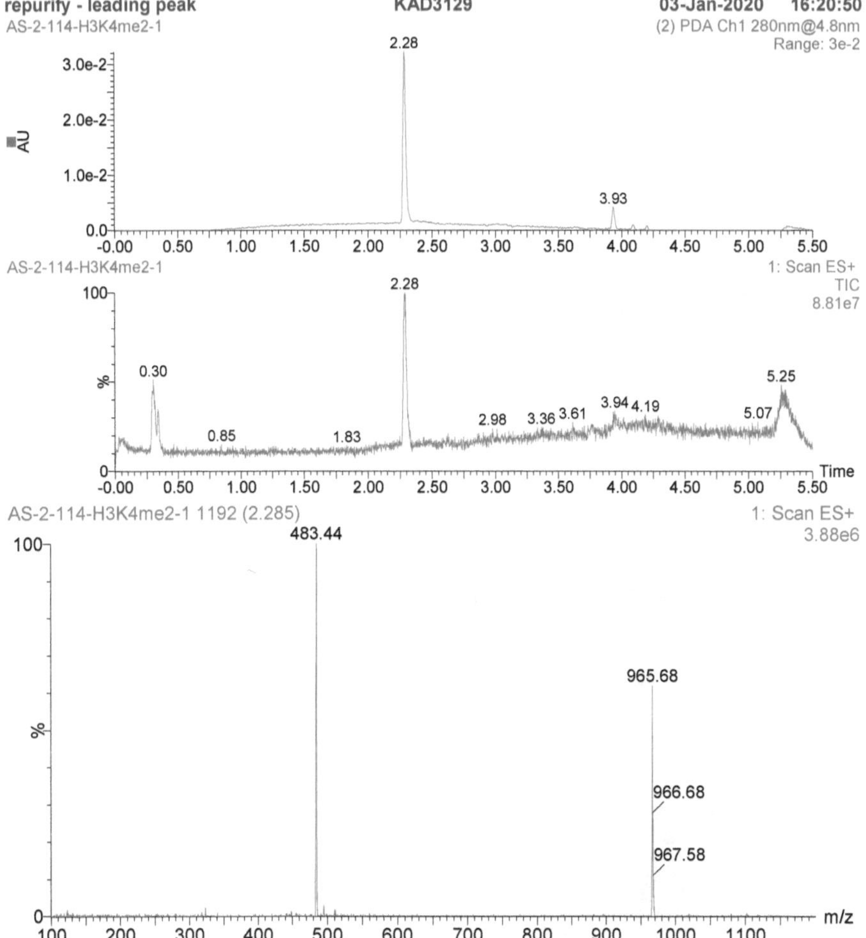

Peptide – K4me1

Sequence (N to C terminus) **– ART(Kme1)QTAY** (N terminus= free amine, C terminus= amide)
Prominent mass spectrocscopy peaks – 951.67 (M+H$^+$), 476.34 (M+2H$^+$), 318.20 (M+3H$^+$),
1065.38 (M+TFA+H$^+$)

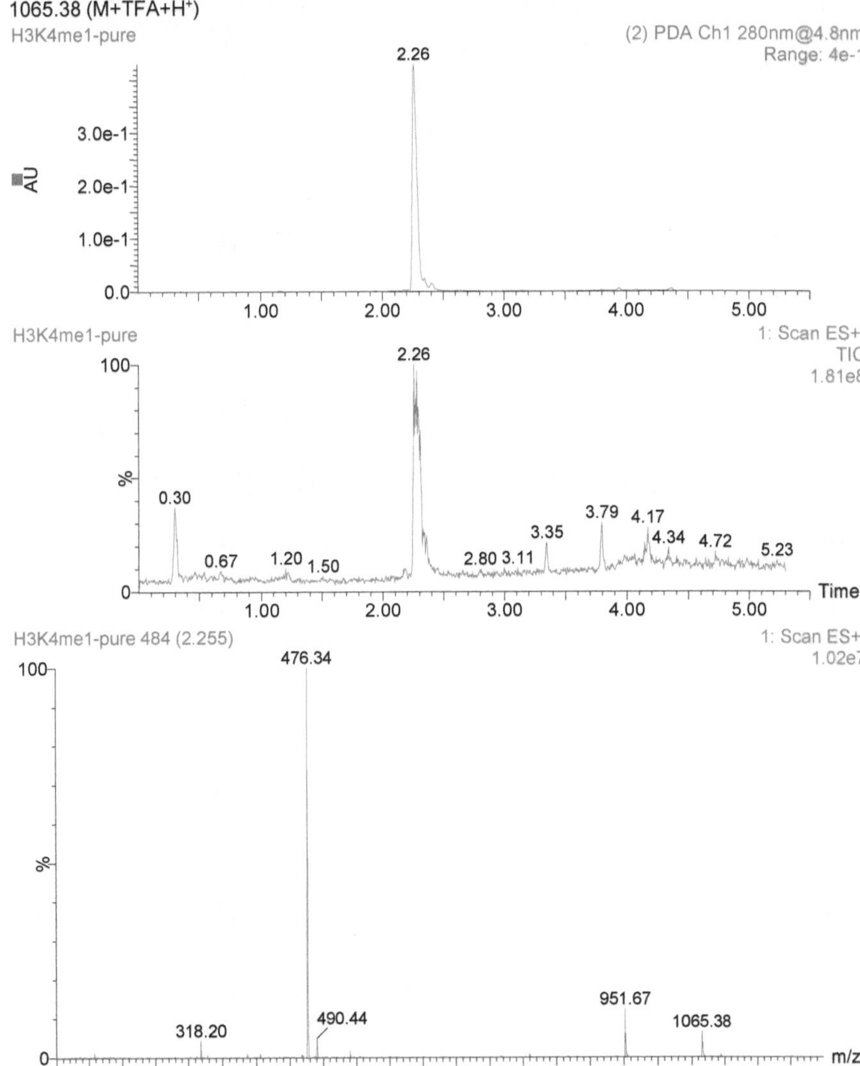

Peptide - H3R2me2a

Sequence(N to C terminus)**– A(Rme2a)TKQTAY** (N terminus= free amine, C terminus= amide)

Prominent mass spectrocscopy peaks – 965.67 (M+H⁺), 483.44 (M+2H⁺)

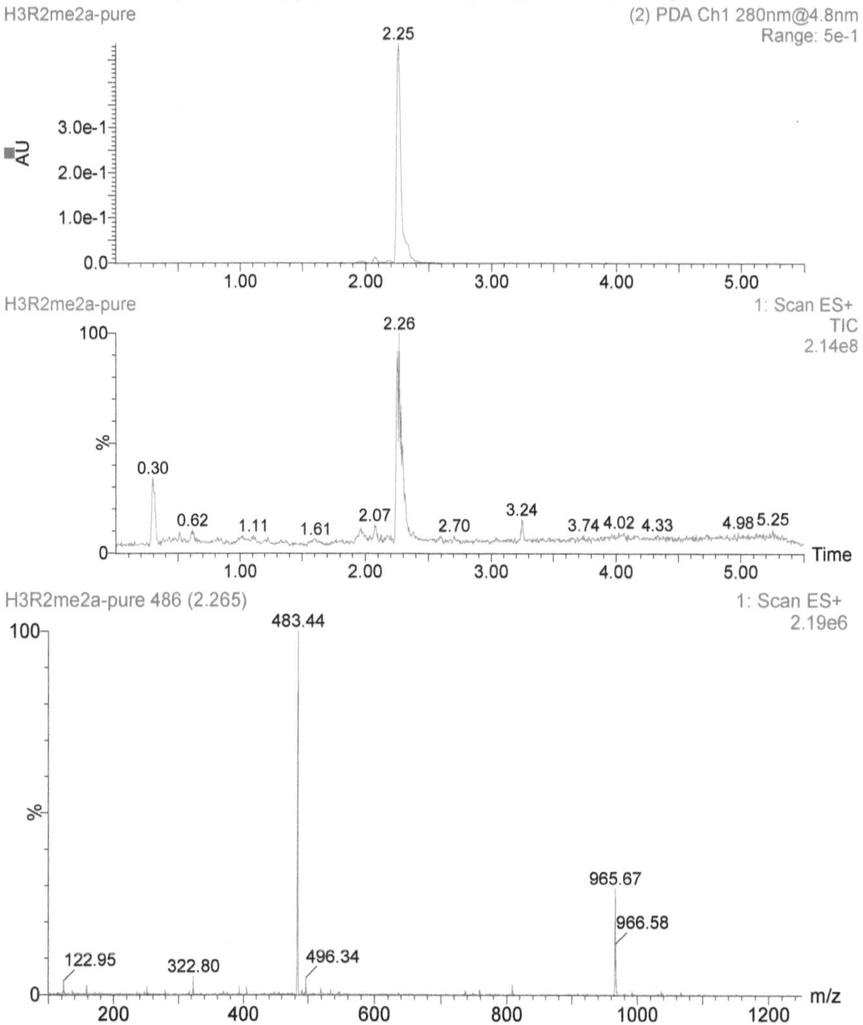

Peptide - H3R2me2s

Sequence(N to C terminus)– **A(Rme2s)TKQTAY** (N terminus= free amine, C terminus= amide)

Prominent mass spectrocscopy peaks – 965.57 (M+H⁺), 483.54 (M+2H⁺)

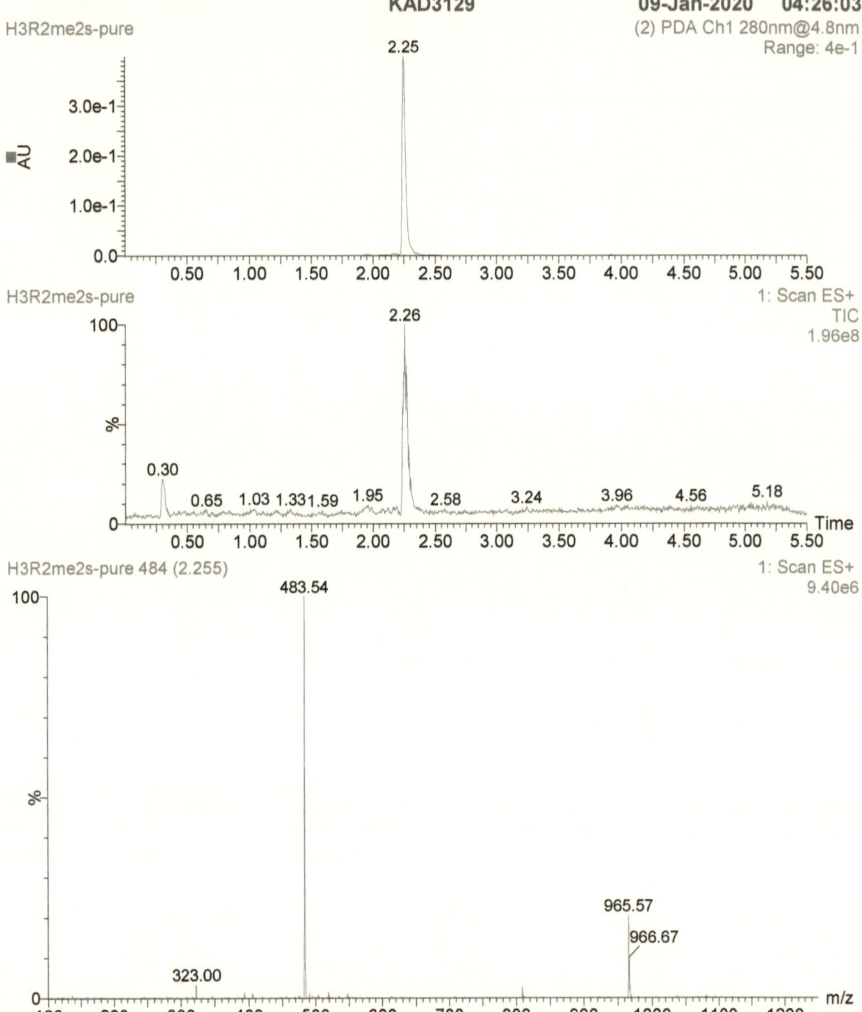

Appendix C – Plots for IDA of all compounds in Chapter 2

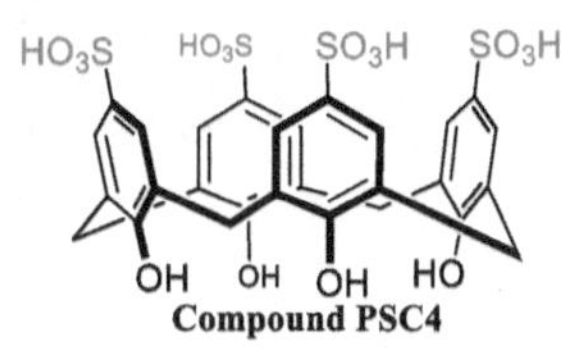

Compound PSC4

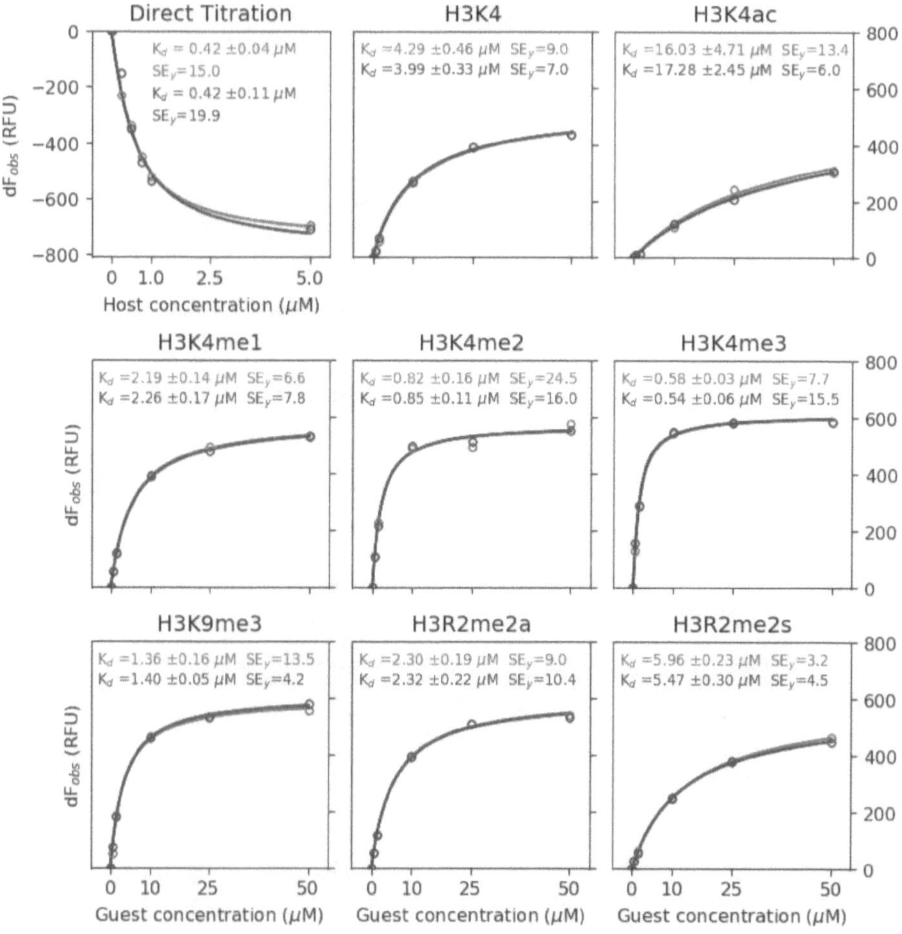

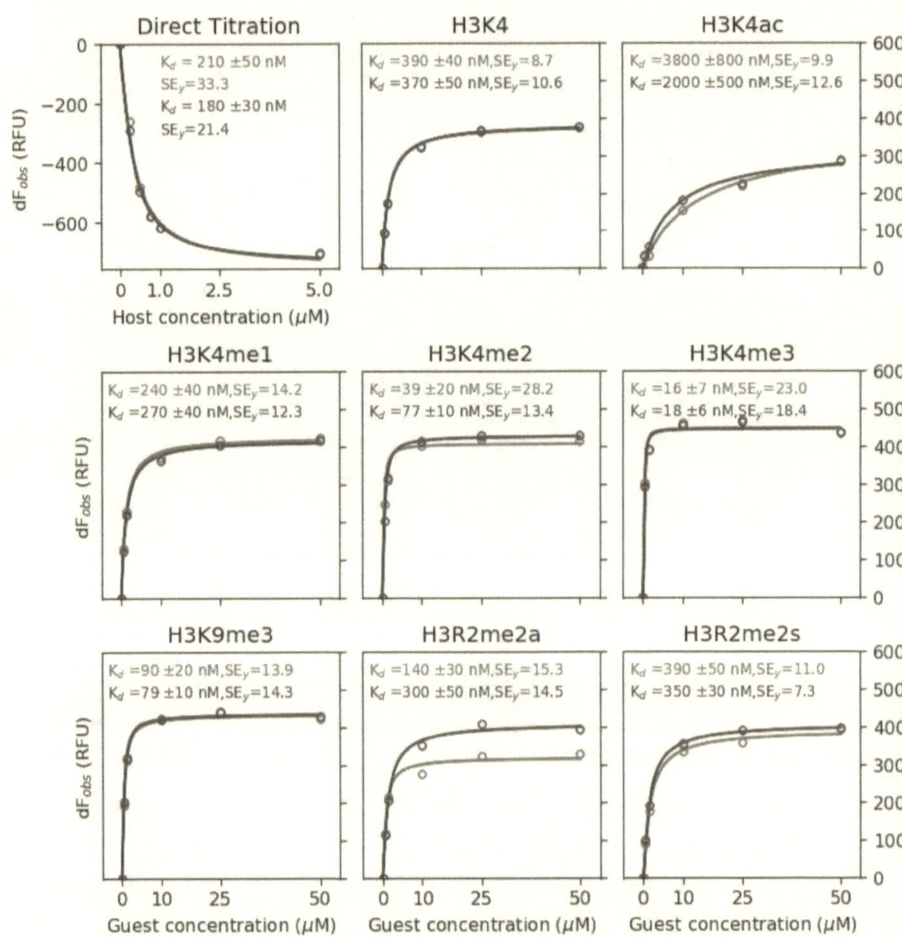
HO3S
SO3H SO3H
SO3H SO3H
SO3H
OH OH OH OH OH HO
Compound PSC6

Direct Titration
K_d = 210 ±50 nM
SE_y=33.3
K_d = 180 ±30 nM
SE_y=21.4
dF_obs (RFU)
0
-200
-400
-600
0 1.0 2.5 5.0
Host concentration (μM)

H3K4
K_d =390 ±40 nM,SE_y=8.7
K_d =370 ±50 nM,SE_y=10.6

H3K4ac
K_d =3800 ±800 nM,SE_y=9.9
K_d =2000 ±500 nM,SE_y=12.6
600
500
400
300
200
100
0

H3K4me1
K_d =240 ±40 nM,SE_y=14.2
K_d =270 ±40 nM,SE_y=12.3
dF_obs (RFU)

H3K4me2
K_d =39 ±20 nM,SE_y=28.2
K_d =77 ±10 nM,SE_y=13.4

H3K4me3
K_d =16 ±7 nM,SE_y=23.0
K_d =18 ±6 nM,SE_y=18.4
600
500
400
300
200
100
0

H3K9me3
K_d =90 ±20 nM,SE_y=13.9
K_d =79 ±10 nM,SE_y=14.3
dF_obs (RFU)

H3R2me2a
K_d =140 ±30 nM,SE_y=15.3
K_d =300 ±50 nM,SE_y=14.5

H3R2me2s
K_d =390 ±50 nM,SE_y=11.0
K_d =350 ±30 nM,SE_y=7.3
600
500
400
300
200
100
0

0 10 25 50
Guest concentration (μM)
0 10 25 50
Guest concentration (μM)
0 10 25 50
Guest concentration (μM)

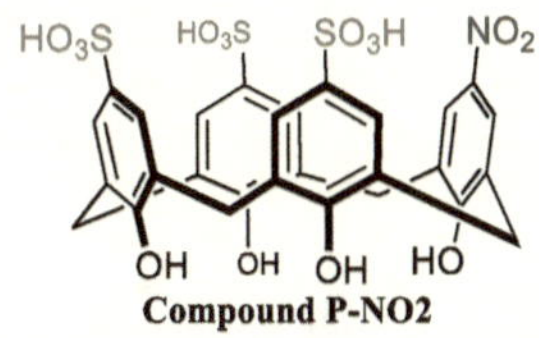

HO3S
HO3S
SO3H
NO2
OH
OH
OH
HO
Compound P-NO2

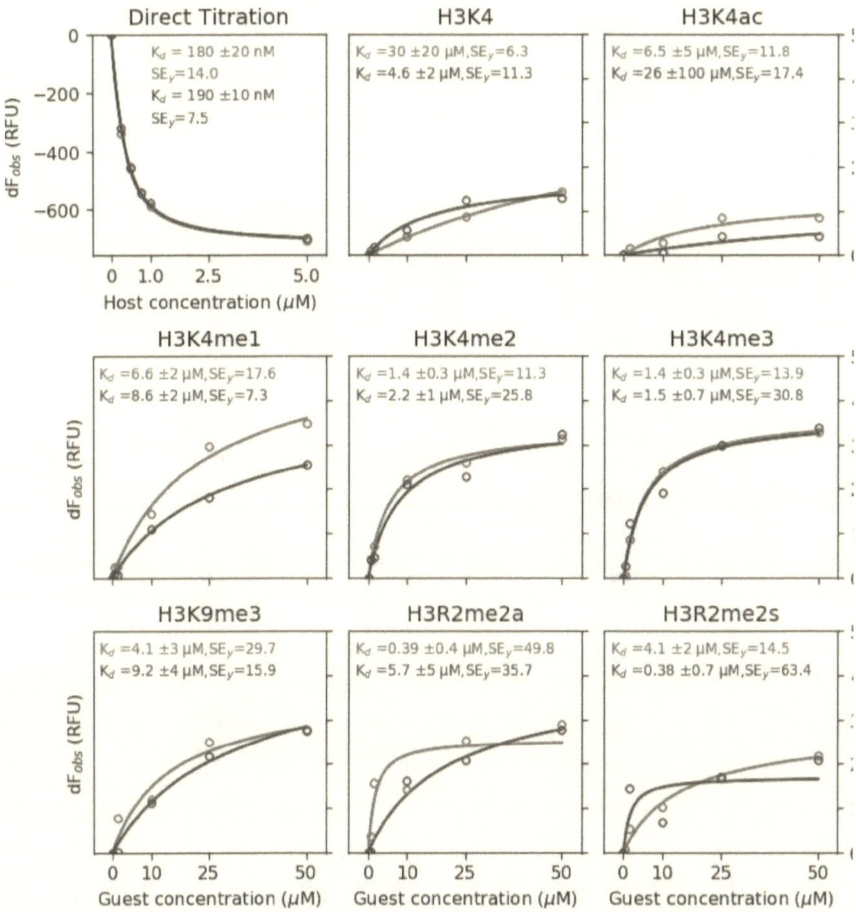

Direct Titration
Kd = 180 ±20 nM
SEy=14.0
Kd = 190 ±10 nM
SEy=7.5
dFobs (RFU)
0
−200
−400
−600
0   1.0   2.5   5.0
Host concentration (µM)

H3K4
Kd =30 ±20 µM,SEy=6.3
Kd =4.6 ±2 µM,SEy=11.3

H3K4ac
Kd =6.5 ±5 µM,SEy=11.8
Kd =26 ±100 µM,SEy=17.4

H3K4me1
Kd =6.6 ±2 µM,SEy=17.6
Kd =8.6 ±2 µM,SEy=7.3
dFobs (RFU)

H3K4me2
Kd =1.4 ±0.3 µM,SEy=11.3
Kd =2.2 ±1 µM,SEy=25.8

H3K4me3
Kd =1.4 ±0.3 µM,SEy=13.9
Kd =1.5 ±0.7 µM,SEy=30.8

H3K9me3
Kd =4.1 ±3 µM,SEy=29.7
Kd =9.2 ±4 µM,SEy=15.9
dFobs (RFU)
0   10   25   50
Guest concentration (µM)

H3R2me2a
Kd =0.39 ±0.4 µM,SEy=49.8
Kd =5.7 ±5 µM,SEy=35.7
0   10   25   50
Guest concentration (µM)

H3R2me2s
Kd =4.1 ±2 µM,SEy=14.5
Kd =0.38 ±0.7 µM,SEy=63.4
0   10   25   50
Guest concentration (µM)

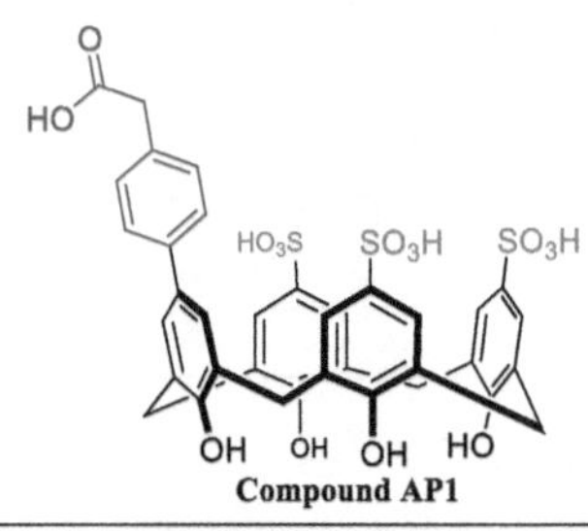

Direct Titration

K_d = 0.15 ±0.03 μM
SE_y=8.2
K_d = 0.14 ±0.04 μM
SE_y=10.8

dF_{obs} (RFU) vs Host concentration (μM)

H3K4

K_d =3.3 ±0.4 μM, SE_y=6.4
K_d =3.1 ±0.4 μM, SE_y=6.8

H3K4ac

K_d =8.9 ±1 μM, SE_y=4.5
K_d =8.9 ±1 μM, SE_y=5.2

H3K4me1

K_d =1.1 ±0.09 μM, SE_y=7.4
K_d =0.96 ±0.3 μM, SE_y=32.0

H3K4me2

K_d =0.27 ±0.03 μM, SE_y=12.7
K_d =0.27 ±0.03 μM, SE_y=10.6

H3K4me3

K_d =0.10 ±0.02 μM, SE_y=20.2
K_d =0.11 ±0.02 μM, SE_y=16.1

H3K9me3

K_d =0.29 ±0.02 μM, SE_y=8.6
K_d =0.26 ±0.04 μM, SE_y=17.2

H3R2me2a

K_d =0.44 ±0.05 μM, SE_y=12.1
K_d =0.45 ±0.05 μM, SE_y=11.1

H3R2me2s

K_d =1.6 ±0.1 μM, SE_y=5.6
K_d =1.4 ±0.1 μM, SE_y=7.6

dF_{obs} (RFU) vs Guest concentration (μM)

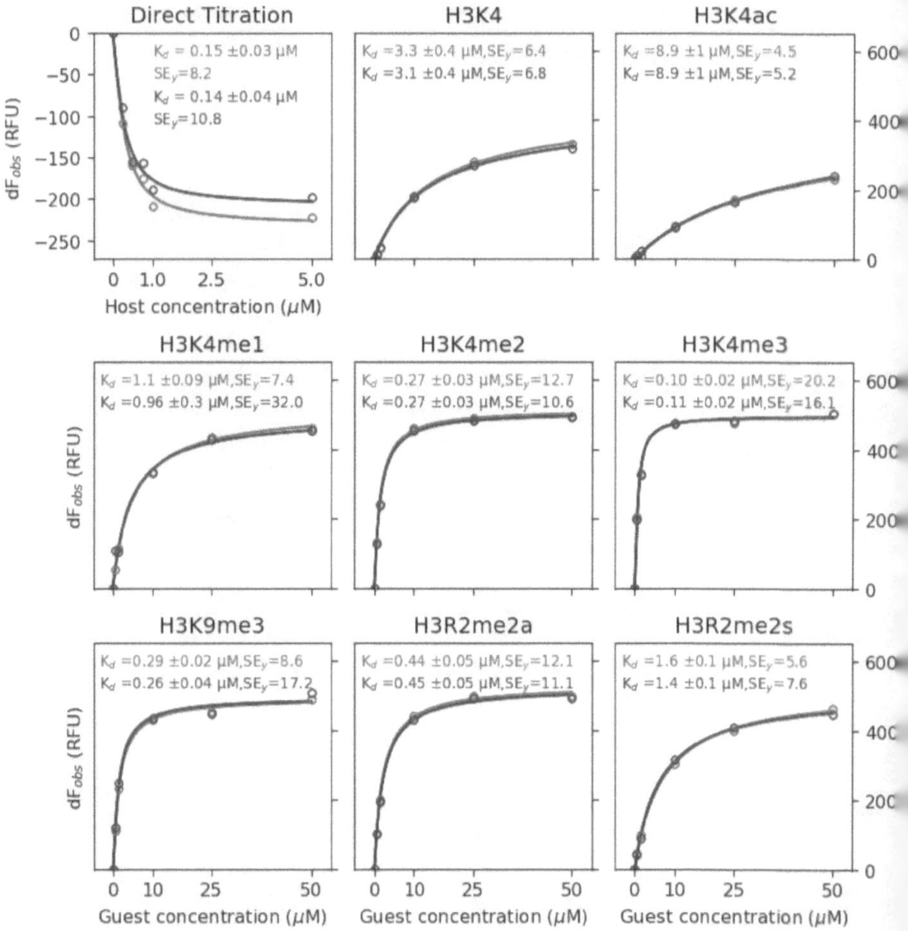

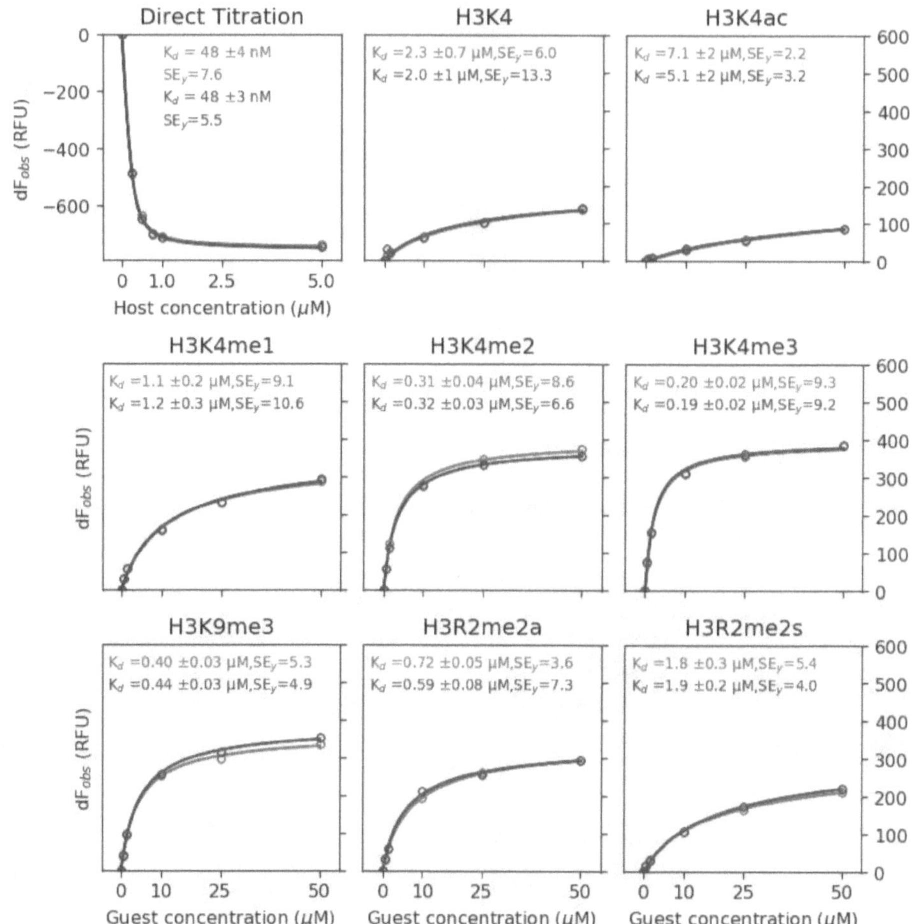
Compound AP3

Direct Titration
K_d = 48 ±4 nM
SE_y=7.6
K_d = 48 ±3 nM
SE_y=5.5
dF_obs (RFU)
Host concentration (μM)

H3K4
K_d =2.3 ±0.7 μM,SE_y=6.0
K_d =2.0 ±1 μM,SE_y=13.3

H3K4ac
K_d =7.1 ±2 μM,SE_y=2.2
K_d =5.1 ±2 μM,SE_y=3.2

H3K4me1
K_d =1.1 ±0.2 μM,SE_y=9.1
K_d =1.2 ±0.3 μM,SE_y=10.6
dF_obs (RFU)

H3K4me2
K_d =0.31 ±0.04 μM,SE_y=8.6
K_d =0.32 ±0.03 μM,SE_y=6.6

H3K4me3
K_d =0.20 ±0.02 μM,SE_y=9.3
K_d =0.19 ±0.02 μM,SE_y=9.2

H3K9me3
K_d =0.40 ±0.03 μM,SE_y=5.3
K_d =0.44 ±0.03 μM,SE_y=4.9
dF_obs (RFU)

H3R2me2a
K_d =0.72 ±0.05 μM,SE_y=3.6
K_d =0.59 ±0.08 μM,SE_y=7.3

H3R2me2s
K_d =1.8 ±0.3 μM,SE_y=5.4
K_d =1.9 ±0.2 μM,SE_y=4.0

Guest concentration (μM)
Guest concentration (μM)
Guest concentration (μM)

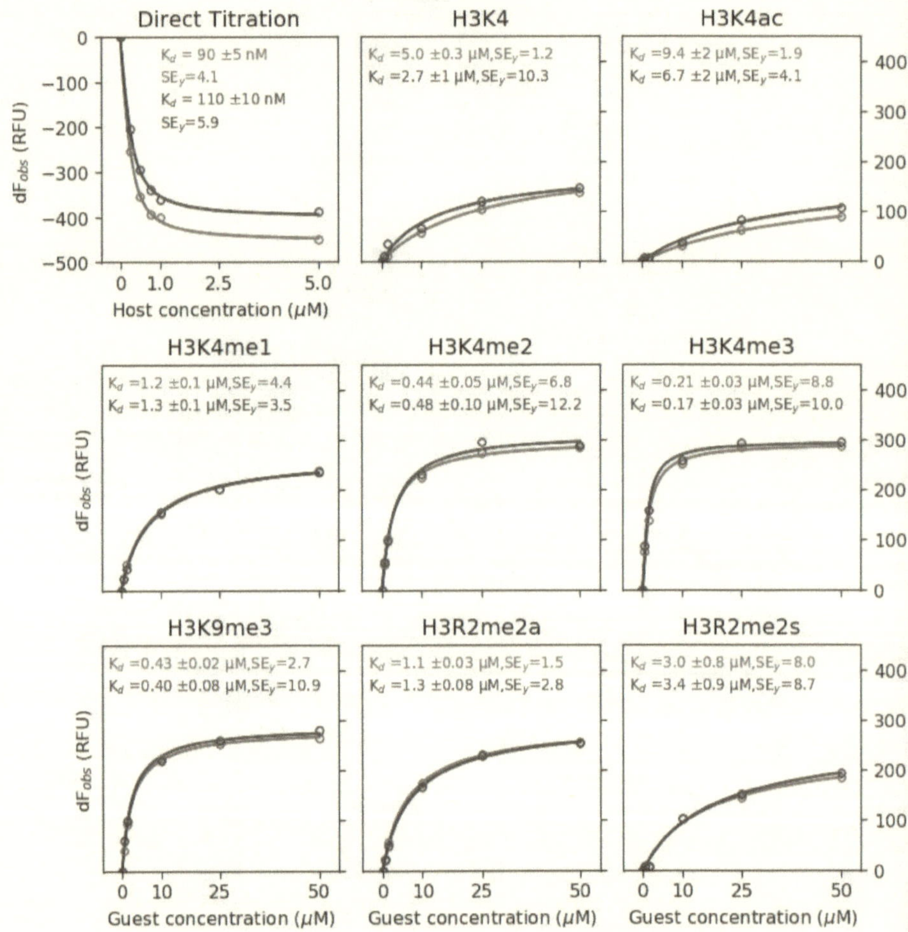
Compound AP4
Direct Titration
K_d = 90 ±5 nM
SE_y=4.1
K_d = 110 ±10 nM
SE_y=5.9
dF_obs (RFU)
Host concentration (μM)
H3K4
K_d =5.0 ±0.3 μM,SE_y=1.2
K_d =2.7 ±1 μM,SE_y=10.3
H3K4ac
K_d =9.4 ±2 μM,SE_y=1.9
K_d =6.7 ±2 μM,SE_y=4.1
H3K4me1
K_d =1.2 ±0.1 μM,SE_y=4.4
K_d =1.3 ±0.1 μM,SE_y=3.5
H3K4me2
K_d =0.44 ±0.05 μM,SE_y=6.8
K_d =0.48 ±0.10 μM,SE_y=12.2
H3K4me3
K_d =0.21 ±0.03 μM,SE_y=8.8
K_d =0.17 ±0.03 μM,SE_y=10.0
dF_obs (RFU)
H3K9me3
K_d =0.43 ±0.02 μM,SE_y=2.7
K_d =0.40 ±0.08 μM,SE_y=10.9
H3R2me2a
K_d =1.1 ±0.03 μM,SE_y=1.5
K_d =1.3 ±0.08 μM,SE_y=2.8
H3R2me2s
K_d =3.0 ±0.8 μM,SE_y=8.0
K_d =3.4 ±0.9 μM,SE_y=8.7
dF_obs (RFU)
Guest concentration (μM)
Guest concentration (μM)
Guest concentration (μM)

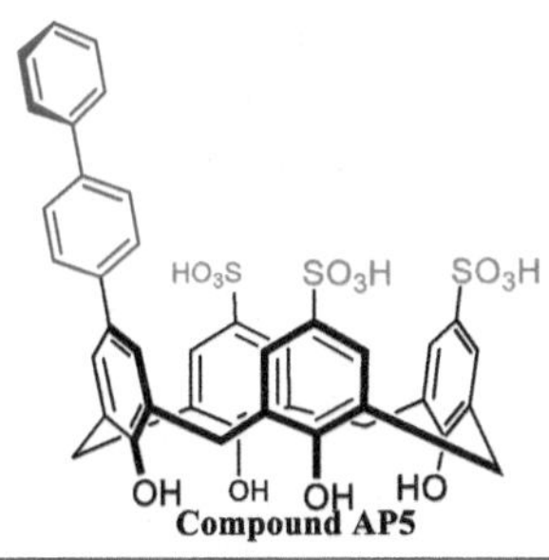

HO3S
SO3H
SO3H
OH
OH
OH
HO
Compound AP5

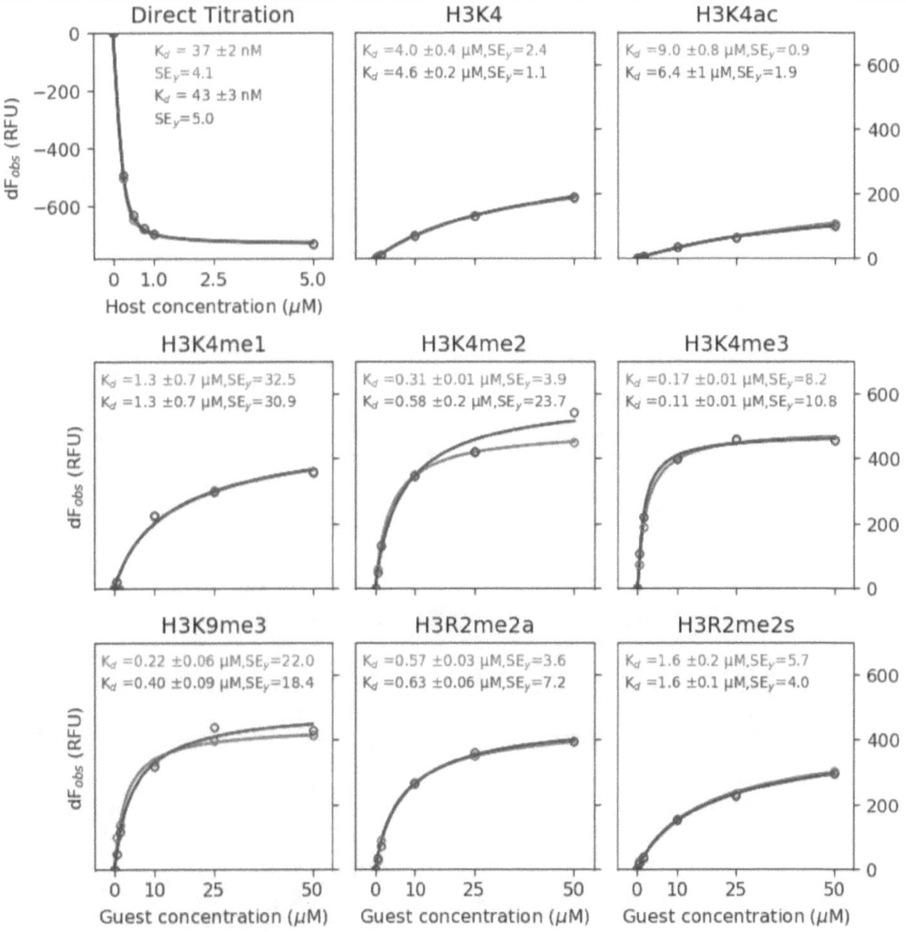

Direct Titration
K_d = 37 ±2 nM
SE_y=4.1
K_d = 43 ±3 nM
SE_y=5.0
dF$_{obs}$ (RFU)
Host concentration (μM)

H3K4
K_d =4.0 ±0.4 μM,SE_y=2.4
K_d =4.6 ±0.2 μM,SE_y=1.1

H3K4ac
K_d =9.0 ±0.8 μM,SE_y=0.9
K_d =6.4 ±1 μM,SE_y=1.9

H3K4me1
K_d =1.3 ±0.7 μM,SE_y=32.5
K_d =1.3 ±0.7 μM,SE_y=30.9

H3K4me2
K_d =0.31 ±0.01 μM,SE_y=3.9
K_d =0.58 ±0.2 μM,SE_y=23.7

H3K4me3
K_d =0.17 ±0.01 μM,SE_y=8.2
K_d =0.11 ±0.01 μM,SE_y=10.8

H3K9me3
K_d =0.22 ±0.06 μM,SE_y=22.0
K_d =0.40 ±0.09 μM,SE_y=18.4

H3R2me2a
K_d =0.57 ±0.03 μM,SE_y=3.6
K_d =0.63 ±0.06 μM,SE_y=7.2

H3R2me2s
K_d =1.6 ±0.2 μM,SE_y=5.7
K_d =1.6 ±0.1 μM,SE_y=4.0

dF$_{obs}$ (RFU)
Guest concentration (μM)
Guest concentration (μM)
Guest concentration (μM)

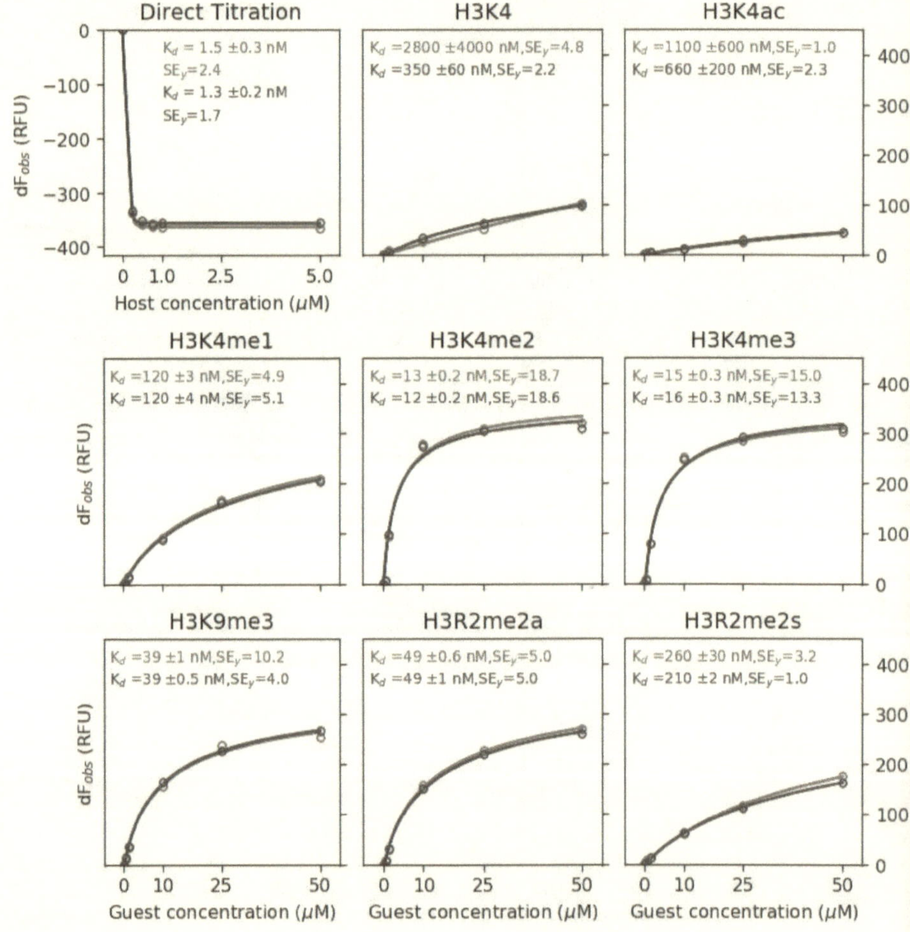

Compound AP6

Direct Titration
K_d = 1.5 ±0.3 nM
SE_y=2.4
K_d = 1.3 ±0.2 nM
SE_y=1.7
dF$_{obs}$ (RFU)
Host concentration (μM)

H3K4
K_d =2800 ±4000 nM,SE_y=4.8
K_d =350 ±60 nM,SE_y=2.2

H3K4ac
K_d =1100 ±600 nM,SE_y=1.0
K_d =660 ±200 nM,SE_y=2.3

H3K4me1
K_d =120 ±3 nM,SE_y=4.9
K_d =120 ±4 nM,SE_y=5.1
dF$_{obs}$ (RFU)

H3K4me2
K_d =13 ±0.2 nM,SE_y=18.7
K_d =12 ±0.2 nM,SE_y=18.6

H3K4me3
K_d =15 ±0.3 nM,SE_y=15.0
K_d =16 ±0.3 nM,SE_y=13.3

H3K9me3
K_d =39 ±1 nM,SE_y=10.2
K_d =39 ±0.5 nM,SE_y=4.0
dF$_{obs}$ (RFU)

H3R2me2a
K_d =49 ±0.6 nM,SE_y=5.0
K_d =49 ±1 nM,SE_y=5.0

H3R2me2s
K_d =260 ±30 nM,SE_y=3.2
K_d =210 ±2 nM,SE_y=1.0

Guest concentration (μM)
Guest concentration (μM)
Guest concentration (μM)

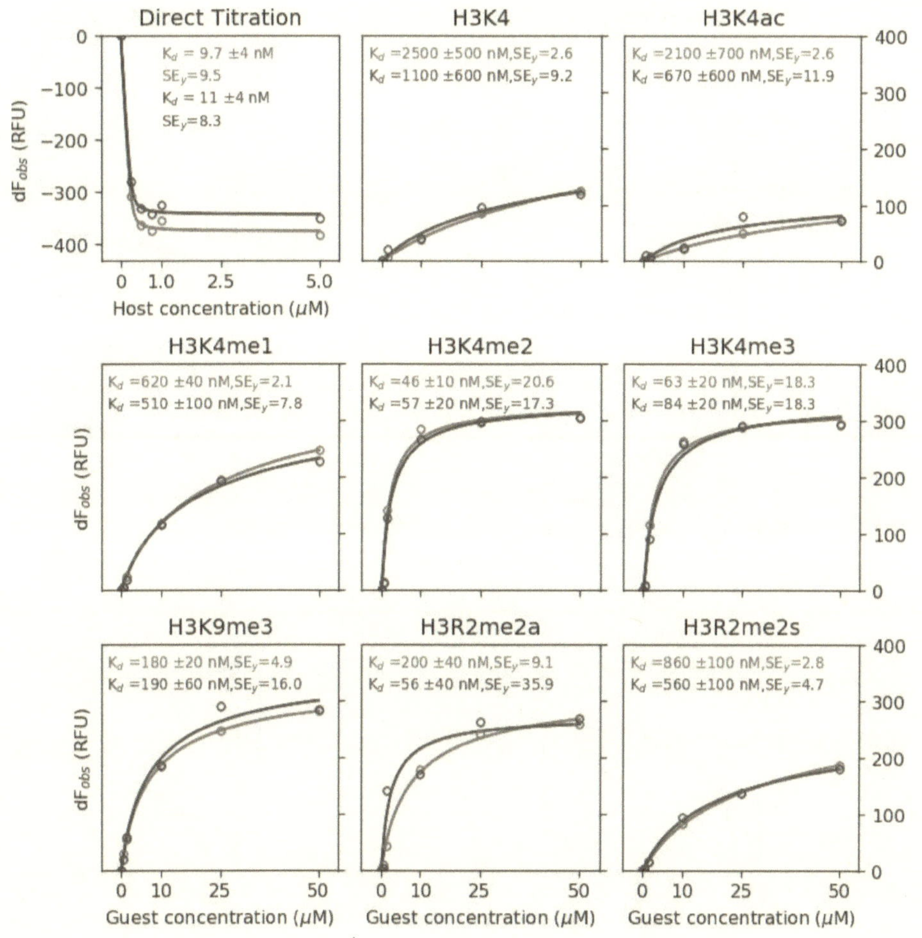
H2N
H
O
O
OH
HO3S
SO3H
SO3H
OH
OH
OH
HO
Compound AP7

Direct Titration
Kd = 9.7 ±4 nM
SEy=9.5
Kd = 11 ±4 nM
SEy=8.3
dFobs (RFU)
0
-100
-200
-300
-400
0   1.0   2.5   5.0
Host concentration (μM)

H3K4
Kd =2500 ±500 nM,SEy=2.6
Kd =1100 ±600 nM,SEy=9.2

H3K4ac
Kd =2100 ±700 nM,SEy=2.6
Kd =670 ±600 nM,SEy=11.9
400
300
200
100
0

H3K4me1
Kd =620 ±40 nM,SEy=2.1
Kd =510 ±100 nM,SEy=7.8
dFobs (RFU)

H3K4me2
Kd =46 ±10 nM,SEy=20.6
Kd =57 ±20 nM,SEy=17.3

H3K4me3
Kd =63 ±20 nM,SEy=18.3
Kd =84 ±20 nM,SEy=18.3
400
300
200
100
0

H3K9me3
Kd =180 ±20 nM,SEy=4.9
Kd =190 ±60 nM,SEy=16.0
dFobs (RFU)
0   10   25   50
Guest concentration (μM)

H3R2me2a
Kd =200 ±40 nM,SEy=9.1
Kd =56 ±40 nM,SEy=35.9
0   10   25   50
Guest concentration (μM)

H3R2me2s
Kd =860 ±100 nM,SEy=2.8
Kd =560 ±100 nM,SEy=4.7
400
300
200
100
0
0   10   25   50
Guest concentration (μM)

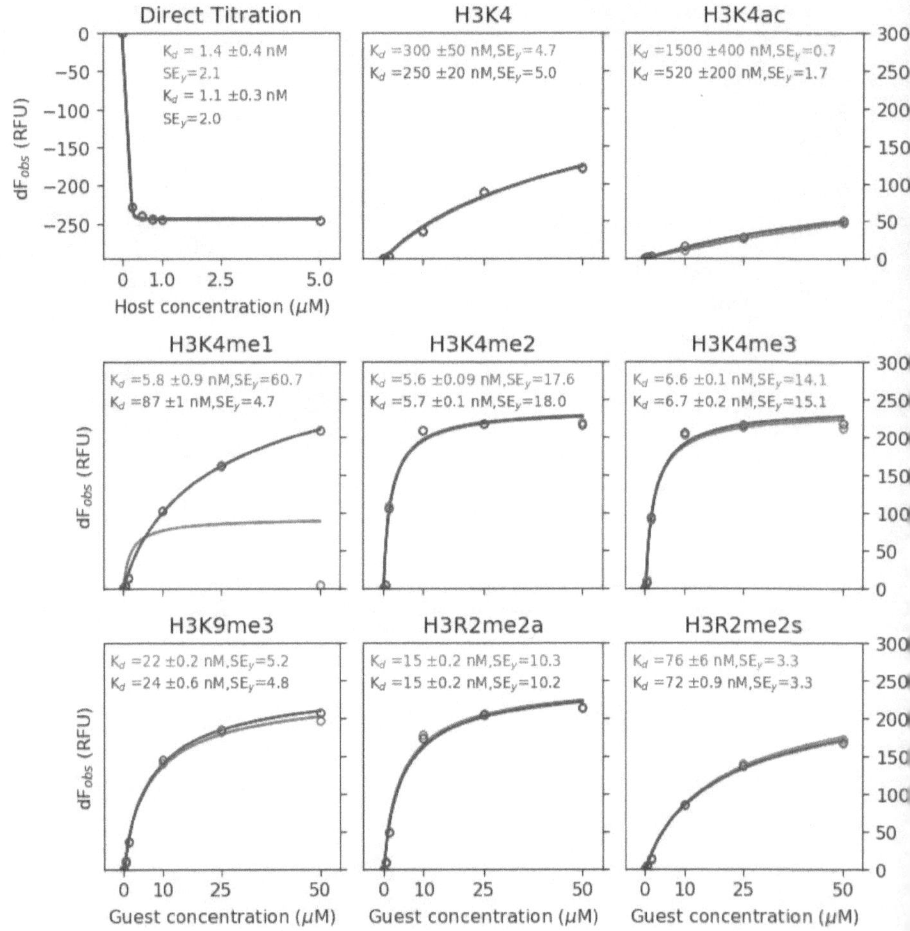

Compound AP8
HO3S
SO3H
SO3H
HO
OH
OH
OH
HO
Direct Titration
Kd = 1.4 ±0.4 nM
SEy=2.1
Kd = 1.1 ±0.3 nM
SEy=2.0
dFobs (RFU)
Host concentration (μM)
H3K4
Kd =300 ±50 nM,SEy=4.7
Kd =250 ±20 nM,SEy=5.0
H3K4ac
Kd =1500 ±400 nM,SEy=0.7
Kd =520 ±200 nM,SEy=1.7
H3K4me1
Kd =5.8 ±0.9 nM,SEy=60.7
Kd =87 ±1 nM,SEy=4.7
H3K4me2
Kd =5.6 ±0.09 nM,SEy=17.6
Kd =5.7 ±0.1 nM,SEy=18.0
H3K4me3
Kd =6.6 ±0.1 nM,SEy=14.1
Kd =6.7 ±0.2 nM,SEy=15.1
dFobs (RFU)
H3K9me3
Kd =22 ±0.2 nM,SEy=5.2
Kd =24 ±0.6 nM,SEy=4.8
H3R2me2a
Kd =15 ±0.2 nM,SEy=10.3
Kd =15 ±0.2 nM,SEy=10.2
H3R2me2s
Kd =76 ±6 nM,SEy=3.3
Kd =72 ±0.9 nM,SEy=3.3
dFobs (RFU)
Guest concentration (μM)
Guest concentration (μM)
Guest concentration (μM)

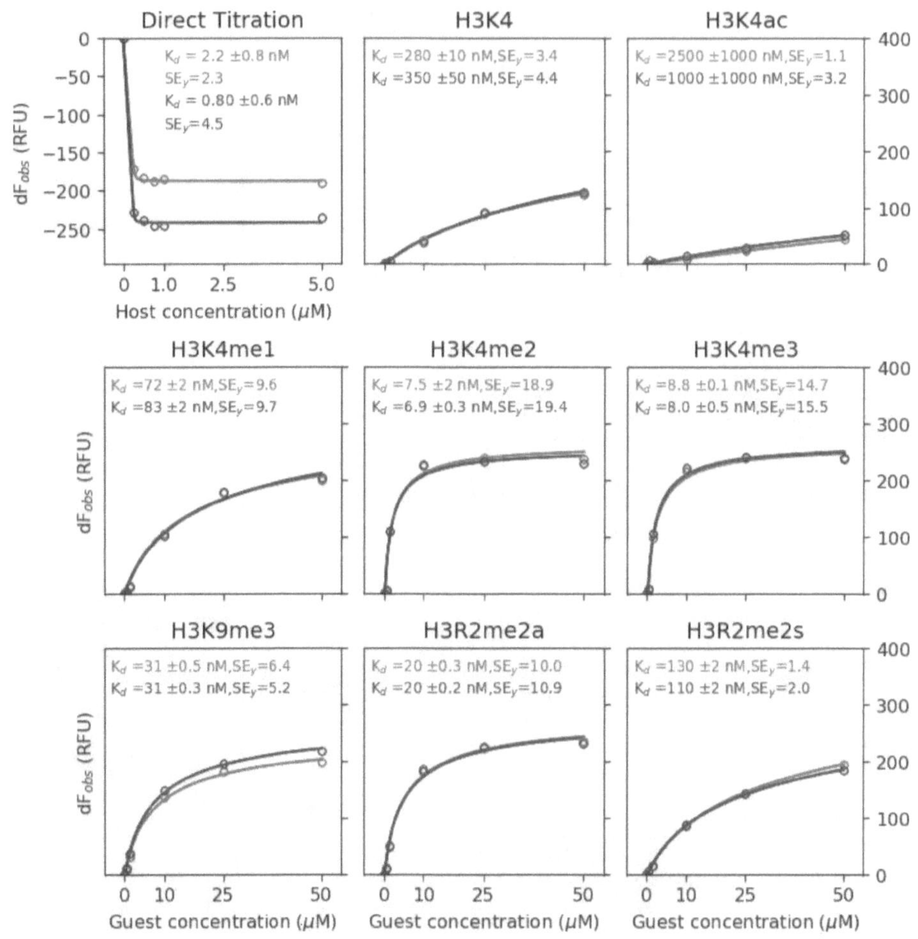
Compound AP9
Direct Titration
K_d = 2.2 ±0.8 nM
SE_y=2.3
K_d = 0.80 ±0.6 nM
SE_y=4.5
dF_obs (RFU)
Host concentration (μM)
H3K4
K_d =280 ±10 nM,SE_y=3.4
K_d =350 ±50 nM,SE_y=4.4
H3K4ac
K_d =2500 ±1000 nM,SE_y=1.1
K_d =1000 ±1000 nM,SE_y=3.2
H3K4me1
K_d =72 ±2 nM,SE_y=9.6
K_d =83 ±2 nM,SE_y=9.7
H3K4me2
K_d =7.5 ±2 nM,SE_y=18.9
K_d =6.9 ±0.3 nM,SE_y=19.4
H3K4me3
K_d =8.8 ±0.1 nM,SE_y=14.7
K_d =8.0 ±0.5 nM,SE_y=15.5
H3K9me3
K_d =31 ±0.5 nM,SE_y=6.4
K_d =31 ±0.3 nM,SE_y=5.2
H3R2me2a
K_d =20 ±0.3 nM,SE_y=10.0
K_d =20 ±0.2 nM,SE_y=10.9
H3R2me2s
K_d =130 ±2 nM,SE_y=1.4
K_d =110 ±2 nM,SE_y=2.0
dF_obs (RFU)
Guest concentration (μM)

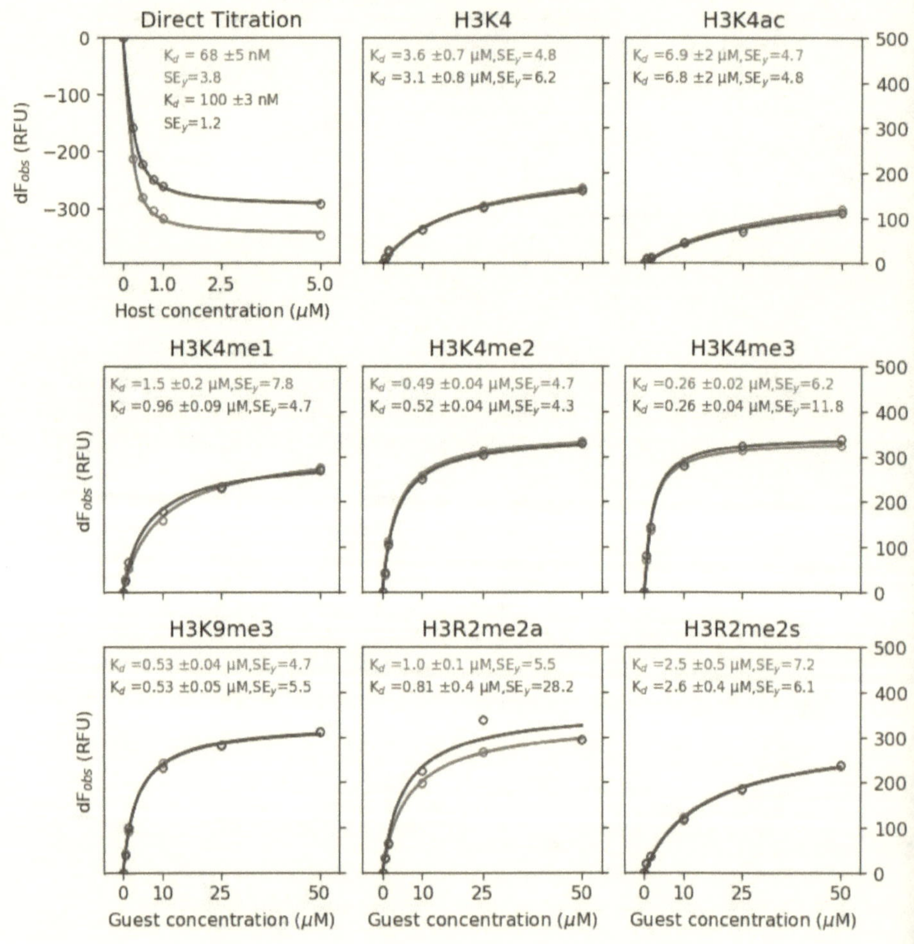

Compound AM1

Direct Titration
K_d = 68 ±5 nM
SE_y=3.8
K_d = 100 ±3 nM
SE_y=1.2
dF$_{obs}$ (RFU)
0
−100
−200
−300
0   1.0   2.5   5.0
Host concentration (μM)

H3K4
K_d =3.6 ±0.7 μM,SE_y=4.8
K_d =3.1 ±0.8 μM,SE_y=6.2

H3K4ac
K_d =6.9 ±2 μM,SE_y=4.7
K_d =6.8 ±2 μM,SE_y=4.8
500
400
300
200
100
0

H3K4me1
K_d =1.5 ±0.2 μM,SE_y=7.8
K_d =0.96 ±0.09 μM,SE_y=4.7
dF$_{obs}$ (RFU)

H3K4me2
K_d =0.49 ±0.04 μM,SE_y=4.7
K_d =0.52 ±0.04 μM,SE_y=4.3

H3K4me3
K_d =0.26 ±0.02 μM,SE_y=6.2
K_d =0.26 ±0.04 μM,SE_y=11.8
500
400
300
200
100
0

H3K9me3
K_d =0.53 ±0.04 μM,SE_y=4.7
K_d =0.53 ±0.05 μM,SE_y=5.5
dF$_{obs}$ (RFU)
0   10   25   50
Guest concentration (μM)

H3R2me2a
K_d =1.0 ±0.1 μM,SE_y=5.5
K_d =0.81 ±0.4 μM,SE_y=28.2
0   10   25   50
Guest concentration (μM)

H3R2me2s
K_d =2.5 ±0.5 μM,SE_y=7.2
K_d =2.6 ±0.4 μM,SE_y=6.1
500
400
300
200
100
0
0   10   25   50
Guest concentration (μM)

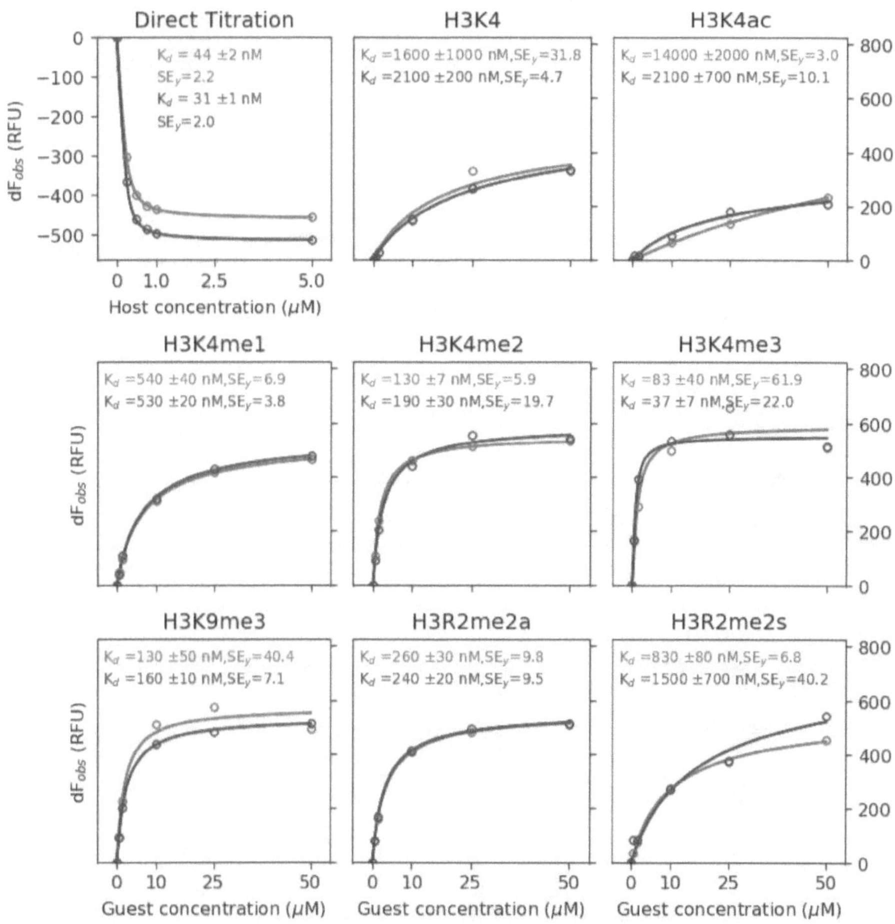

HO₃S   SO₃H   SO₃H
HO OC
OH   OH   OH   HO
Compound AM2

Direct Titration
K_d = 44 ±2 nM
SE_y=2.2
K_d = 31 ±1 nM
SE_y=2.0
Host concentration (µM)
dF_obs (RFU)

H3K4
K_d =1600 ±1000 nM,SE_y=31.8
K_d =2100 ±200 nM,SE_y=4.7

H3K4ac
K_d =14000 ±2000 nM,SE_y=3.0
K_d =2100 ±700 nM,SE_y=10.1

H3K4me1
K_d =540 ±40 nM,SE_y=6.9
K_d =530 ±20 nM,SE_y=3.8

H3K4me2
K_d =130 ±7 nM,SE_y=5.9
K_d =190 ±30 nM,SE_y=19.7

H3K4me3
K_d =83 ±40 nM,SE_y=61.9
K_d =37 ±7 nM,SE_y=22.0

H3K9me3
K_d =130 ±50 nM,SE_y=40.4
K_d =160 ±10 nM,SE_y=7.1

H3R2me2a
K_d =260 ±30 nM,SE_y=9.8
K_d =240 ±20 nM,SE_y=9.5

H3R2me2s
K_d =830 ±80 nM,SE_y=6.8
K_d =1500 ±700 nM,SE_y=40.2

Guest concentration (µM)   Guest concentration (µM)   Guest concentration (µM)

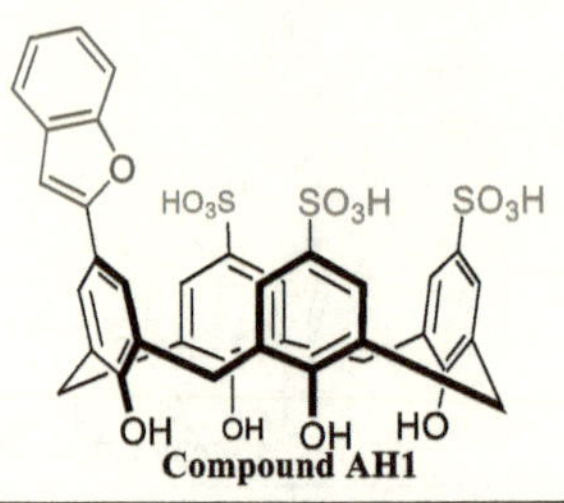

HO3S   SO3H   SO3H
OH   OH   OH   HO
Compound AH1

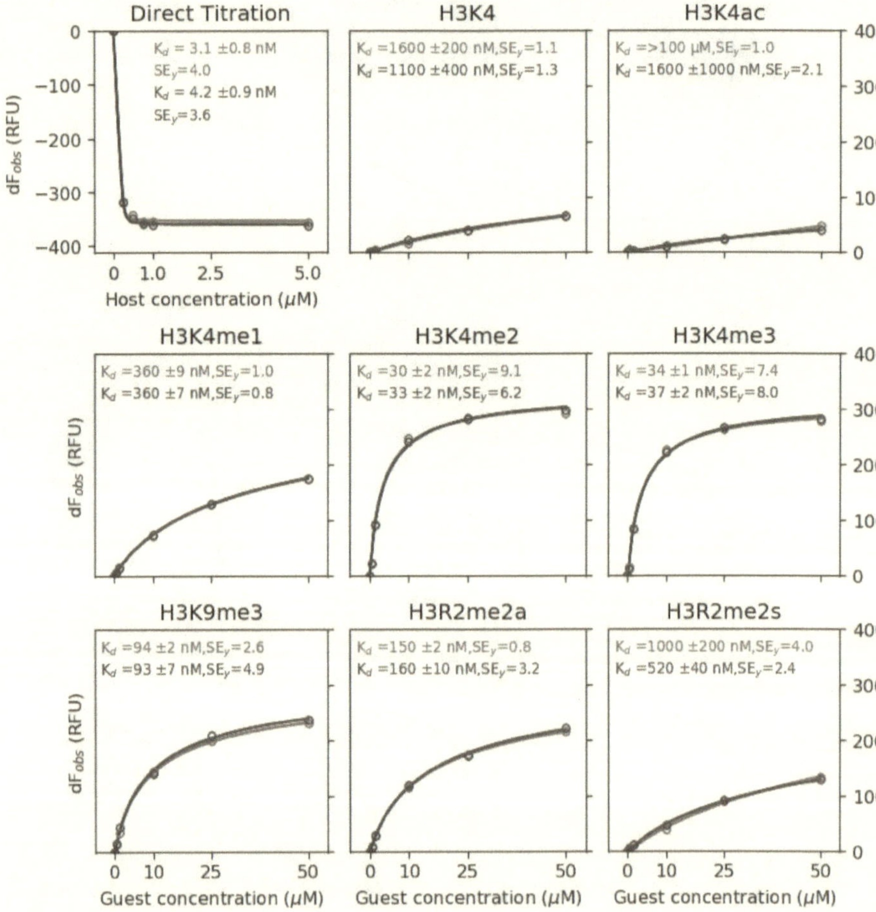

Direct Titration
K_d = 3.1 ±0.8 nM
SE_y=4.0
K_d = 4.2 ±0.9 nM
SE_y=3.6
dF_{obs} (RFU)
Host concentration (μM)

H3K4
K_d =1600 ±200 nM,SE_y=1.1
K_d =1100 ±400 nM,SE_y=1.3

H3K4ac
K_d =>100 μM,SE_y=1.0
K_d =1600 ±1000 nM,SE_y=2.1

H3K4me1
K_d =360 ±9 nM,SE_y=1.0
K_d =360 ±7 nM,SE_y=0.8
dF_{obs} (RFU)

H3K4me2
K_d =30 ±2 nM,SE_y=9.1
K_d =33 ±2 nM,SE_y=6.2

H3K4me3
K_d =34 ±1 nM,SE_y=7.4
K_d =37 ±2 nM,SE_y=8.0

H3K9me3
K_d =94 ±2 nM,SE_y=2.6
K_d =93 ±7 nM,SE_y=4.9
dF_{obs} (RFU)
Guest concentration (μM)

H3R2me2a
K_d =150 ±2 nM,SE_y=0.8
K_d =160 ±10 nM,SE_y=3.2
Guest concentration (μM)

H3R2me2s
K_d =1000 ±200 nM,SE_y=4.0
K_d =520 ±40 nM,SE_y=2.4
Guest concentration (μM)

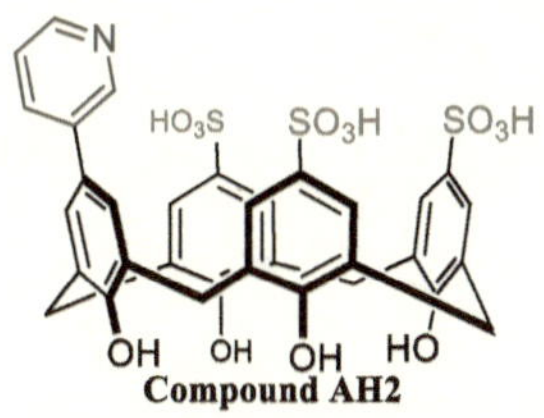

Compound AH2
Direct Titration
Kd = 600 ±30 nM
SEy=3.8
Kd = 740 ±200 nM
SEy=20.0
dFobs (RFU)
Host concentration (µM)
H3K4
Kd =1100 ±300 nM,SEy=3.6
Kd =1300 ±400 nM,SEy=4.4
H3K4ac
Kd =3200 ±800 nM,SEy=3.2
Kd =2000 ±300 nM,SEy=2.5
H3K4me1
Kd =1100 ±400 nM,SEy=14.4
Kd =940 ±400 nM,SEy=15.4
H3K4me2
Kd =320 ±100 nM,SEy=13.6
Kd =480 ±90 nM,SEy=8.5
H3K4me3
Kd =370 ±60 nM,SEy=8.2
Kd =390 ±100 nM,SEy=15.9
H3K9me3
Kd =560 ±200 nM,SEy=14.3
Kd =720 ±200 nM,SEy=11.3
H3R2me2a
Kd =1400 ±500 nM,SEy=13.8
Kd =970 ±300 nM,SEy=11.5
H3R2me2s
Kd =1300 ±400 nM,SEy=8.3
Kd =1400 ±900 nM,SEy=13.9
dFobs (RFU)
Guest concentration (µM)
Guest concentration (µM)
Guest concentration (µM)

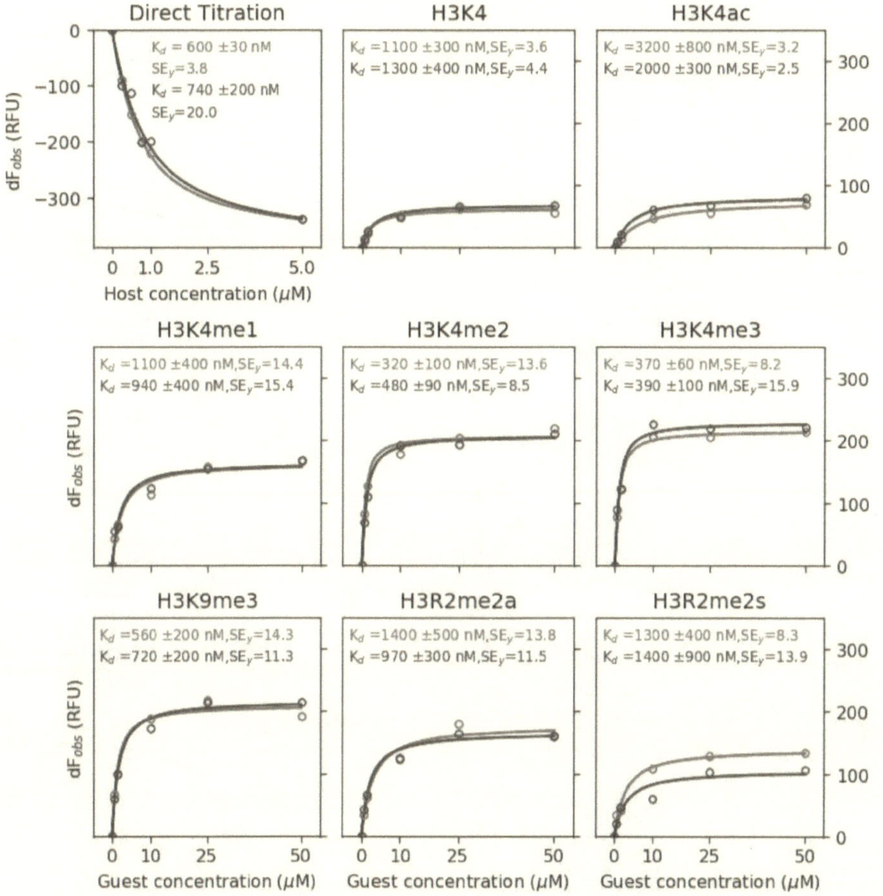

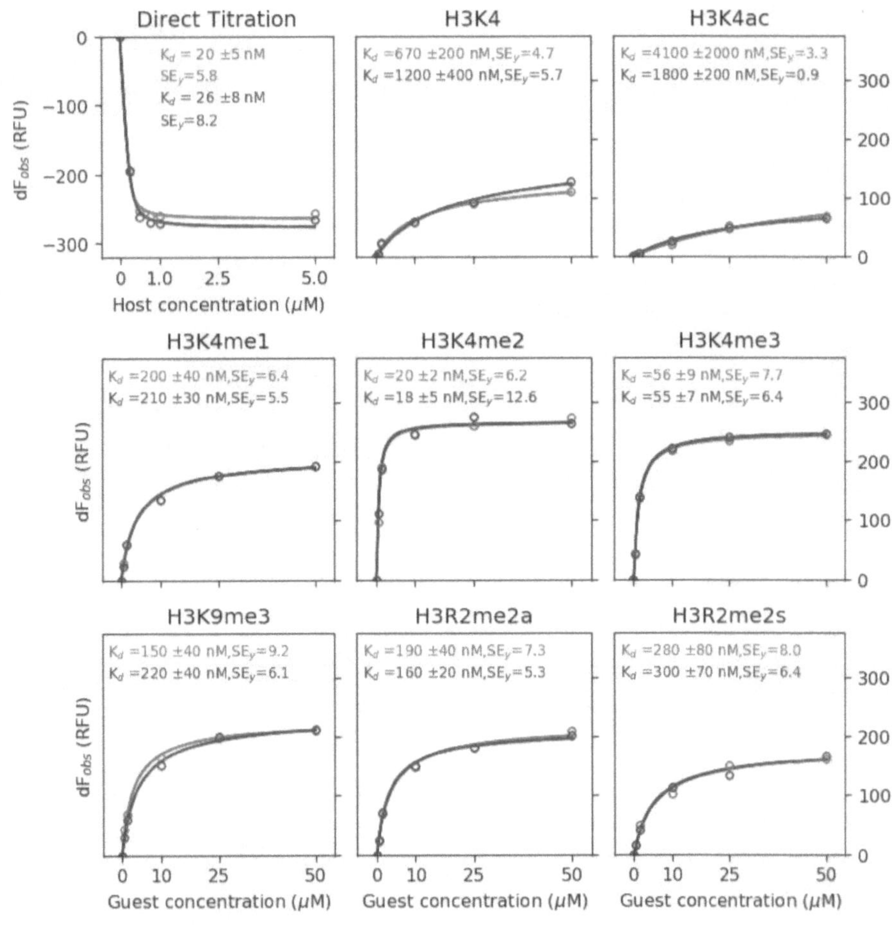

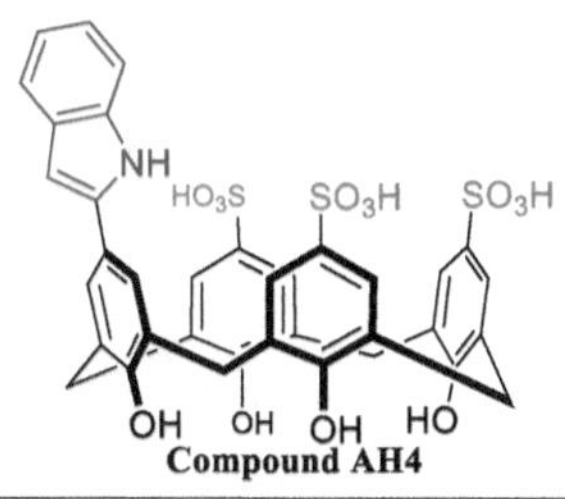

NH
HO3S
SO3H
SO3H
OH
OH
OH
HO
Compound AH4

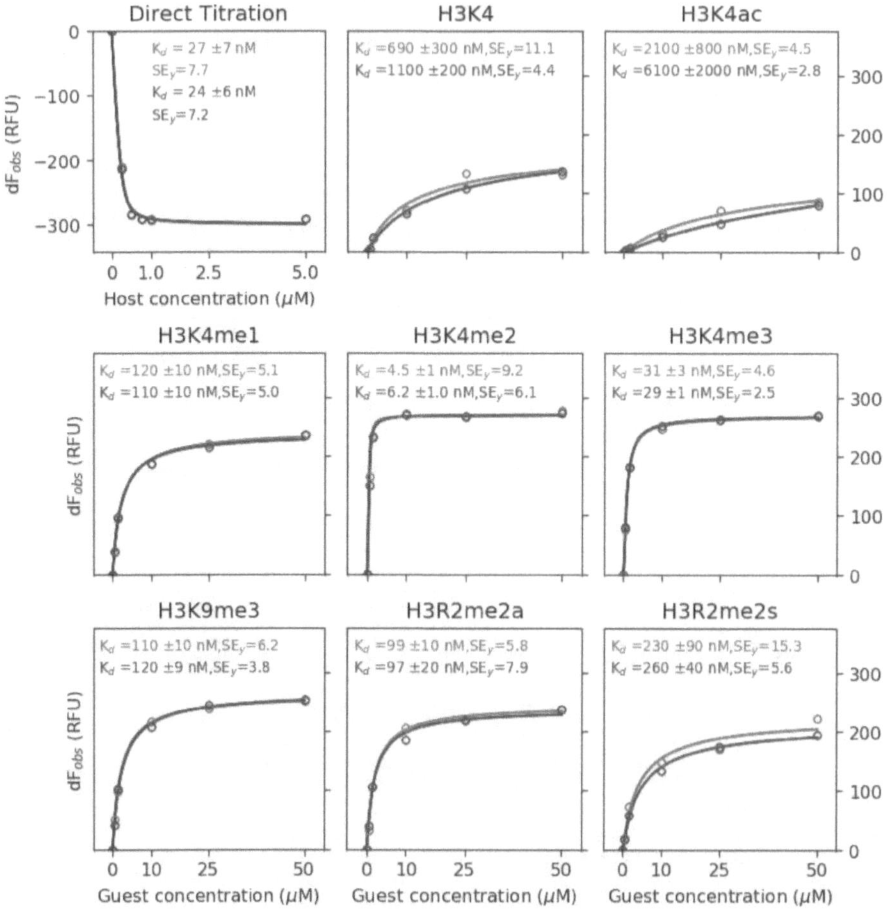

Direct Titration
K_d = 27 ±7 nM
SE_y=7.7
K_d = 24 ±6 nM
SE_y=7.2
dF$_{obs}$ (RFU)

H3K4
K_d =690 ±300 nM,SE_y=11.1
K_d =1100 ±200 nM,SE_y=4.4

H3K4ac
K_d =2100 ±800 nM,SE_y=4.5
K_d =6100 ±2000 nM,SE_y=2.8

Host concentration (μM)

H3K4me1
K_d =120 ±10 nM,SE_y=5.1
K_d =110 ±10 nM,SE_y=5.0

H3K4me2
K_d =4.5 ±1 nM,SE_y=9.2
K_d =6.2 ±1.0 nM,SE_y=6.1

H3K4me3
K_d =31 ±3 nM,SE_y=4.6
K_d =29 ±1 nM,SE_y=2.5

H3K9me3
K_d =110 ±10 nM,SE_y=6.2
K_d =120 ±9 nM,SE_y=3.8

H3R2me2a
K_d =99 ±10 nM,SE_y=5.8
K_d =97 ±20 nM,SE_y=7.9

H3R2me2s
K_d =230 ±90 nM,SE_y=15.3
K_d =260 ±40 nM,SE_y=5.6

Guest concentration (μM)
Guest concentration (μM)
Guest concentration (μM)

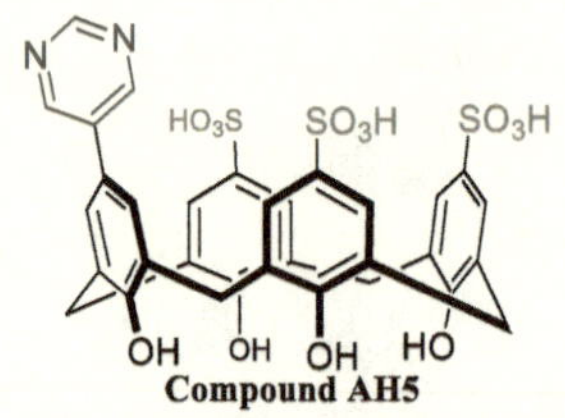

Compound AH5

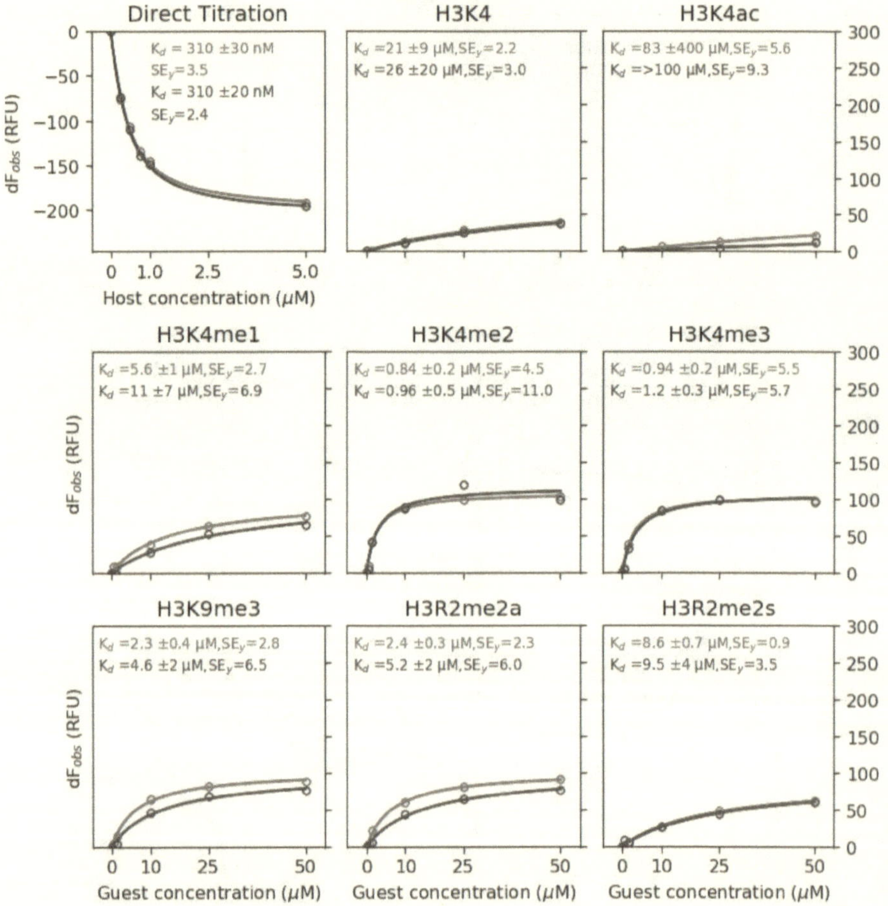

Direct Titration

K_d = 310 ±30 nM
SE_y=3.5
K_d = 310 ±20 nM
SE_y=2.4

dF$_{obs}$ (RFU)

Host concentration (μM)

H3K4

K_d =21 ±9 µM, SE_y=2.2
K_d =26 ±20 µM, SE_y=3.0

H3K4ac

K_d =83 ±400 µM, SE_y=5.6
K_d =>100 µM, SE_y=9.3

H3K4me1

K_d =5.6 ±1 µM, SE_y=2.7
K_d =11 ±7 µM, SE_y=6.9

H3K4me2

K_d =0.84 ±0.2 µM, SE_y=4.5
K_d =0.96 ±0.5 µM, SE_y=11.0

H3K4me3

K_d =0.94 ±0.2 µM, SE_y=5.5
K_d =1.2 ±0.3 µM, SE_y=5.7

H3K9me3

K_d =2.3 ±0.4 µM, SE_y=2.8
K_d =4.6 ±2 µM, SE_y=6.5

H3R2me2a

K_d =2.4 ±0.3 µM, SE_y=2.3
K_d =5.2 ±2 µM, SE_y=6.0

H3R2me2s

K_d =8.6 ±0.7 µM, SE_y=0.9
K_d =9.5 ±4 µM, SE_y=3.5

dF$_{obs}$ (RFU)

Guest concentration (μM)

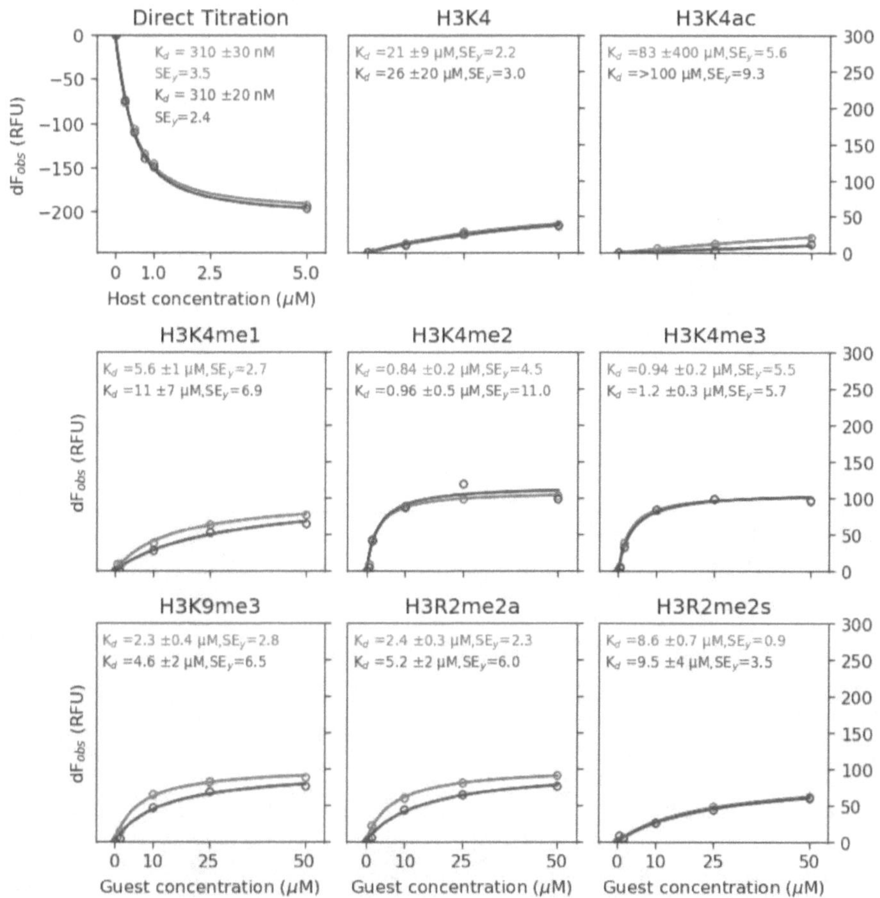

Compound AH6
Direct Titration
K_d = 310 ±30 nM
SE_y=3.5
K_d = 310 ±20 nM
SE_y=2.4
dF_obs (RFU)
Host concentration (µM)
H3K4
K_d =21 ±9 µM,SE_y=2.2
K_d =26 ±20 µM,SE_y=3.0
H3K4ac
K_d =83 ±400 µM,SE_y=5.6
K_d =>100 µM,SE_y=9.3
H3K4me1
K_d =5.6 ±1 µM,SE_y=2.7
K_d =11 ±7 µM,SE_y=6.9
H3K4me2
K_d =0.84 ±0.2 µM,SE_y=4.5
K_d =0.96 ±0.5 µM,SE_y=11.0
H3K4me3
K_d =0.94 ±0.2 µM,SE_y=5.5
K_d =1.2 ±0.3 µM,SE_y=5.7
H3K9me3
K_d =2.3 ±0.4 µM,SE_y=2.8
K_d =4.6 ±2 µM,SE_y=6.5
H3R2me2a
K_d =2.4 ±0.3 µM,SE_y=2.3
K_d =5.2 ±2 µM,SE_y=6.0
H3R2me2s
K_d =8.6 ±0.7 µM,SE_y=0.9
K_d =9.5 ±4 µM,SE_y=3.5
dF_obs (RFU)
Guest concentration (µM)

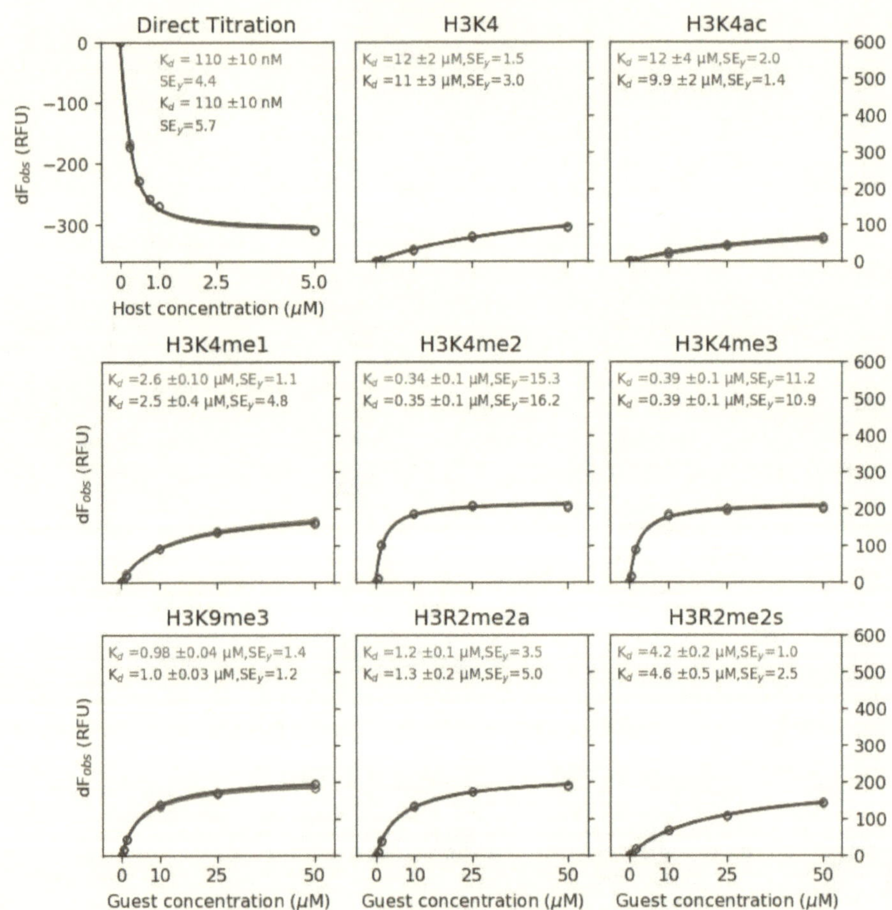

N
HO₃S   SO₃H   SO₃H
OH   OH   OH   HO
Compound AH7

Direct Titration
Kd = 110 ±10 nM
SEy=4.4
Kd = 110 ±10 nM
SEy=5.7
dFobs (RFU)
0
−100
−200
−300
0   1.0   2.5   5.0
Host concentration (μM)

H3K4
Kd =12 ±2 μM,SEy=1.5
Kd =11 ±3 μM,SEy=3.0

H3K4ac
Kd =12 ±4 μM,SEy=2.0
Kd =9.9 ±2 μM,SEy=1.4
600
500
400
300
200
100
0

H3K4me1
Kd =2.6 ±0.10 μM,SEy=1.1
Kd =2.5 ±0.4 μM,SEy=4.8
dFobs (RFU)

H3K4me2
Kd =0.34 ±0.1 μM,SEy=15.3
Kd =0.35 ±0.1 μM,SEy=16.2

H3K4me3
Kd =0.39 ±0.1 μM,SEy=11.2
Kd =0.39 ±0.1 μM,SEy=10.9
600
500
400
300
200
100
0

H3K9me3
Kd =0.98 ±0.04 μM,SEy=1.4
Kd =1.0 ±0.03 μM,SEy=1.2
dFobs (RFU)
0   10   25   50
Guest concentration (μM)

H3R2me2a
Kd =1.2 ±0.1 μM,SEy=3.5
Kd =1.3 ±0.2 μM,SEy=5.0
0   10   25   50
Guest concentration (μM)

H3R2me2s
Kd =4.2 ±0.2 μM,SEy=1.0
Kd =4.6 ±0.5 μM,SEy=2.5
600
500
400
300
200
100
0
0   10   25   50
Guest concentration (μM)

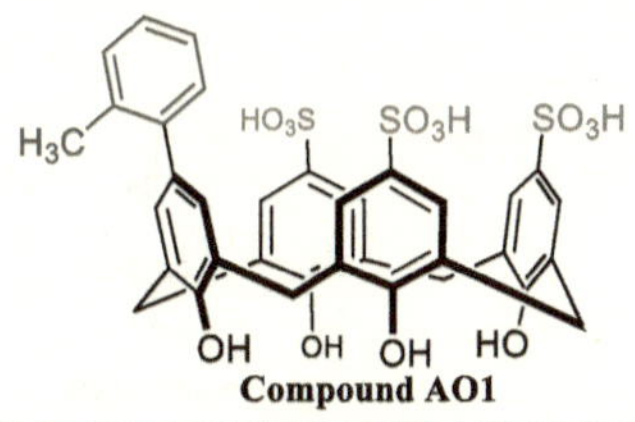

Direct Titration

H3K4

H3K4ac

H3K4me1

H3K4me2

H3K4me3

H3K9me3

H3R2me2a

H3R2me2s

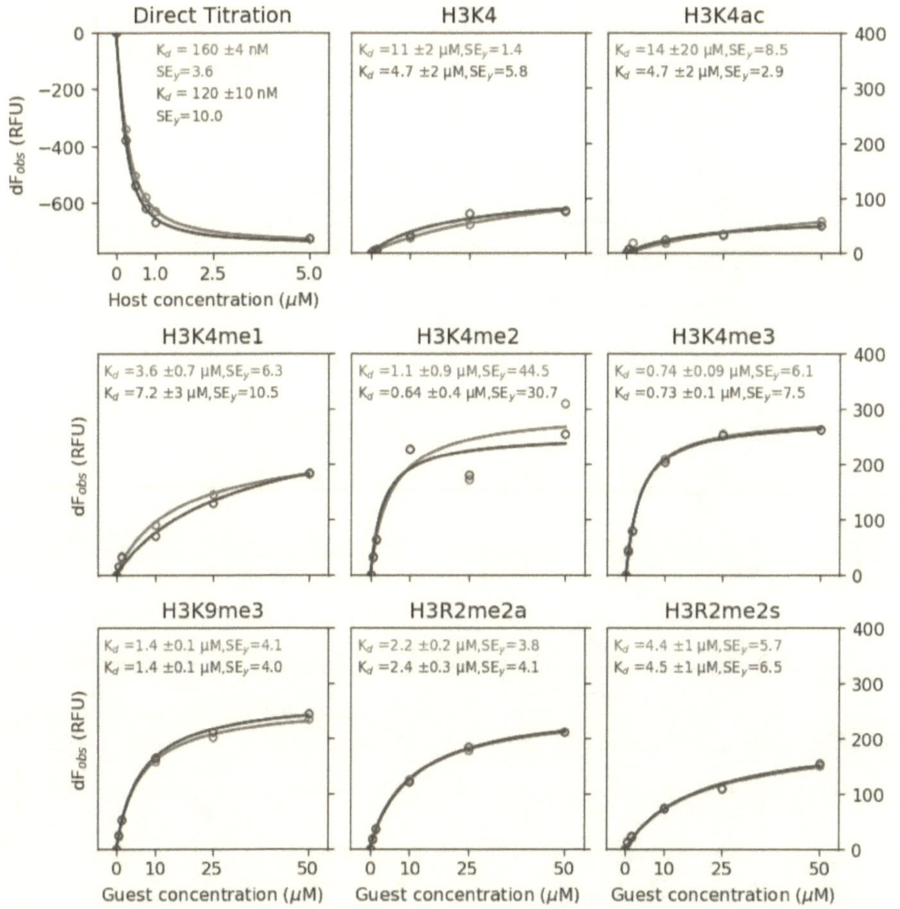

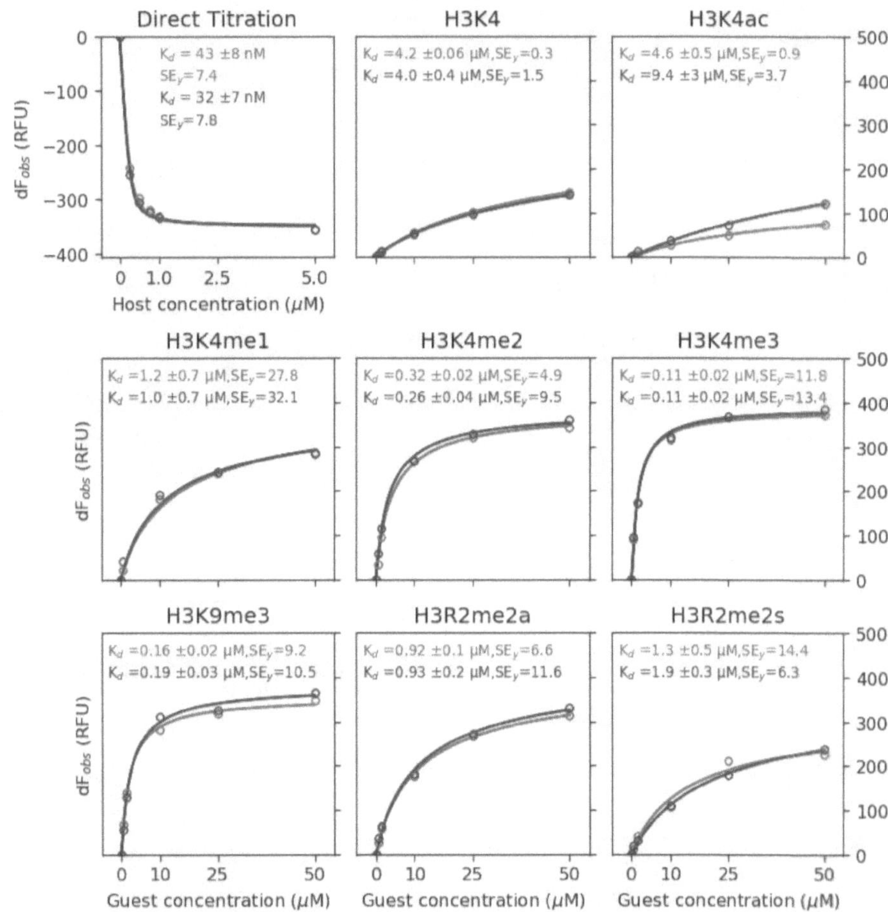

H3C—O
HO3S
SO3H
SO3H
OH
OH
OH
HO
Compound AO2
Direct Titration
Kd = 43 ±8 nM
SEy=7.4
Kd = 32 ±7 nM
SEy=7.8
dFobs (RFU)
0
−100
−200
−300
−400
0   1.0   2.5   5.0
Host concentration (µM)
H3K4
Kd =4.2 ±0.06 µM,SEy=0.3
Kd =4.0 ±0.4 µM,SEy=1.5
H3K4ac
Kd =4.6 ±0.5 µM,SEy=0.9
Kd =9.4 ±3 µM,SEy=3.7
500
400
300
200
100
0
H3K4me1
Kd =1.2 ±0.7 µM,SEy=27.8
Kd =1.0 ±0.7 µM,SEy=32.1
H3K4me2
Kd =0.32 ±0.02 µM,SEy=4.9
Kd =0.26 ±0.04 µM,SEy=9.5
H3K4me3
Kd =0.11 ±0.02 µM,SEy=11.8
Kd =0.11 ±0.02 µM,SEy=13.4
500
400
300
200
100
0
dFobs (RFU)
H3K9me3
Kd =0.16 ±0.02 µM,SEy=9.2
Kd =0.19 ±0.03 µM,SEy=10.5
H3R2me2a
Kd =0.92 ±0.1 µM,SEy=6.6
Kd =0.93 ±0.2 µM,SEy=11.6
H3R2me2s
Kd =1.3 ±0.5 µM,SEy=14.4
Kd =1.9 ±0.3 µM,SEy=6.3
500
400
300
200
100
0
dFobs (RFU)
0   10   25   50
Guest concentration (µM)
0   10   25   50
Guest concentration (µM)
0   10   25   50
Guest concentration (µM)

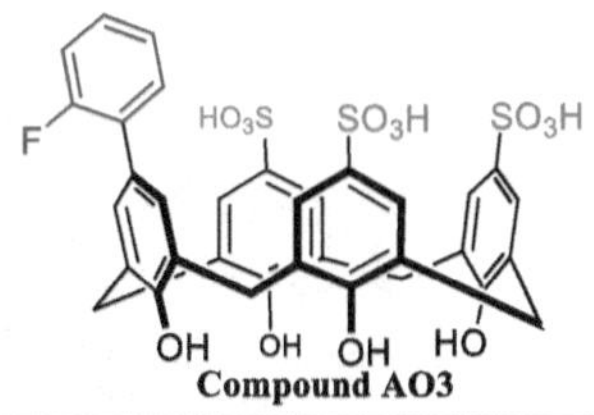

HO3S   SO3H   SO3H
F
OH   OH   OH   HO
Compound AO3

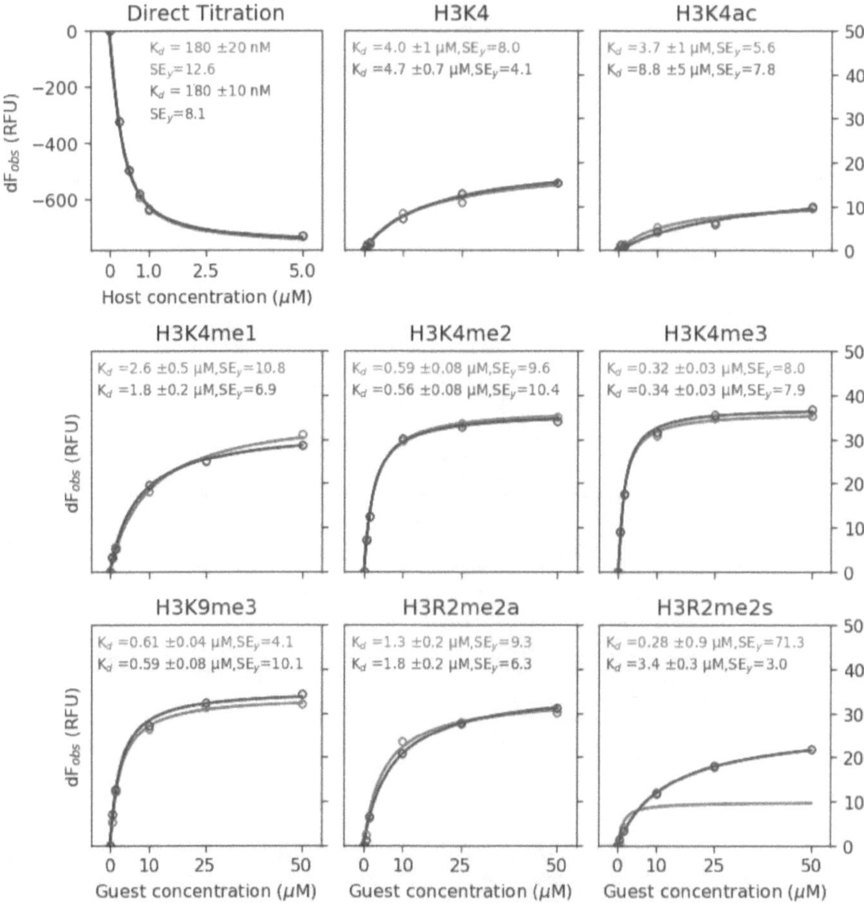

Direct Titration
K_d = 180 ±20 nM
SE_y=12.6
K_d = 180 ±10 nM
SE_y=8.1
dF$_{obs}$ (RFU)
Host concentration (μM)

H3K4
K_d =4.0 ±1 μM,SE_y=8.0
K_d =4.7 ±0.7 μM,SE_y=4.1

H3K4ac
K_d =3.7 ±1 μM,SE_y=5.6
K_d =8.8 ±5 μM,SE_y=7.8

H3K4me1
K_d =2.6 ±0.5 μM,SE_y=10.8
K_d =1.8 ±0.2 μM,SE_y=6.9

H3K4me2
K_d =0.59 ±0.08 μM,SE_y=9.6
K_d =0.56 ±0.08 μM,SE_y=10.4

H3K4me3
K_d =0.32 ±0.03 μM,SE_y=8.0
K_d =0.34 ±0.03 μM,SE_y=7.9

H3K9me3
K_d =0.61 ±0.04 μM,SE_y=4.1
K_d =0.59 ±0.08 μM,SE_y=10.1

H3R2me2a
K_d =1.3 ±0.2 μM,SE_y=9.3
K_d =1.8 ±0.2 μM,SE_y=6.3

H3R2me2s
K_d =0.28 ±0.9 μM,SE_y=71.3
K_d =3.4 ±0.3 μM,SE_y=3.0

dF$_{obs}$ (RFU)
Guest concentration (μM)

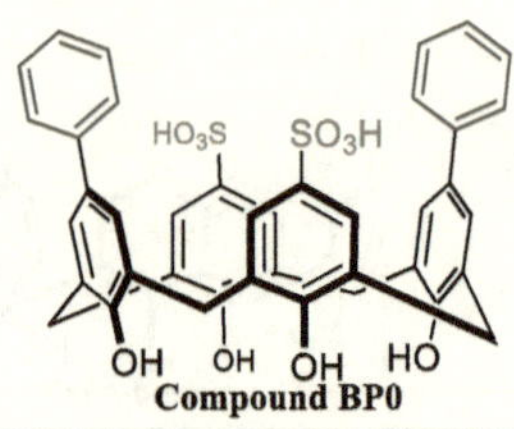

Direct Titration

dF$_{obs}$ (RFU)

Host concentration (μM)

H3K4

H3K4ac

H3K4me1

dF$_{obs}$ (RFU)

H3K4me2

H3K4me3

H3K9me3

dF$_{obs}$ (RFU)

Guest concentration (μM)

H3R2me2a

Guest concentration (μM)

H3R2me2s

Guest concentration (μM)

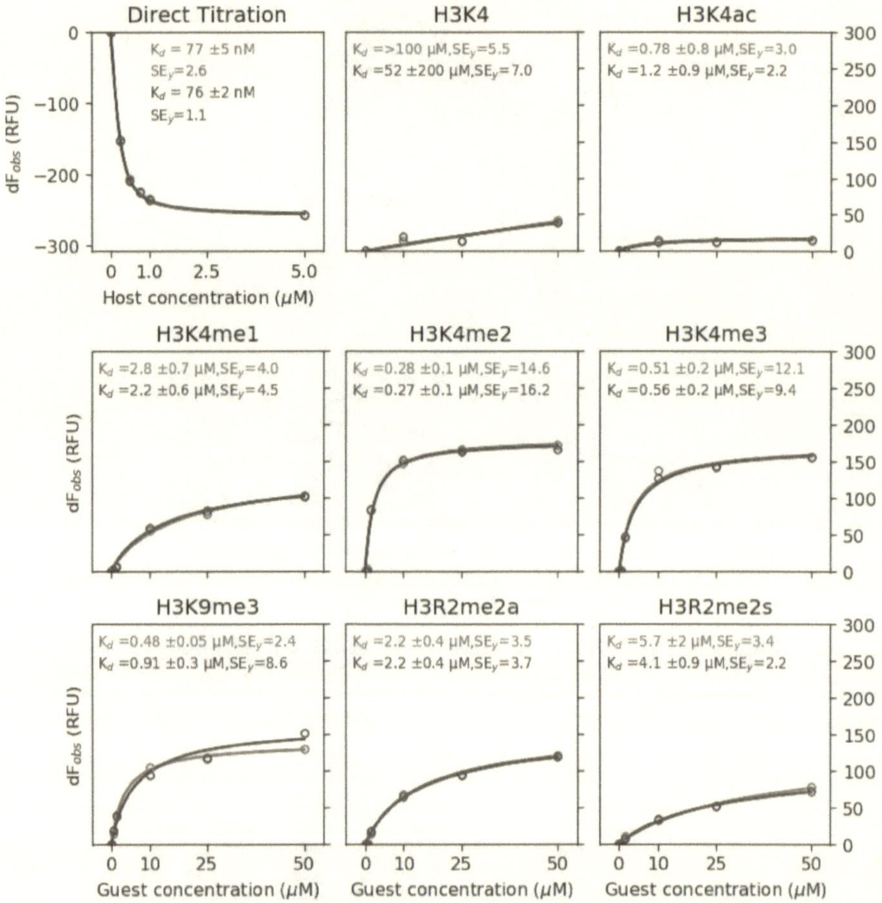

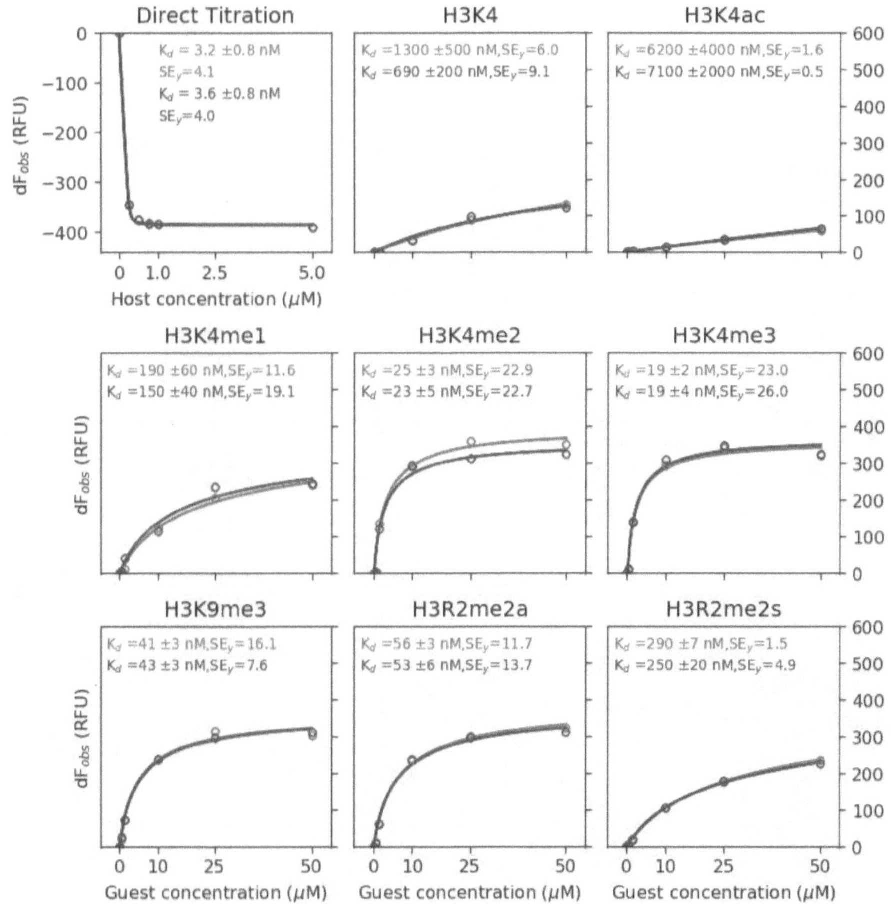

HO
O
OH
HO
O
OH
HO₃S
SO₃H
OH
OH
OH
HO
Compound BP1
Direct Titration
K_d = 3.2 ±0.8 nM
SE_y=4.1
K_d = 3.6 ±0.8 nM
SE_y=4.0
dF$_{obs}$ (RFU)
0
−100
−200
−300
−400
0 1.0 2.5 5.0
Host concentration (μM)
H3K4
K_d =1300 ±500 nM,SE_y=6.0
K_d =690 ±200 nM,SE_y=9.1
H3K4ac
K_d =6200 ±4000 nM,SE_y=1.6
K_d =7100 ±2000 nM,SE_y=0.5
600
500
400
300
200
100
0
H3K4me1
K_d =190 ±60 nM,SE_y=11.6
K_d =150 ±40 nM,SE_y=19.1
H3K4me2
K_d =25 ±3 nM,SE_y=22.9
K_d =23 ±5 nM,SE_y=22.7
H3K4me3
K_d =19 ±2 nM,SE_y=23.0
K_d =19 ±4 nM,SE_y=26.0
600
500
400
300
200
100
0
dF$_{obs}$ (RFU)
H3K9me3
K_d =41 ±3 nM,SE_y=16.1
K_d =43 ±3 nM,SE_y=7.6
H3R2me2a
K_d =56 ±3 nM,SE_y=11.7
K_d =53 ±6 nM,SE_y=13.7
H3R2me2s
K_d =290 ±7 nM,SE_y=1.5
K_d =250 ±20 nM,SE_y=4.9
600
500
400
300
200
100
0
dF$_{obs}$ (RFU)
0 10 25 50
Guest concentration (μM)
0 10 25 50
Guest concentration (μM)
0 10 25 50
Guest concentration (μM)

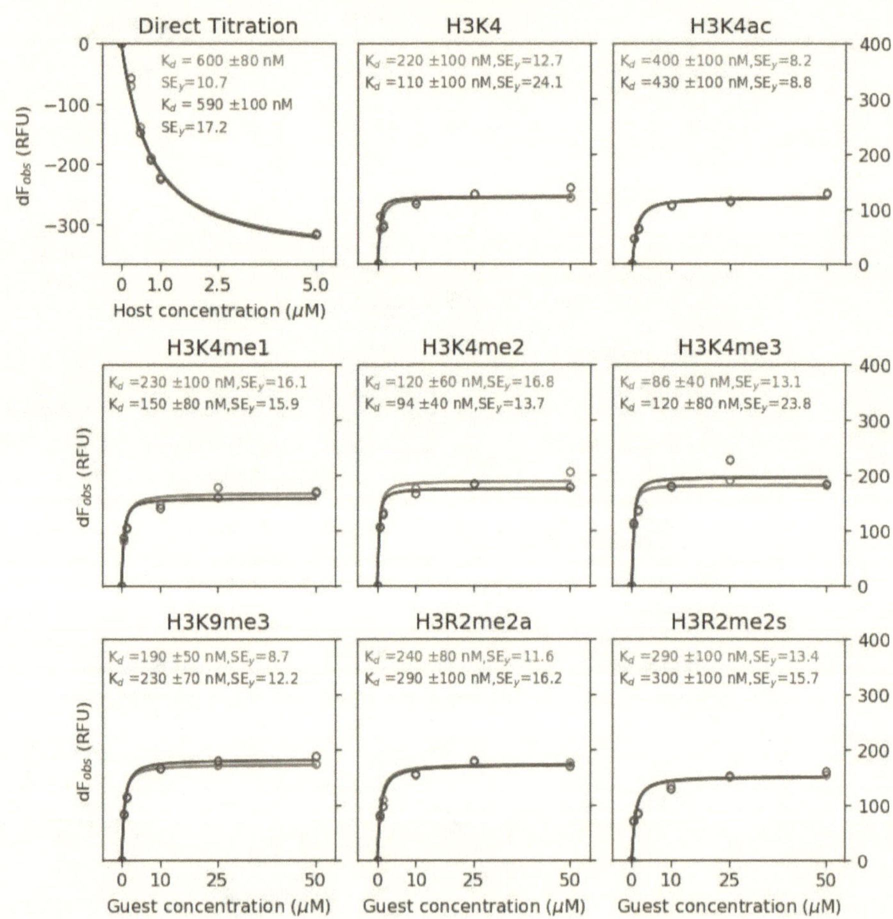

Compound BM1
Direct Titration
K_d = 600 ±80 nM
SE_y=10.7
K_d = 590 ±100 nM
SE_y=17.2
H3K4
K_d =220 ±100 nM,SE_y=12.7
K_d =110 ±100 nM,SE_y=24.1
H3K4ac
K_d =400 ±100 nM,SE_y=8.2
K_d =430 ±100 nM,SE_y=8.8
H3K4me1
K_d =230 ±100 nM,SE_y=16.1
K_d =150 ±80 nM,SE_y=15.9
H3K4me2
K_d =120 ±60 nM,SE_y=16.8
K_d =94 ±40 nM,SE_y=13.7
H3K4me3
K_d =86 ±40 nM,SE_y=13.1
K_d =120 ±80 nM,SE_y=23.8
H3K9me3
K_d =190 ±50 nM,SE_y=8.7
K_d =230 ±70 nM,SE_y=12.2
H3R2me2a
K_d =240 ±80 nM,SE_y=11.6
K_d =290 ±100 nM,SE_y=16.2
H3R2me2s
K_d =290 ±100 nM,SE_y=13.4
K_d =300 ±100 nM,SE_y=15.7
dF_{obs} (RFU)
Host concentration (μM)
Guest concentration (μM)

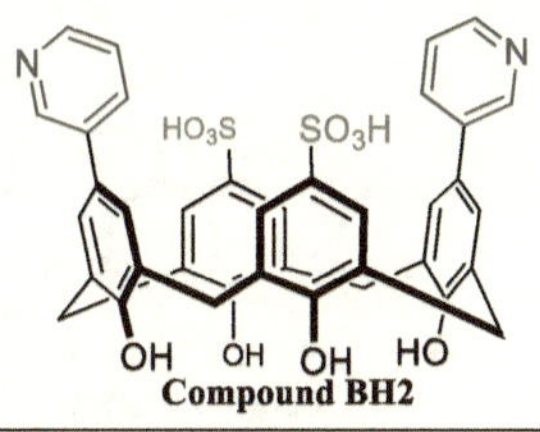

HO3S   SO3H
N
N
OH   OH   OH   HO
Compound BH2

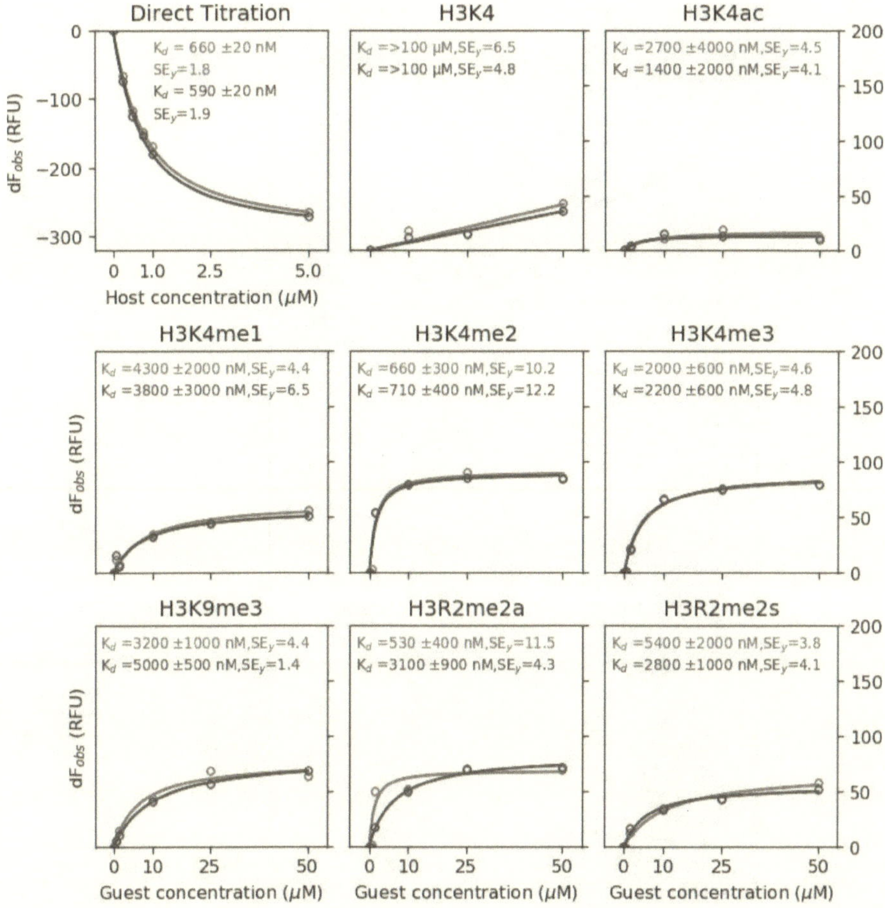

Direct Titration
K_d = 660 ±20 nM
SE_y=1.8
K_d = 590 ±20 nM
SE_y=1.9
dF$_{obs}$ (RFU)
0
−100
−200
−300
0   1.0   2.5   5.0
Host concentration (μM)

H3K4
K_d =>100 μM,SE_y=6.5
K_d =>100 μM,SE_y=4.8

H3K4ac
K_d =2700 ±4000 nM,SE_y=4.5
K_d =1400 ±2000 nM,SE_y=4.1
200
150
100
50
0

H3K4me1
K_d =4300 ±2000 nM,SE_y=4.4
K_d =3800 ±3000 nM,SE_y=6.5
dF$_{obs}$ (RFU)

H3K4me2
K_d =660 ±300 nM,SE_y=10.2
K_d =710 ±400 nM,SE_y=12.2

H3K4me3
K_d =2000 ±600 nM,SE_y=4.6
K_d =2200 ±600 nM,SE_y=4.8
200
150
100
50
0

H3K9me3
K_d =3200 ±1000 nM,SE_y=4.4
K_d =5000 ±500 nM,SE_y=1.4
dF$_{obs}$ (RFU)
0   10   25   50
Guest concentration (μM)

H3R2me2a
K_d =530 ±400 nM,SE_y=11.5
K_d =3100 ±900 nM,SE_y=4.3
0   10   25   50
Guest concentration (μM)

H3R2me2s
K_d =5400 ±2000 nM,SE_y=3.8
K_d =2800 ±1000 nM,SE_y=4.1
200
150
100
50
0
0   10   25   50
Guest concentration (μM)

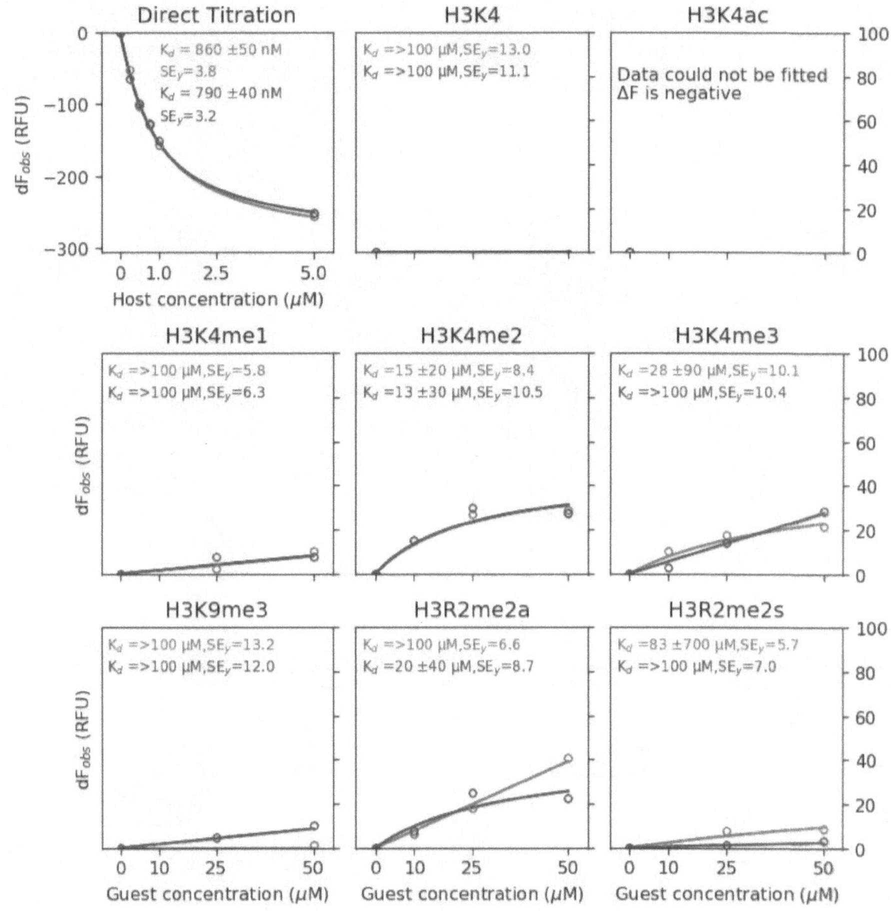

Compound BH5

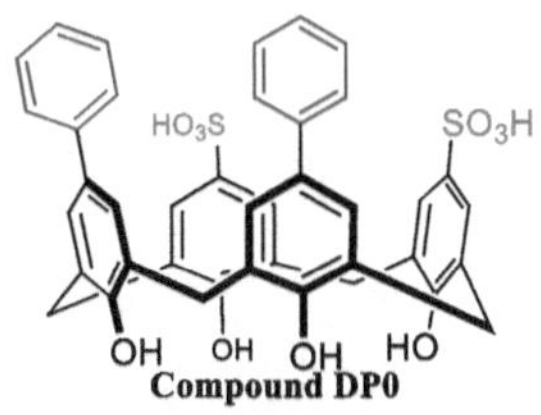

Direct Titration

dF$_{obs}$ (RFU)

Host concentration (μM)

H3K4

H3K4ac

H3K4me1

dF$_{obs}$ (RFU)

H3K4me2

H3K4me3

H3K9me3

dF$_{obs}$ (RFU)

Guest concentration (μM)

H3R2me2a

Guest concentration (μM)

H3R2me2s

Guest concentration (μM)

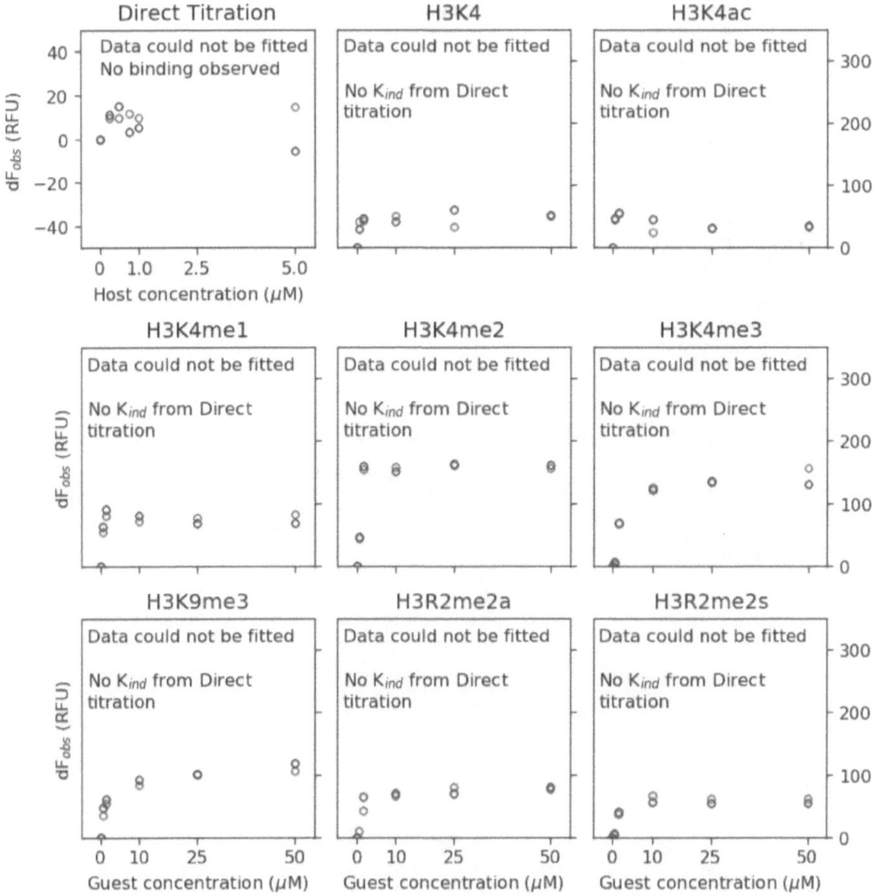

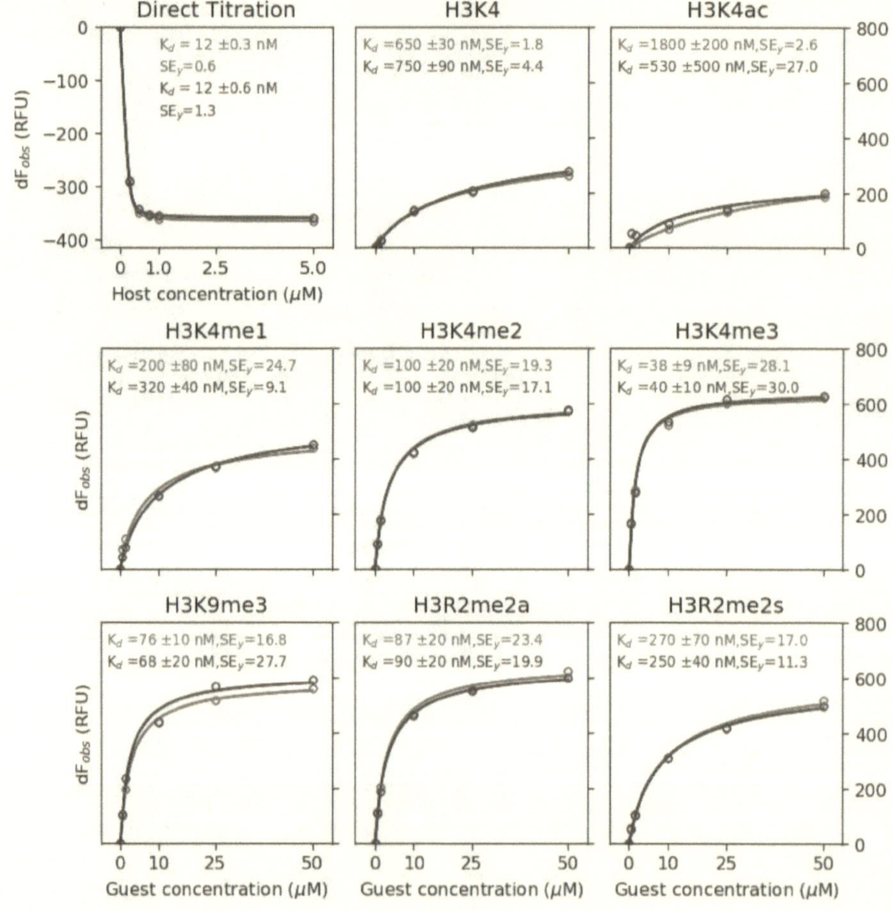
Compound DP2
Direct Titration
Kd = 12 ±0.3 nM
SEy=0.6
Kd = 12 ±0.6 nM
SEy=1.3
dFobs (RFU)
Host concentration (µM)
H3K4
Kd =650 ±30 nM,SEy=1.8
Kd =750 ±90 nM,SEy=4.4
H3K4ac
Kd =1800 ±200 nM,SEy=2.6
Kd =530 ±500 nM,SEy=27.0
H3K4me1
Kd =200 ±80 nM,SEy=24.7
Kd =320 ±40 nM,SEy=9.1
H3K4me2
Kd =100 ±20 nM,SEy=19.3
Kd =100 ±20 nM,SEy=17.1
H3K4me3
Kd =38 ±9 nM,SEy=28.1
Kd =40 ±10 nM,SEy=30.0
H3K9me3
Kd =76 ±10 nM,SEy=16.8
Kd =68 ±20 nM,SEy=27.7
H3R2me2a
Kd =87 ±20 nM,SEy=23.4
Kd =90 ±20 nM,SEy=19.9
H3R2me2s
Kd =270 ±70 nM,SEy=17.0
Kd =250 ±40 nM,SEy=11.3
dFobs (RFU)
Guest concentration (µM)

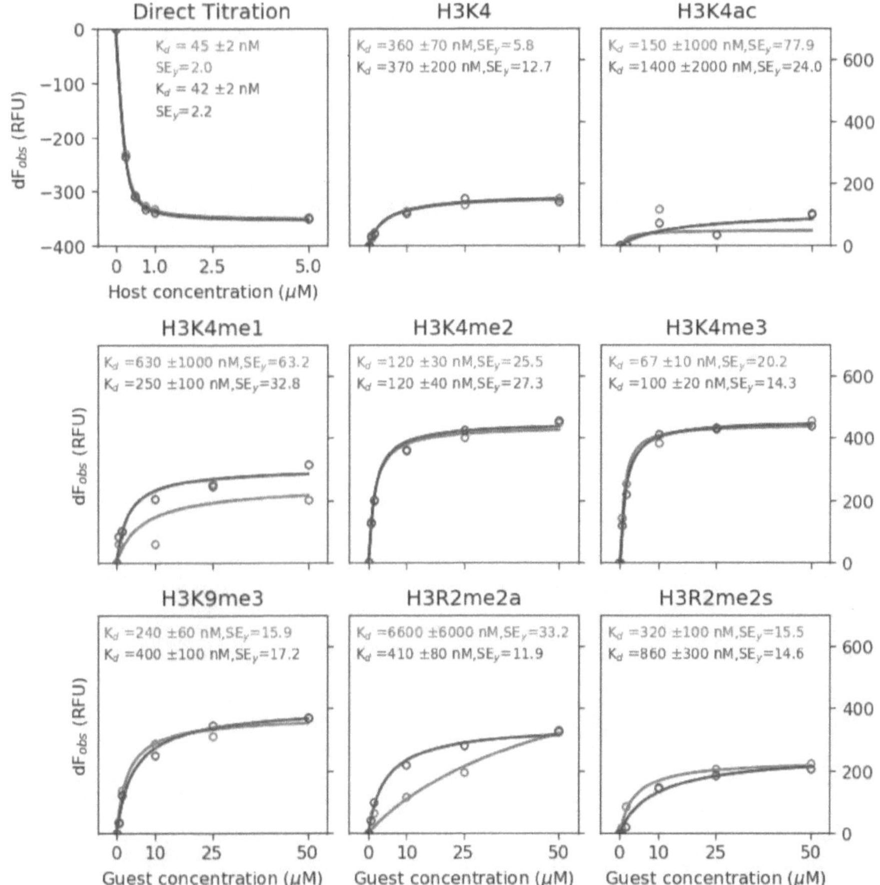
HO3S
SO3H
H3C
CH3
O
O
OH
OH
OH
HO
Compound DM1
Direct Titration
Kd = 45 ±2 nM
SEy=2.0
Kd = 42 ±2 nM
SEy=2.2
dFobs (RFU)
0
−100
−200
−300
−400
0 1.0 2.5 5.0
Host concentration (μM)
H3K4
Kd =360 ±70 nM,SEy=5.8
Kd =370 ±200 nM,SEy=12.7
H3K4ac
Kd =150 ±1000 nM,SEy=77.9
Kd =1400 ±2000 nM,SEy=24.0
600
400
200
0
H3K4me1
Kd =630 ±1000 nM,SEy=63.2
Kd =250 ±100 nM,SEy=32.8
dFobs (RFU)
H3K4me2
Kd =120 ±30 nM,SEy=25.5
Kd =120 ±40 nM,SEy=27.3
H3K4me3
Kd =67 ±10 nM,SEy=20.2
Kd =100 ±20 nM,SEy=14.3
600
400
200
0
H3K9me3
Kd =240 ±60 nM,SEy=15.9
Kd =400 ±100 nM,SEy=17.2
dFobs (RFU)
H3R2me2a
Kd =6600 ±6000 nM,SEy=33.2
Kd =410 ±80 nM,SEy=11.9
H3R2me2s
Kd =320 ±100 nM,SEy=15.5
Kd =860 ±300 nM,SEy=14.6
600
400
200
0
0 10 25 50
Guest concentration (μM)
0 10 25 50
Guest concentration (μM)
0 10 25 50
Guest concentration (μM)

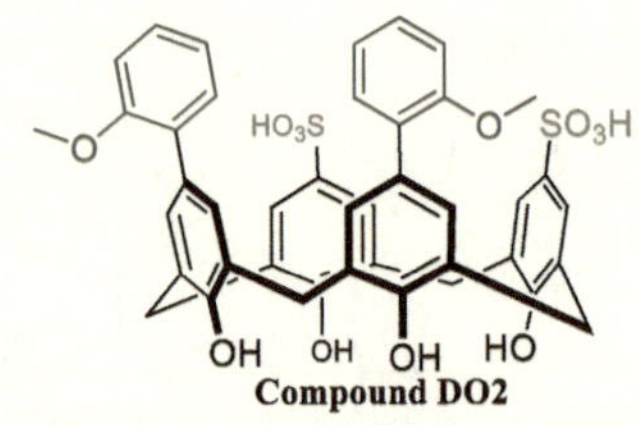

HO₃S
SO₃H
OH OH OH HO
Compound DO2

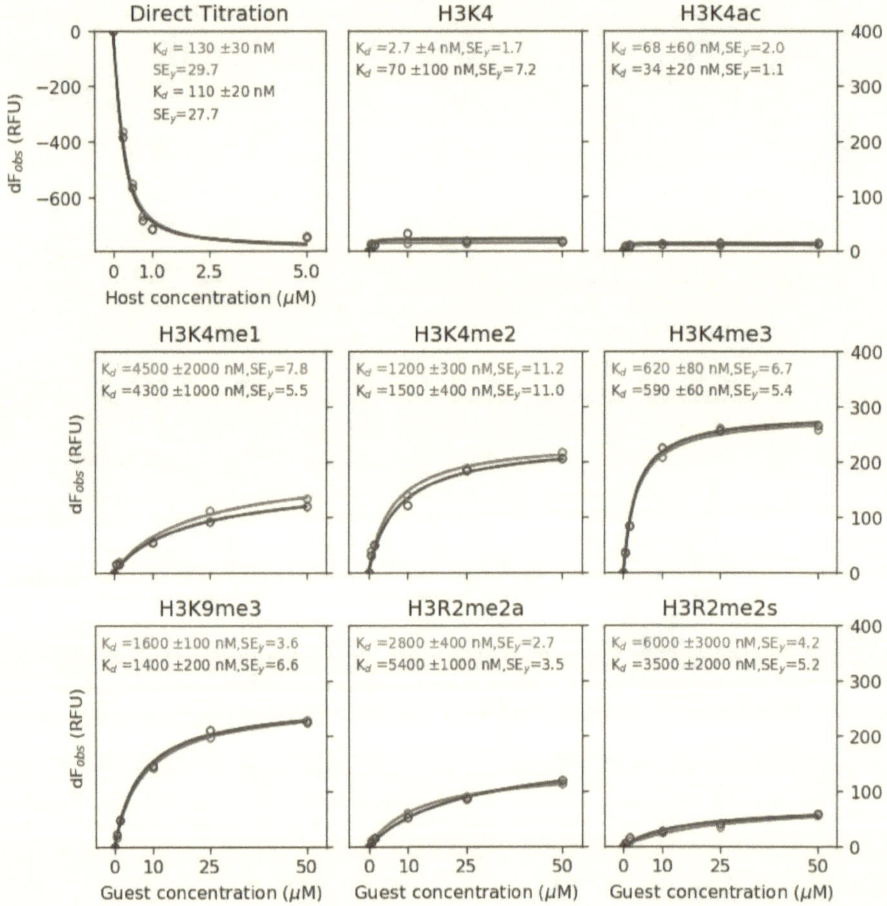

Direct Titration
K_d = 130 ±30 nM
SE_y=29.7
K_d = 110 ±20 nM
SE_y=27.7
dF_{obs} (RFU)
0
-200
-400
-600
0   1.0   2.5   5.0
Host concentration (µM)

H3K4
K_d =2.7 ±4 nM,SE_y=1.7
K_d =70 ±100 nM,SE_y=7.2

H3K4ac
K_d =68 ±60 nM,SE_y=2.0
K_d =34 ±20 nM,SE_y=1.1
400
300
200
100
0

H3K4me1
K_d =4500 ±2000 nM,SE_y=7.8
K_d =4300 ±1000 nM,SE_y=5.5
dF_{obs} (RFU)

H3K4me2
K_d =1200 ±300 nM,SE_y=11.2
K_d =1500 ±400 nM,SE_y=11.0

H3K4me3
K_d =620 ±80 nM,SE_y=6.7
K_d =590 ±60 nM,SE_y=5.4
400
300
200
100
0

H3K9me3
K_d =1600 ±100 nM,SE_y=3.6
K_d =1400 ±200 nM,SE_y=6.6
dF_{obs} (RFU)
0   10   25   50
Guest concentration (µM)

H3R2me2a
K_d =2800 ±400 nM,SE_y=2.7
K_d =5400 ±1000 nM,SE_y=3.5
0   10   25   50
Guest concentration (µM)

H3R2me2s
K_d =6000 ±3000 nM,SE_y=4.2
K_d =3500 ±2000 nM,SE_y=5.2
400
300
200
100
0
0   10   25   50
Guest concentration (µM)

F
HO3S
F
SO3H
OH
OH
OH
HO

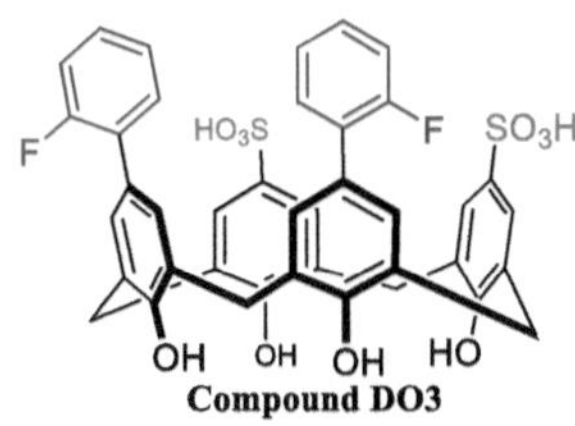

Compound DO3

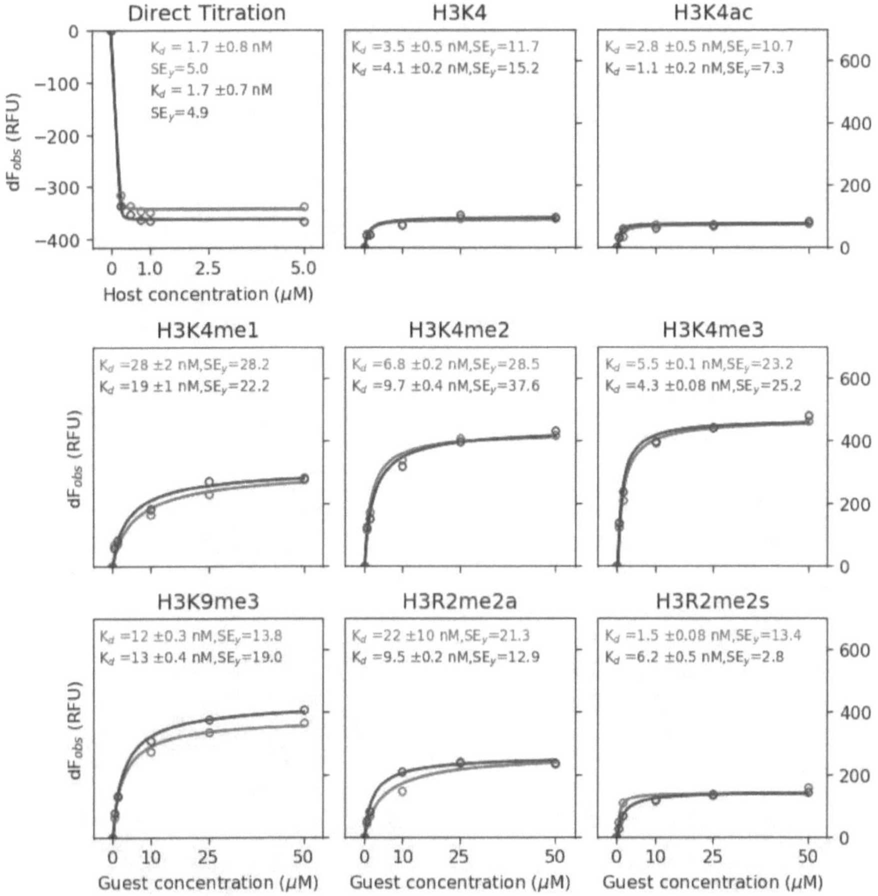

Direct Titration
Kd = 1.7 ±0.8 nM
SEy=5.0
Kd = 1.7 ±0.7 nM
SEy=4.9
dFobs (RFU)
Host concentration (μM)
H3K4
Kd =3.5 ±0.5 nM,SEy=11.7
Kd =4.1 ±0.2 nM,SEy=15.2
H3K4ac
Kd =2.8 ±0.5 nM,SEy=10.7
Kd =1.1 ±0.2 nM,SEy=7.3
H3K4me1
Kd =28 ±2 nM,SEy=28.2
Kd =19 ±1 nM,SEy=22.2
H3K4me2
Kd =6.8 ±0.2 nM,SEy=28.5
Kd =9.7 ±0.4 nM,SEy=37.6
H3K4me3
Kd =5.5 ±0.1 nM,SEy=23.2
Kd =4.3 ±0.08 nM,SEy=25.2
H3K9me3
Kd =12 ±0.3 nM,SEy=13.8
Kd =13 ±0.4 nM,SEy=19.0
H3R2me2a
Kd =22 ±10 nM,SEy=21.3
Kd =9.5 ±0.2 nM,SEy=12.9
H3R2me2s
Kd =1.5 ±0.08 nM,SEy=13.4
Kd =6.2 ±0.5 nM,SEy=2.8
Guest concentration (μM)

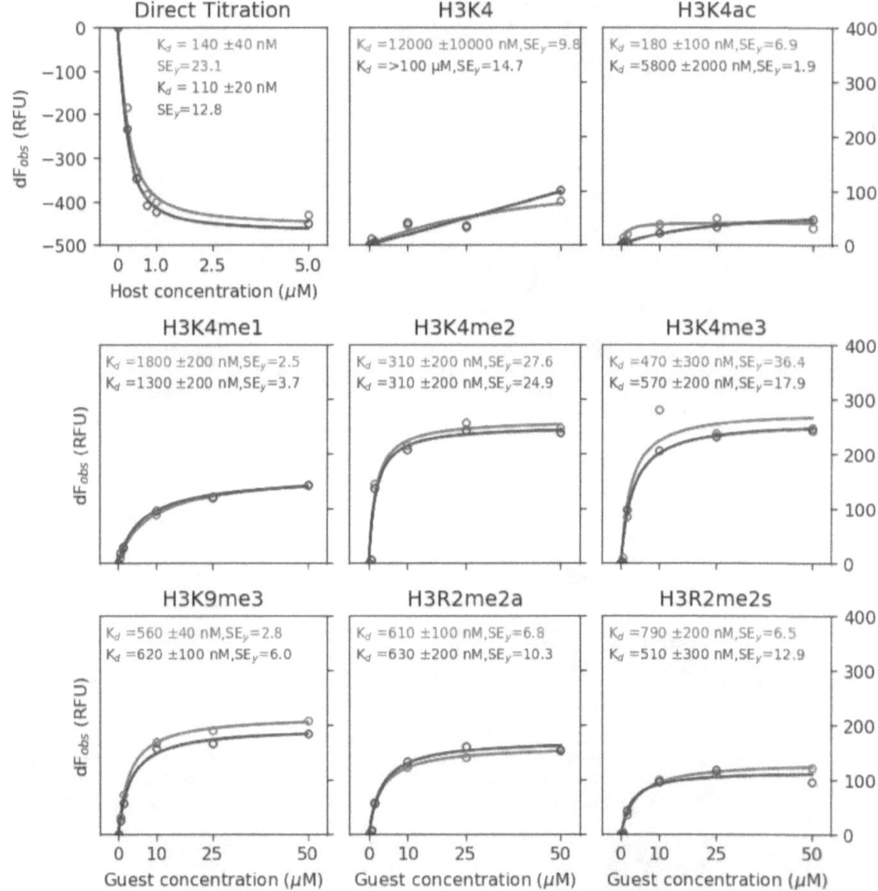
Compound CP1
Direct Titration
H3K4
H3K4ac
dF_obs (RFU)
0
−100
−200
−300
−400
−500
K_d = 140 ±40 nM
SE_y=23.1
K_d = 110 ±20 nM
SE_y=12.8
0 1.0 2.5 5.0
Host concentration (μM)
K_d =12000 ±10000 nM,SE_y=9.8
K_d =>100 μM,SE_y=14.7
K_d =180 ±100 nM,SE_y=6.9
K_d =5800 ±2000 nM,SE_y=1.9
400
300
200
100
0
H3K4me1
H3K4me2
H3K4me3
K_d =1800 ±200 nM,SE_y=2.5
K_d =1300 ±200 nM,SE_y=3.7
K_d =310 ±200 nM,SE_y=27.6
K_d =310 ±200 nM,SE_y=24.9
K_d =470 ±300 nM,SE_y=36.4
K_d =570 ±200 nM,SE_y=17.9
400
300
200
100
0
H3K9me3
H3R2me2a
H3R2me2s
K_d =560 ±40 nM,SE_y=2.8
K_d =620 ±100 nM,SE_y=6.0
K_d =610 ±100 nM,SE_y=6.8
K_d =630 ±200 nM,SE_y=10.3
K_d =790 ±200 nM,SE_y=6.5
K_d =510 ±300 nM,SE_y=12.9
400
300
200
100
0
0 10 25 50
Guest concentration (μM)
0 10 25 50
Guest concentration (μM)
0 10 25 50
Guest concentration (μM)

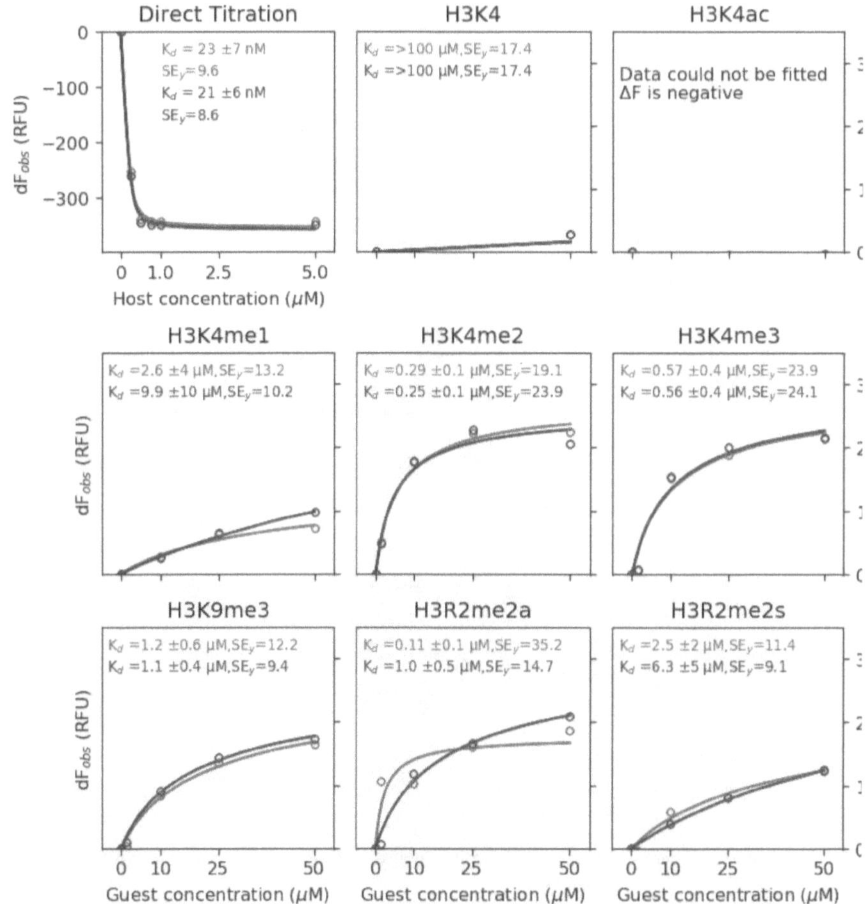
HO3S
Compound CP2
Direct Titration
$K_d = 23 \pm 7$ nM
$SE_y = 9.6$
$K_d = 21 \pm 6$ nM
$SE_y = 8.6$
dF$_{obs}$ (RFU)
0
−100
−200
−300
0   1.0   2.5   5.0
Host concentration (μM)
H3K4
$K_d => 100$ μM, $SE_y = 17.4$
$K_d => 100$ μM, $SE_y = 17.4$
H3K4ac
Data could not be fitted
ΔF is negative
H3K4me1
$K_d = 2.6 \pm 4$ μM, $SE_y = 13.2$
$K_d = 9.9 \pm 10$ μM, $SE_y = 10.2$
dF$_{obs}$ (RFU)
H3K4me2
$K_d = 0.29 \pm 0.1$ μM, $SE_y = 19.1$
$K_d = 0.25 \pm 0.1$ μM, $SE_y = 23.9$
H3K4me3
$K_d = 0.57 \pm 0.4$ μM, $SE_y = 23.9$
$K_d = 0.56 \pm 0.4$ μM, $SE_y = 24.1$
H3K9me3
$K_d = 1.2 \pm 0.6$ μM, $SE_y = 12.2$
$K_d = 1.1 \pm 0.4$ μM, $SE_y = 9.4$
dF$_{obs}$ (RFU)
H3R2me2a
$K_d = 0.11 \pm 0.1$ μM, $SE_y = 35.2$
$K_d = 1.0 \pm 0.5$ μM, $SE_y = 14.7$
H3R2me2s
$K_d = 2.5 \pm 2$ μM, $SE_y = 11.4$
$K_d = 6.3 \pm 5$ μM, $SE_y = 9.1$
0   10   25   50
Guest concentration (μM)
0   10   25   50
Guest concentration (μM)
0   10   25   50
Guest concentration (μM)

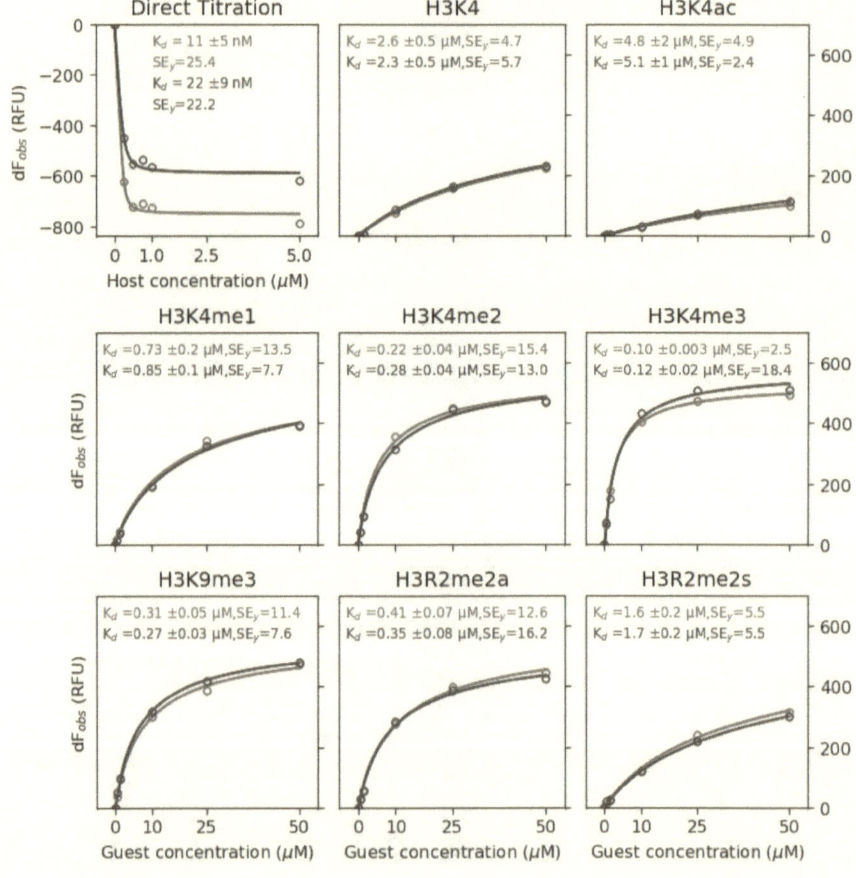

O
OH
O
S
O
NH
HO3S
SO3H
SO3H
OH
OH
OH
HO
Compound E1
Direct Titration
Kd = 11 ±5 nM
SEy=25.4
Kd = 22 ±9 nM
SEy=22.2
dFobs (RFU)
0
−200
−400
−600
−800
0   1.0   2.5   5.0
Host concentration (µM)
H3K4
Kd =2.6 ±0.5 µM,SEy=4.7
Kd =2.3 ±0.5 µM,SEy=5.7
H3K4ac
Kd =4.8 ±2 µM,SEy=4.9
Kd =5.1 ±1 µM,SEy=2.4
600
400
200
0
H3K4me1
Kd =0.73 ±0.2 µM,SEy=13.5
Kd =0.85 ±0.1 µM,SEy=7.7
H3K4me2
Kd =0.22 ±0.04 µM,SEy=15.4
Kd =0.28 ±0.04 µM,SEy=13.0
H3K4me3
Kd =0.10 ±0.003 µM,SEy=2.5
Kd =0.12 ±0.02 µM,SEy=18.4
600
400
200
0
dFobs (RFU)
H3K9me3
Kd =0.31 ±0.05 µM,SEy=11.4
Kd =0.27 ±0.03 µM,SEy=7.6
H3R2me2a
Kd =0.41 ±0.07 µM,SEy=12.6
Kd =0.35 ±0.08 µM,SEy=16.2
H3R2me2s
Kd =1.6 ±0.2 µM,SEy=5.5
Kd =1.7 ±0.2 µM,SEy=5.5
600
400
200
0
dFobs (RFU)
0   10   25   50
Guest concentration (µM)
0   10   25   50
Guest concentration (µM)
0   10   25   50
Guest concentration (µM)

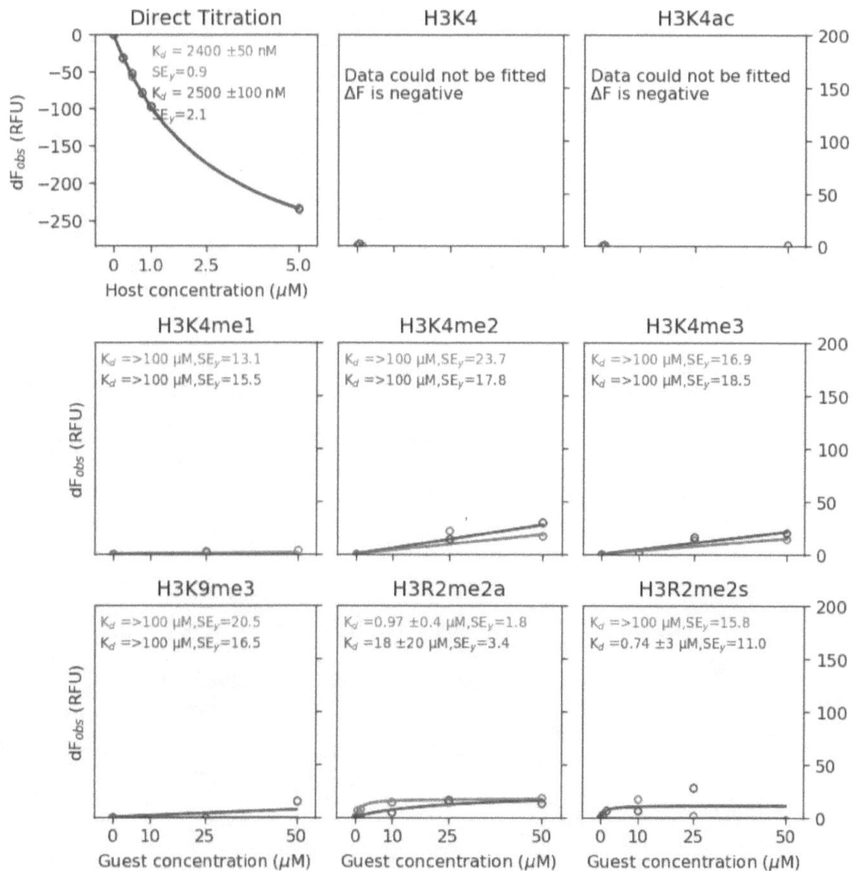

Compound E2
Direct Titration
K_d = 2400 ±50 nM
SE_y=0.9
K_d = 2500 ±100 nM
SE_y=2.1
dF$_{obs}$ (RFU)
Host concentration (μM)
H3K4
Data could not be fitted
ΔF is negative
H3K4ac
Data could not be fitted
ΔF is negative
H3K4me1
K_d =>100 μM,SE_y=13.1
K_d =>100 μM,SE_y=15.5
H3K4me2
K_d =>100 μM,SE_y=23.7
K_d =>100 μM,SE_y=17.8
H3K4me3
K_d =>100 μM,SE_y=16.9
K_d =>100 μM,SE_y=18.5
H3K9me3
K_d =>100 μM,SE_y=20.5
K_d =>100 μM,SE_y=16.5
H3R2me2a
K_d =0.97 ±0.4 μM,SE_y=1.8
K_d =18 ±20 μM,SE_y=3.4
H3R2me2s
K_d =>100 μM,SE_y=15.8
K_d =0.74 ±3 μM,SE_y=11.0
dF$_{obs}$ (RFU)
Guest concentration (μM)
Guest concentration (μM)
Guest concentration (μM)

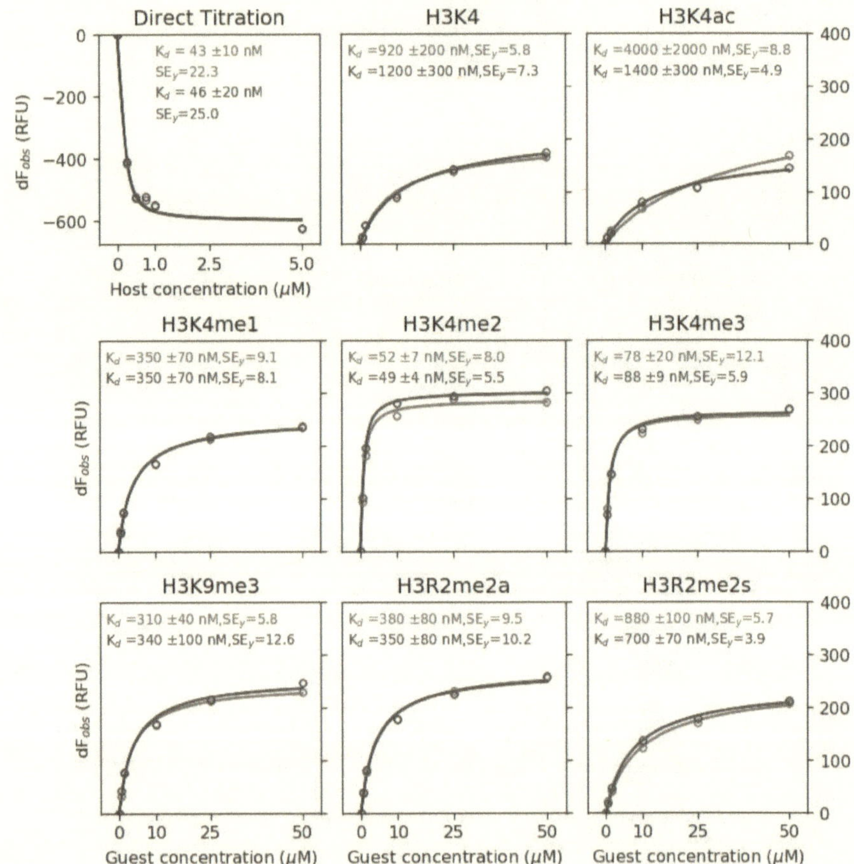

HO3S
SO3H
SO3H
NH
O
OH
OH
OH
HO
Compound E3

Direct Titration
Kd = 43 ±10 nM
SEy=22.3
Kd = 46 ±20 nM
SEy=25.0
dFobs (RFU)
0
−200
−400
−600
0   1.0   2.5   5.0
Host concentration (µM)

H3K4
Kd =920 ±200 nM,SEy=5.8
Kd =1200 ±300 nM,SEy=7.3

H3K4ac
Kd =4000 ±2000 nM,SEy=8.8
Kd =1400 ±300 nM,SEy=4.9
400
300
200
100
0

H3K4me1
Kd =350 ±70 nM,SEy=9.1
Kd =350 ±70 nM,SEy=8.1
dFobs (RFU)

H3K4me2
Kd =52 ±7 nM,SEy=8.0
Kd =49 ±4 nM,SEy=5.5

H3K4me3
Kd =78 ±20 nM,SEy=12.1
Kd =88 ±9 nM,SEy=5.9
400
300
200
100
0

H3K9me3
Kd =310 ±40 nM,SEy=5.8
Kd =340 ±100 nM,SEy=12.6
dFobs (RFU)

H3R2me2a
Kd =380 ±80 nM,SEy=9.5
Kd =350 ±80 nM,SEy=10.2

H3R2me2s
Kd =880 ±100 nM,SEy=5.7
Kd =700 ±70 nM,SEy=3.9
400
300
200
100
0

0   10   25   50
Guest concentration (µM)
0   10   25   50
Guest concentration (µM)
0   10   25   50
Guest concentration (µM)

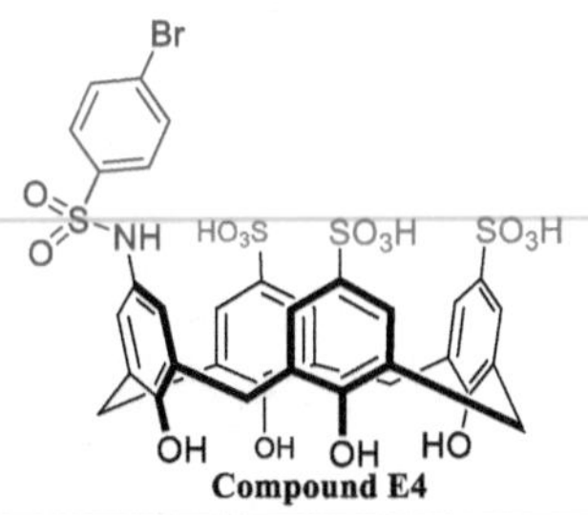

Br
O=S=O
NH
HO_3S
SO_3H
SO_3H
OH OH OH HO
Compound E4

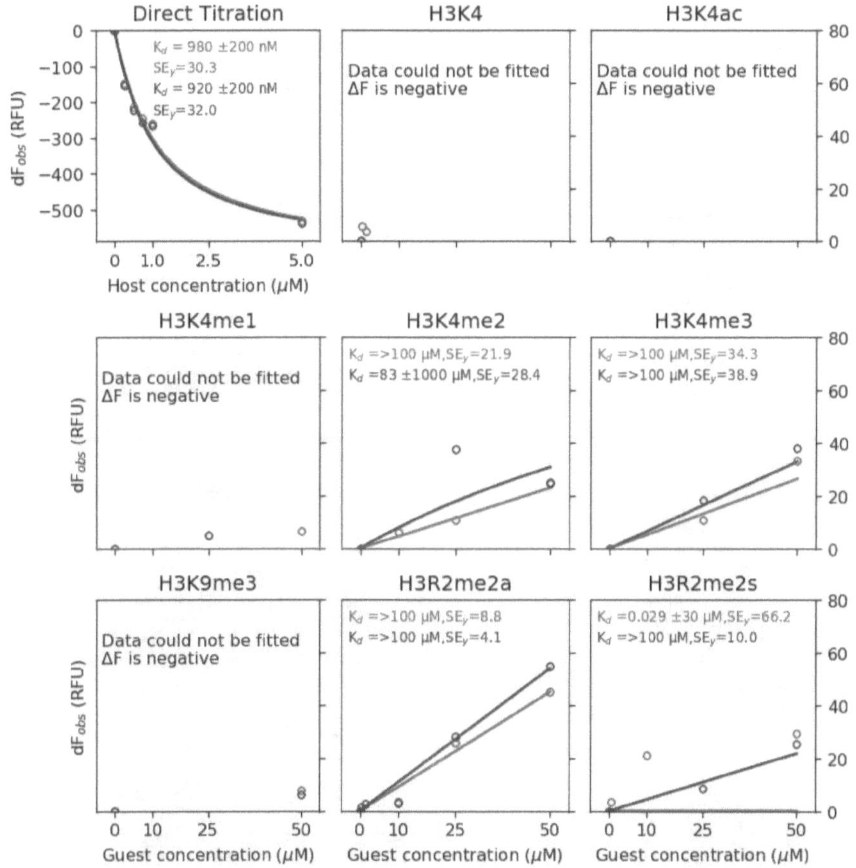

Direct Titration
K_d = 980 ±200 nM
SE_y=30.3
K_d = 920 ±200 nM
SE_y=32.0
dF_{obs} (RFU)
0
−100
−200
−300
−400
−500
0 1.0 2.5 5.0
Host concentration (μM)

H3K4
Data could not be fitted
ΔF is negative
80
60
40
20
0

H3K4ac
Data could not be fitted
ΔF is negative

H3K4me1
Data could not be fitted
ΔF is negative
dF_{obs} (RFU)

H3K4me2
K_d =>100 μM,SE_y=21.9
K_d =83 ±1000 μM,SE_y=28.4
80
60
40
20
0

H3K4me3
K_d =>100 μM,SE_y=34.3
K_d =>100 μM,SE_y=38.9

H3K9me3
Data could not be fitted
ΔF is negative
dF_{obs} (RFU)

H3R2me2a
K_d =>100 μM,SE_y=8.8
K_d =>100 μM,SE_y=4.1
80
60
40
20
0

H3R2me2s
K_d =0.029 ±30 μM,SE_y=66.2
K_d =>100 μM,SE_y=10.0

0 10 25 50
Guest concentration (μM)
0 10 25 50
Guest concentration (μM)
0 10 25 50
Guest concentration (μM)

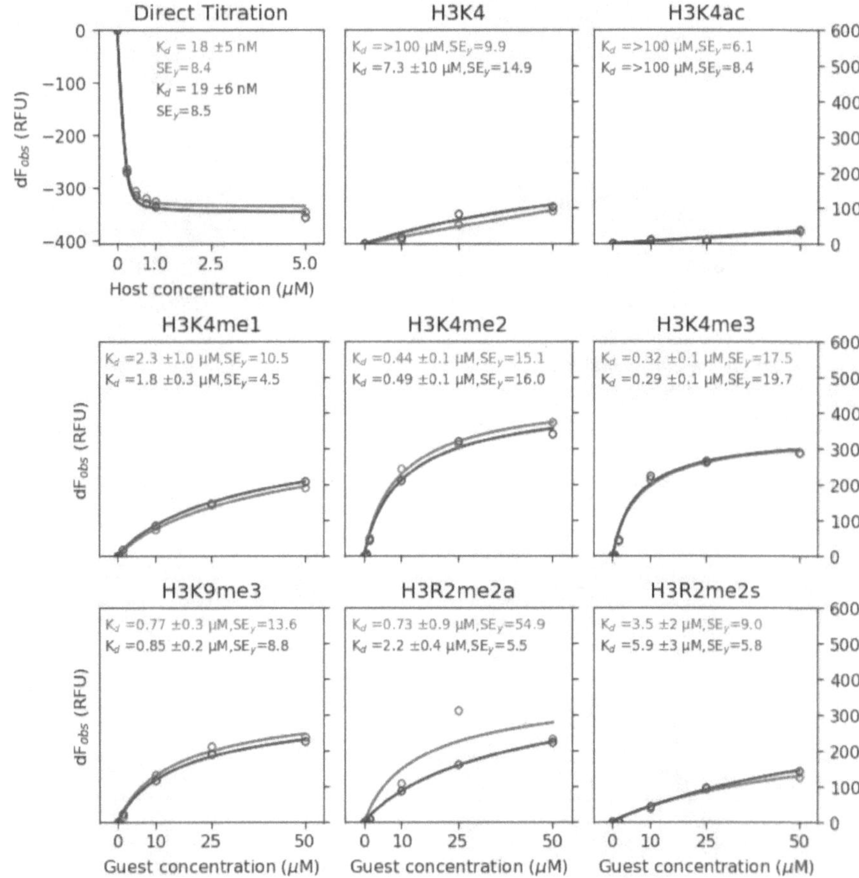

Br
HO3S   SO3H   SO3H
NH
O
OH   OH   OH   HO
Compound E5

Direct Titration
Kd = 18 ±5 nM
SEy=8.4
Kd = 19 ±6 nM
SEy=8.5

H3K4
Kd =>100 µM,SEy=9.9
Kd =7.3 ±10 µM,SEy=14.9

H3K4ac
Kd =>100 µM,SEy=6.1
Kd =>100 µM,SEy=8.4

H3K4me1
Kd =2.3 ±1.0 µM,SEy=10.5
Kd =1.8 ±0.3 µM,SEy=4.5

H3K4me2
Kd =0.44 ±0.1 µM,SEy=15.1
Kd =0.49 ±0.1 µM,SEy=16.0

H3K4me3
Kd =0.32 ±0.1 µM,SEy=17.5
Kd =0.29 ±0.1 µM,SEy=19.7

H3K9me3
Kd =0.77 ±0.3 µM,SEy=13.6
Kd =0.85 ±0.2 µM,SEy=8.8

H3R2me2a
Kd =0.73 ±0.9 µM,SEy=54.9
Kd =2.2 ±0.4 µM,SEy=5.5

H3R2me2s
Kd =3.5 ±2 µM,SEy=9.0
Kd =5.9 ±3 µM,SEy=5.8

dFobs (RFU)
Host concentration (µM)
Guest concentration (µM)

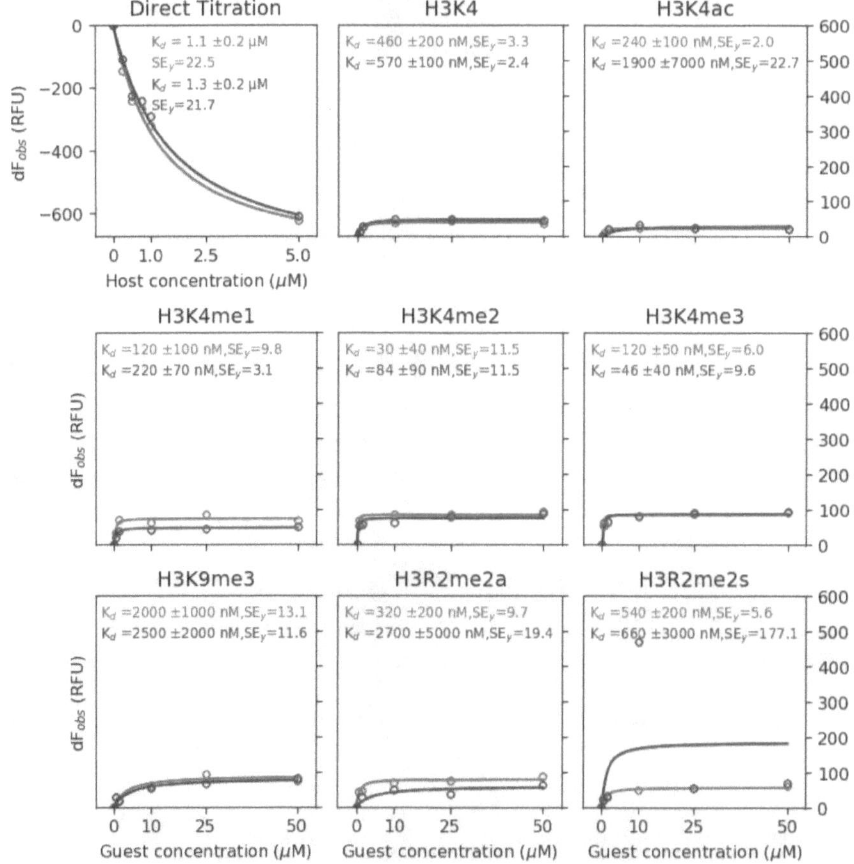
Compound E6
Direct Titration
$K_d = 1.1 \pm 0.2\ \mu M$
$SE_y = 22.5$
$K_d = 1.3 \pm 0.2\ \mu M$
$SE_y = 21.7$
Host concentration (μM)
H3K4
$K_d = 460 \pm 200$ nM, $SE_y = 3.3$
$K_d = 570 \pm 100$ nM, $SE_y = 2.4$
H3K4ac
$K_d = 240 \pm 100$ nM, $SE_y = 2.0$
$K_d = 1900 \pm 7000$ nM, $SE_y = 22.7$
H3K4me1
$K_d = 120 \pm 100$ nM, $SE_y = 9.8$
$K_d = 220 \pm 70$ nM, $SE_y = 3.1$
H3K4me2
$K_d = 30 \pm 40$ nM, $SE_y = 11.5$
$K_d = 84 \pm 90$ nM, $SE_y = 11.5$
H3K4me3
$K_d = 120 \pm 50$ nM, $SE_y = 6.0$
$K_d = 46 \pm 40$ nM, $SE_y = 9.6$
H3K9me3
$K_d = 2000 \pm 1000$ nM, $SE_y = 13.1$
$K_d = 2500 \pm 2000$ nM, $SE_y = 11.6$
H3R2me2a
$K_d = 320 \pm 200$ nM, $SE_y = 9.7$
$K_d = 2700 \pm 5000$ nM, $SE_y = 19.4$
H3R2me2s
$K_d = 540 \pm 200$ nM, $SE_y = 5.6$
$K_d = 660 \pm 3000$ nM, $SE_y = 177.1$
dF$_{obs}$ (RFU)
Guest concentration (μM)

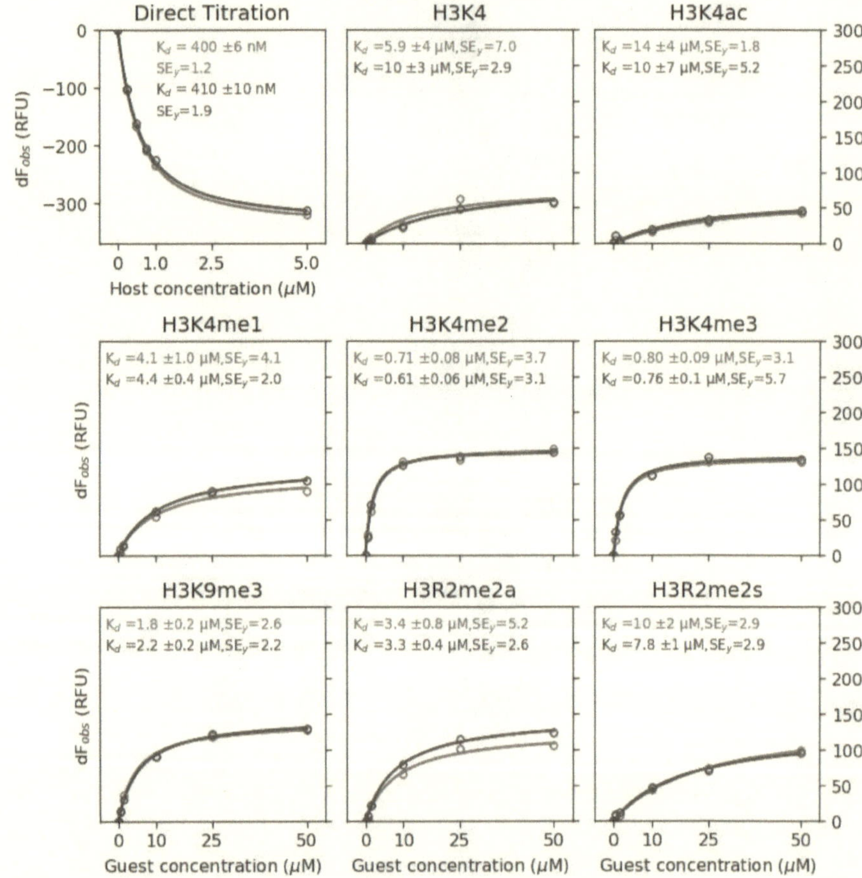

Compound E7
Direct Titration
K_d = 400 ±6 nM
SE_y=1.2
K_d = 410 ±10 nM
SE_y=1.9
dF_obs (RFU)
0
-100
-200
-300
0   1.0   2.5   5.0
Host concentration (μM)
H3K4
K_d =5.9 ±4 μM,SE_y=7.0
K_d =10 ±3 μM,SE_y=2.9
H3K4ac
K_d =14 ±4 μM,SE_y=1.8
K_d =10 ±7 μM,SE_y=5.2
300
250
200
150
100
50
0
H3K4me1
K_d =4.1 ±1.0 μM,SE_y=4.1
K_d =4.4 ±0.4 μM,SE_y=2.0
dF_obs (RFU)
H3K4me2
K_d =0.71 ±0.08 μM,SE_y=3.7
K_d =0.61 ±0.06 μM,SE_y=3.1
H3K4me3
K_d =0.80 ±0.09 μM,SE_y=3.1
K_d =0.76 ±0.1 μM,SE_y=5.7
300
250
200
150
100
50
0
H3K9me3
K_d =1.8 ±0.2 μM,SE_y=2.6
K_d =2.2 ±0.2 μM,SE_y=2.2
dF_obs (RFU)
H3R2me2a
K_d =3.4 ±0.8 μM,SE_y=5.2
K_d =3.3 ±0.4 μM,SE_y=2.6
H3R2me2s
K_d =10 ±2 μM,SE_y=2.9
K_d =7.8 ±1 μM,SE_y=2.9
300
250
200
150
100
50
0
0   10   25   50
Guest concentration (μM)
0   10   25   50
Guest concentration (μM)
0   10   25   50
Guest concentration (μM)

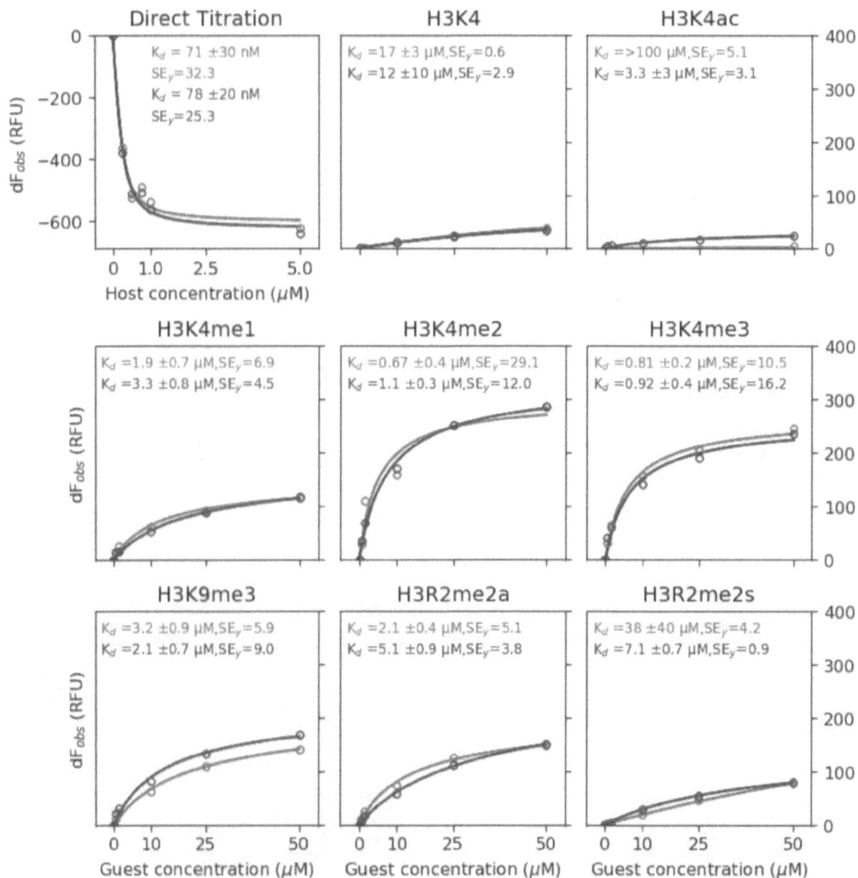

Compound E8
HN
NH
HO3S
SO3H
SO3H
OH
OH
OH
HO

Direct Titration
Kd = 71 ±30 nM
SEy=32.3
Kd = 78 ±20 nM
SEy=25.3
dF_obs (RFU)
0
−200
−400
−600
0   1.0   2.5   5.0
Host concentration (µM)

H3K4
Kd =17 ±3 µM,SEy=0.6
Kd =12 ±10 µM,SEy=2.9

H3K4ac
Kd =>100 µM,SEy=5.1
Kd =3.3 ±3 µM,SEy=3.1
400
300
200
100
0

H3K4me1
Kd =1.9 ±0.7 µM,SEy=6.9
Kd =3.3 ±0.8 µM,SEy=4.5
dF_obs (RFU)

H3K4me2
Kd =0.67 ±0.4 µM,SEy=29.1
Kd =1.1 ±0.3 µM,SEy=12.0

H3K4me3
Kd =0.81 ±0.2 µM,SEy=10.5
Kd =0.92 ±0.4 µM,SEy=16.2
400
300
200
100
0

H3K9me3
Kd =3.2 ±0.9 µM,SEy=5.9
Kd =2.1 ±0.7 µM,SEy=9.0
dF_obs (RFU)
0   10   25   50
Guest concentration (µM)

H3R2me2a
Kd =2.1 ±0.4 µM,SEy=5.1
Kd =5.1 ±0.9 µM,SEy=3.8
0   10   25   50
Guest concentration (µM)

H3R2me2s
Kd =38 ±40 µM,SEy=4.2
Kd =7.1 ±0.7 µM,SEy=0.9
400
300
200
100
0
0   10   25   50
Guest concentration (µM)

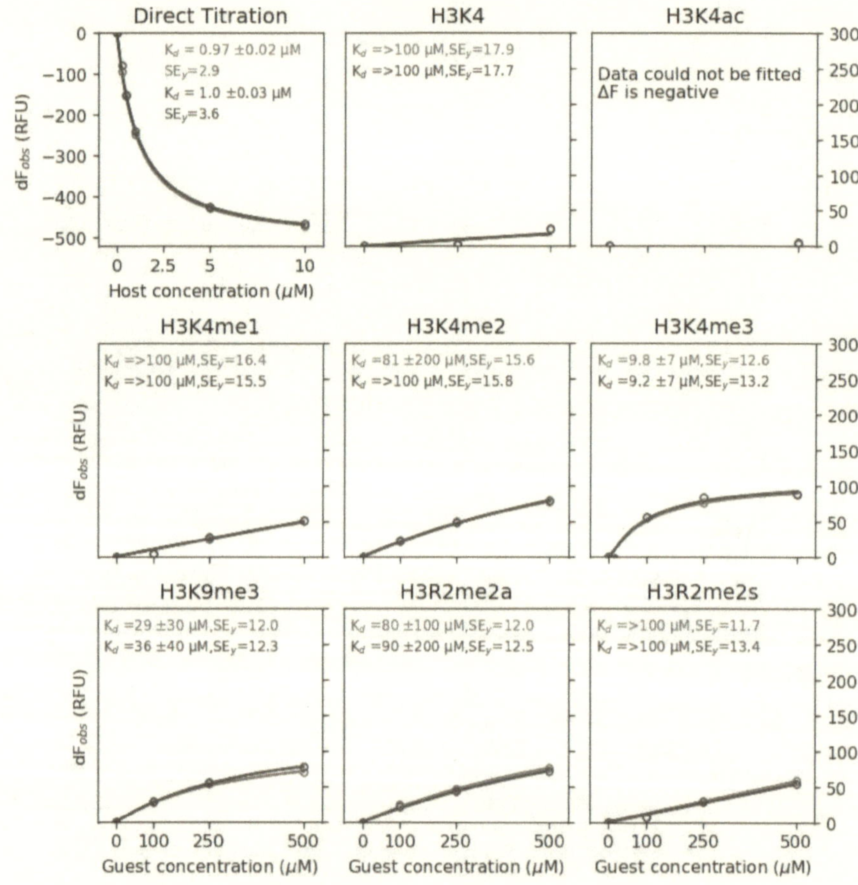

NO2
HO3S   SO3H   SO3H
S
O
O
NH
OH   OH   OH   HO
Compound E9

Direct Titration
Kd = 0.97 ±0.02 µM
SEy=2.9
Kd = 1.0 ±0.03 µM
SEy=3.6
dFobs (RFU)
0
-100
-200
-300
-400
-500
0   2.5   5   10
Host concentration (µM)

H3K4
Kd =>100 µM,SEy=17.9
Kd =>100 µM,SEy=17.7

H3K4ac
Data could not be fitted
ΔF is negative
300
250
200
150
100
50
0

H3K4me1
Kd =>100 µM,SEy=16.4
Kd =>100 µM,SEy=15.5
dFobs (RFU)

H3K4me2
Kd =81 ±200 µM,SEy=15.6
Kd =>100 µM,SEy=15.8

H3K4me3
Kd =9.8 ±7 µM,SEy=12.6
Kd =9.2 ±7 µM,SEy=13.2
300
250
200
150
100
50
0

H3K9me3
Kd =29 ±30 µM,SEy=12.0
Kd =36 ±40 µM,SEy=12.3
dFobs (RFU)
0   100   250   500
Guest concentration (µM)

H3R2me2a
Kd =80 ±100 µM,SEy=12.0
Kd =90 ±200 µM,SEy=12.5
0   100   250   500
Guest concentration (µM)

H3R2me2s
Kd =>100 µM,SEy=11.7
Kd =>100 µM,SEy=13.4
300
250
200
150
100
50
0
0   100   250   500
Guest concentration (µM)

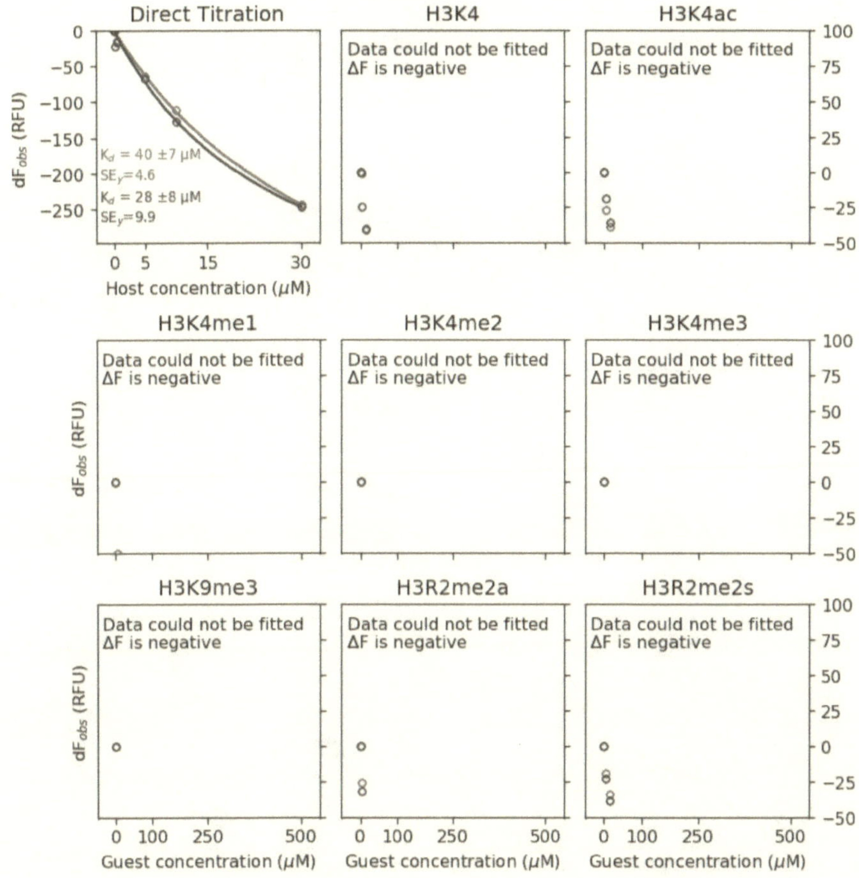
Br
HO3S SO3H SO3H
NH
OH OH OH HO
Compound E10
Direct Titration
H3K4
H3K4ac
Data could not be fitted
ΔF is negative
dF_obs (RFU)
K_d = 40 ±7 μM
SE_y=4.6
K_d = 28 ±8 μM
SE_y=9.9
Host concentration (μM)
H3K4me1
H3K4me2
H3K4me3
Data could not be fitted
ΔF is negative
dF_obs (RFU)
H3K9me3
H3R2me2a
H3R2me2s
Data could not be fitted
ΔF is negative
dF_obs (RFU)
Guest concentration (μM)

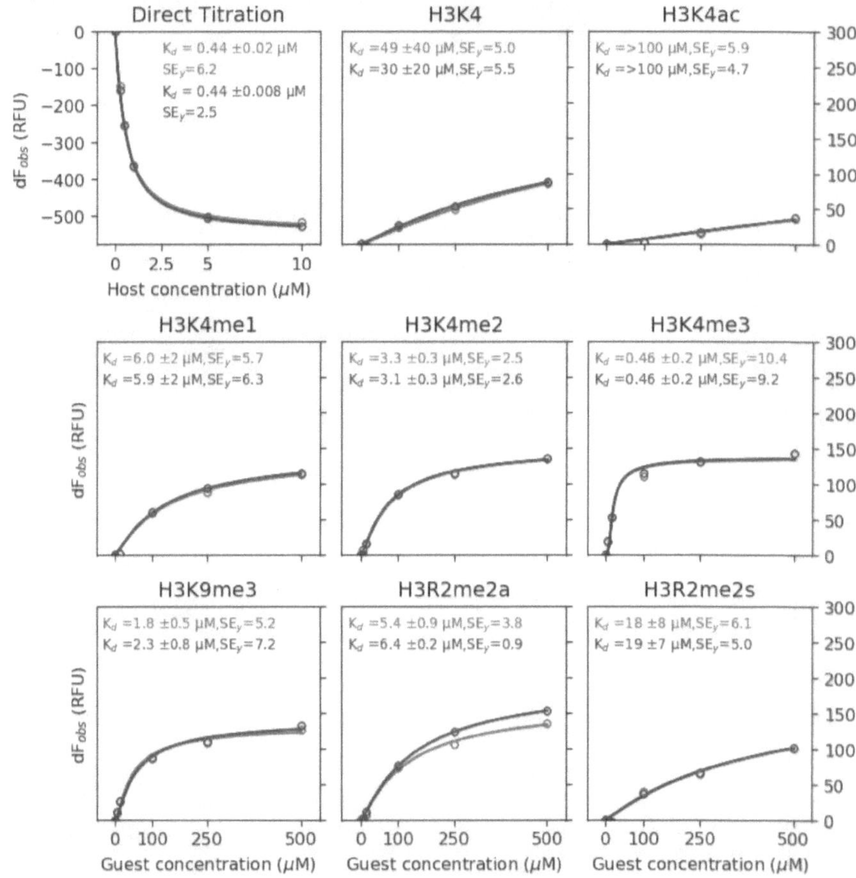
Compound E11
Direct Titration
K_d = 0.44 ±0.02 µM
SE_y=6.2
K_d = 0.44 ±0.008 µM
SE_y=2.5
dF_obs (RFU)
Host concentration (µM)
H3K4
K_d =49 ±40 µM,SE_y=5.0
K_d =30 ±20 µM,SE_y=5.5
H3K4ac
K_d =>100 µM,SE_y=5.9
K_d =>100 µM,SE_y=4.7
H3K4me1
K_d =6.0 ±2 µM,SE_y=5.7
K_d =5.9 ±2 µM,SE_y=6.3
H3K4me2
K_d =3.3 ±0.3 µM,SE_y=2.5
K_d =3.1 ±0.3 µM,SE_y=2.6
H3K4me3
K_d =0.46 ±0.2 µM,SE_y=10.4
K_d =0.46 ±0.2 µM,SE_y=9.2
dF_obs (RFU)
H3K9me3
K_d =1.8 ±0.5 µM,SE_y=5.2
K_d =2.3 ±0.8 µM,SE_y=7.2
H3R2me2a
K_d =5.4 ±0.9 µM,SE_y=3.8
K_d =6.4 ±0.2 µM,SE_y=0.9
H3R2me2s
K_d =18 ±8 µM,SE_y=6.1
K_d =19 ±7 µM,SE_y=5.0
dF_obs (RFU)
Guest concentration (µM)
Guest concentration (µM)
Guest concentration (µM)

www.ingramcontent.com/pod-product-compliance
Lightning Source LLC
LaVergne TN
LVHW040009200726
843493LV00005B/1200